OXFORD IB PREPARED

BIOLOGY

IB DIPLOMA PROGRAMME

Debora M. Primrose

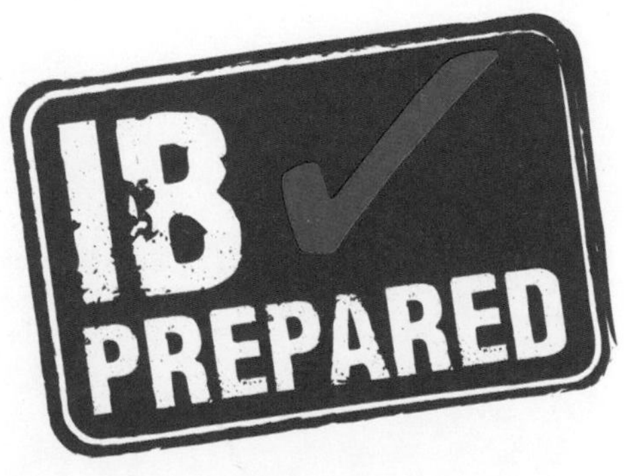

OXFORD
UNIVERSITY PRESS

Great Clarendon Street, Oxford, OX2 6DP, United Kingdom

Oxford University Press is a department of the University of Oxford. It furthers the University's objective of excellence in research, scholarship, and education by publishing worldwide. Oxford is a registered trade mark of Oxford University Press in the UK and in certain other countries

First published in 2019

ISBN 978 0 19 842363 8

Printed in Great Britain by Bell and Bain Ltd. Glasgow

Acknowledgements

Photo credits:

Cover image: James Paterson/PhotoPlus Magazine via Getty Images; **p3**: Peter Hermes Furian/Shutterstock; **p3**: Jeenson; **p3**: Paramjeet; **p5**: MEDIMAGE/SCIENCE PHOTO LIBRARY; **p9**: Sebastian Kaulitzki; **p9**: D. PHILLIPS / SCIENCE PHOTO LIBRARY; **p9**: DR. STANLEY FLEGLER/VISUALS UNLIMITED, INC./SCIENCE PHOTO LIBRARY; **p13**: PR. G GIMENEZ-MARTIN / SCIENCE PHOTO LIBRARY; **p14**: Leonid Andronov/Shutterstock; **p33**: BIOPHOTO ASSOCIATES/SCIENCE PHOTO LIBRARY; **p33**: BIOPHOTO ASSOCIATES/SCIENCE PHOTO LIBRARY; **p40**: Vladimir Melnik/ Shutterstock; **p50**: Zens/Shutterstock; **p59**: Jurgen Ziewe/Shutterstock; **p60**: STEVE GSCHMEISSNER / SCIENCE PHOTO LIBRARY; **p63**: Ttsz/ Istockphoto; **p70**: Steve Cymro/Shutterstock; **p79**: The Biochemist Artist/ Shutterstock; **p90**: Volodymyr Krasyuk/Shutterstock; **p102**: Dimarion/ Shutterstock; **p102**: Dimarion/Shutterstock; **p104**: Gallagher, Aisling; **p118**: DR KEITH WHEELER/SCIENCE PHOTO LIBRARY; **p118**: PRARTHAK; **p134**: Jose Luis Calvo/Shutterstock; **p136**: R. BICK, B. POINDEXTER, UT MEDICAL SCHOOL/SCIENCE PHOTO LIBRARY; **p143**: Wearset Ltd and David Russell Illustration; **p146**: Michal Kowalski/Shutterstock; **p147**: DR JEREMY BURGESS / SCIENCE PHOTO LIBRARY; **p154**: DR GOPAL MURTI/ SCIENCE PHOTO LIBRARY; **p155**: SINCLAIR STAMMERS/SCIENCE PHOTO LIBRARY; **p170**: Mark Herreid/Shutterstock; **p171**: Toeytoey/Shutterstock; **p185**: Cathy Keifer/Shutterstock; **p201**: Ase/Shutterstock; **p222**: Beata Aldridge/Shutterstock; **p234**: Bildagentur Zoonar GmbH/Shutterstock; **p234**: AlessandroZocc/Shutterstock; **p234**: Denis Tabler/Shutterstock; **p234**: TB studio/Shutterstock; **p235**: MARTYN F. CHILLMAID / SCIENCE PHOTO LIBRARY.

Artwork by Aptara Corp. and OUP.

The publisher would like to thank the International Baccalaureate for their kind permission to adapt questions from past examinations and content from the subject guide. The questions adapted for this book, and the corresponding past paper references, are summarized here:

Sample student answer, **p4**: M17 SLP2 TZ2 Q3**(a)**; Sample student answer, **p5**: M17 SLP2 TZ2 Q2**(b)(ii)**; Sample student answer, **p7**: M16 HLP2 Q6**(a)**; Sample student answer, **p9**: M17 HLP2 TZ2 Q3**(c)**; Sample student answer, **p12**: M16 SLP2 Q1**(f)** and **(g)**; Practice problem 5, **p13**: M17 HLP2 TZ2; Sample student answer, **p19**: M16 SLP2 TZ0 Q2; Sample student answer, **p20**: M17 HLP2 TZ2 Q6**(a)**; Sample student answer, **p22**: M17 SLP3 TZ2 Q2; Sample student answer, **p23**: M16 SLP2 TZ0 Q4**(a)**; Sample student answer, **p26**: M17 SLP3 TZ2 Q2; Sample student answer, **p29**: M17 SLP2 TZ2 Q6**(b)**; Practice problem 2, **p29**: M16 SLP2 Q5**(a)**; Sample student answer, **p31**: M17 SLP2 TZ2 Q4**(a)(i)** and **(ii)**; Sample student answer, **p36**: M17 SLP2 TZ2 Q3**(b)**; Sample student answer, **p44**: M17 HLP2 TZ2 Q4**(a)** and **(c)**; Sample student answer, **p45**: M16 HLP2 Q7**(b)**; Sample student answer, **p46**: M16 HLP3 Q1; Sample student answer, **p48**: M16 SLP2 Q3**(b)**; Practice problem 3, **p49**: M17 TZ2 SLP2 Q6**(c)**; Sample student answer, **p51**: M17 TZ2 HLP2 Q7**(a)**; Sample student answer, **p52**: M17 TZ2 SLP2 Q4**(b)(ii)**; Sample student answer, **p55**: M17 TZ2 SLP2 Q4**(b)(i)**; Practice problem 4, **p59**: M16 SLP2 Q4**(b)** and **(c)**; Sample student answer, **p68**: M17 TZ2 SLP2; Example 6.5.2, **p73**: M16 SLP3; Sample student answer, **p76**: M17 SLP2 TZ2 Q2**(a)** and M16 SLP2 Q2**(c)**; Sample student answer, **p76**: M16 HLP2 Q1**(b)**; Example 6.6.1, **p77**: N12 SLP1; Example 6.6.2, **p77**: N17 HLP1; Example 7.2.1, **p85**: M17 TZ2 HLP2 Q8**(c)**; Sample student answer, **p88**: M17 TZ2 HLP2 Q6**(b)**; Sample student answer, **p92**: M17 HLP2 TZ2; Sample student answer, **p97**: M16 HLP2; Sample student answer, **p106**: M17 TZ2 HLP2 and M16 HLP2 Q7**(c)**; Sample student answer, **p109**: M16 HLP3; Sample student answer, **p110**: M16 HLP3 Q3; Practice problem 3, **p127**: M16 HLP2 Q4**(b)**; Sample student answer, **p132**: M17 HLP2 TZ2 Q6**(c)**; Sample student answer, **p139**: M16 HLP2 Q6**(c)**; Sample student answer, **p142**: M17 HLP2 TZ2; Sample student answer, **p152**: Adapted from M17 TZ2 HLP3 Q2; Example A.4.2**(a)**, **p166**: M16 HLP3 Q7**(a)**; Example A.6.1, **p170**: M16 HLP3 Q7**(c)**; Practice problem 3, **p171**: M16 HLP3 Q8; Sample student answer, **p177**: M16 HLP3; Sample student answer, **p187**: Adapted from M17 TZ2 Q14**(c)**; Sample student answer, **p190**: M17 TZ2 Q17; Sample student answers, **p193**: M17 TZ2 Q16**(c)–(e)**; Practice problem 2, **p201**: M17 TZ2 Q15**(a)** to **(e)**; Sample student answer, **p204**: M17 TZ2 HLP3 Q19**(a)**; Example D.1.2, **p205**: **(a)(i)** adapted from M16 SLP3 Q20**(a)**; Sample student answer, **p206**: M16 HLP3 Q19; Sample student answer, **p208**: M17 TZ2 SLP3 Q19**(a)** and **(b)**; Example D.2.2, **p210**: M16 HLP3 Q19; Sample student answer, **p210**: M17 HLP3 TZ2 Q20; Sample student answer, **p211**: M16 Q19; Sample student answer, **p213**: M16 Q19; Sample student answer, **p214**: M17 SLP3 TZ2 Q21; Sample student answer, **p217**: M17 TZ2 HLP3 Q22; Sample student answer, **p220**: M17 TZ2 HLP3.

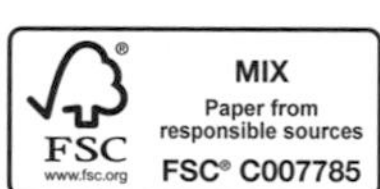

Contents

Answers to questions and exam papers in this book can be found on your free support website. Access the support website here: www.oxfordsecondary.com/ib-prepared-support

INTRODUCTION

This book provides full coverage of the IB diploma syllabus in biology and offers support to students preparing for their examinations. The book will help you revise the study material, learn the essential terms and concepts, strengthen your problem-solving skills and improve your approach to IB examinations. The book is packed with worked examples and exam tips that demonstrate best practices and warn against common errors. All topics are illustrated by annotated student answers to questions from past examinations, which explain why marks may be scored or missed.

A separate section is dedicated to data-based and practical questions, which are the most distinctive feature of the syllabus (first assessment in 2016). Numerous examples show how to tackle unfamiliar situations, interpret and analyse experimental data, and suggest improvements to experimental procedures. Practice problems and a complete set of IB-style examination papers provide further opportunities to check your knowledge and skills, boost your confidence and monitor the progress of your studies. Full solutions to all problems and examination papers are given online at **www.oxfordsecondary.com/ib-prepared-support.**

As any study guide, this book is not intended to replace your course materials, such as textbooks, laboratory manuals, past papers and markschemes, the IB Biology syllabus and your own notes. To succeed in the examination, you will need to use a broad range of resources, many of which are available online. This book will navigate you through this critical part of your studies, making your preparation for the exam less stressful and more efficient.

DP Biology assessment

All standard level (SL) and higher level (HL) students must complete the internal assessment and take three papers as part of their external assessment. Papers 1 and 2 are usually sat on one day and Paper 3 a day or two later. The internal and external assessment marks are combined as shown in the table at the top of page V to give your overall DP Biology grade, from 1 (lowest) to 7 (highest).

Overview of the book structure

The book is divided into several sections that cover the internal assessment, core SL and additional higher level (AHL) topics, data-based and practical questions, the four options (A–D) and a complete set of practice examination papers.

The largest section of the book, **core topics**, follows the structure of the IB diploma biology syllabus (for first assessment 2016) and covers all *understandings* and *applications and skills* assessment statements. Topics 1–6 contain common material for SL and HL students while topics 7–11 are intended for HL students only. The *nature of science* concepts are also discussed where applicable.

The **data-based and practical questions** section (chapter 12) provides a detailed analysis of problems and laboratory experiments that often appear in section A of paper 3. Similar to core topics, the discussion is illustrated by worked examples and sample scripts, followed by IB-style practice problems.

The **options** section reviews the material assessed in the second part of paper 3. Each of the four options is presented as a series of SL and AHL subtopics.

The **internal assessment** section outlines the nature of the investigation that you will have to carry out and explains how to select a suitable topic, collect and process experimental data, draw conclusions and present your report in a suitable format to satisfy the marking criteria and achieve the highest grade.

The final section contains IB-style **practice examination papers 1, 2 and 3**, written exclusively for this book. These papers will give you an opportunity to test yourself before the actual exam and at the same time provide additional practice problems for every topic of core and options material.

The answers and solutions to all practice problems and examination papers are given online at **www.oxfordsecondary.com/ib-prepared-support.** Blank answer sheets for examination papers are also available at the same address.

Assessment overview

Assessment	Description	Topics	SL		HL	
			marks	weight	marks	weight
Internal	Experimental work with a written report	—	24	20%	24	20%
Paper 1	Multiple-choice questions	Core: 1–6 (SL) 1–11 (AHL)	30	20%	40	20%
Paper 2	Short- and extended-response questions		50	40%	72	36%
Paper 3	*Section A:* data-based and practical questions	—	15	20%	15	24%
	Section B: short- and extended-response questions	Option of your choice	20		30	

The final IB diploma score is calculated by combining grades for six subjects with up to three additional points from *theory of knowledge* and *extended essay* components.

Command terms

Command terms are pre-defined words and phrases used in all IB Biology questions and problems. Each command term specifies the type and depth of the response expected from you in a particular question. For example, the command terms *state, outline, explain* and *discuss* require answers with increasingly higher levels of detail, from a single word, short sentence or numerical value ("state") to comprehensive analysis ("discuss"), as shown in the next table.

Question	Possible answer
State the effect of increasing temperature on the reaction rate.	Rate increases.
Outline how an increase in temperature affects the reaction rate.	For most reactions, the rate approximately doubles when temperature increases by 10 degrees.
Explain why an increase in temperature increases the reaction rate.	As temperature increases, the average speed and thus kinetic energy of particles also increase. The particles collide with one another more frequently and with a greater force. As a result, the frequency of successful collisions increases, so the rate increases.
Discuss the effects of increasing temperature and the presence of an enzyme on the reaction rate.	Both factors increase the rate by increasing the frequency of successful collisions. However, an increase in temperature increases the frequency and intensity of all collisions (successful and unsuccessful) but has no effect on the activation energy. In contrast, an enzyme has no effect on the frequency or intensity of collisions but lowers the activation energy by providing an alternative reaction pathway and thus allowing slow-moving particles to collide successfully. Thus, the same macroscopic effect is achieved by different microscopic changes.

A list of commonly used command terms in biology examination questions is given in the following table. Understanding the exact meaning of frequently used command terms is essential for your success in the examination. Therefore, you should explore this table and use it regularly as a reference when answering questions in this book.

Command term	Definition
Analyse	Break down in order to bring out the essential elements or structure
Annotate	Add brief notes to a diagram or graph
Calculate	Obtain a numerical answer showing your working
Comment	Give a judgment based on a given statement or result of a calculation
Compare	Give an account of the similarities between two or more items
Compare and contrast	Give an account of similarities and differences between two or more items
Construct	Present information in a diagrammatic or logical form
Deduce	Reach a conclusion from the information given
Define	Give the precise meaning of a word, phrase, concept or physical quantity
Describe	Give a detailed account
Design	Produce a plan, simulation or model
Determine	Obtain the only possible answer
Discuss	Offer a considered and balanced review that includes a range of arguments, factors or hypotheses

Continued on page VI

Command term	Definition
Distinguish	Make clear the differences between two or more items
Draw	Represent by a labelled, accurate diagram or graph, drawn to scale, with plotted points (if appropriate) joined in a straight line or smooth curve
Estimate	Obtain an approximate value
Explain	Give a detailed account including reasons or causes
Identify	Provide an answer from a number of possibilities
Label	Add labels to a diagram
List	Give a sequence of brief answers with no explanation
Outline	Give a brief account or summary
Predict	Give an expected result
Sketch	Represent by means of a diagram or graph (labelled as appropriate), giving a general idea of the required shape or relationship
State	Give a specific name, value or other brief answer without explanation
Suggest	Propose a solution, hypothesis or other possible answer

A complete list of command terms is available in the subject guide.

Preparation and exam strategies

In addition to the above suggestions, there are some simple rules you should follow during your preparation study and the exam itself.

1. **Get ready for study.** Have enough sleep, eat well, drink plenty of water and reduce your stress by positive thinking and physical exercise. A good night's sleep is particularly important before the exam day, as it can improve your score.
2. **Organize your study environment.** Find a comfortable place with adequate lighting, temperature and ventilation. Avoid distractions. Keep your papers and computer files organized. Bookmark useful online and offline material.
3. **Plan your studies.** Make a list of your tasks and arrange them by importance. Break up large tasks into smaller, easily manageable parts. Create an agenda for your studying time and make sure that you can complete each task before the deadline.
4. **Use this book as your first point of reference.** Work your way through the topics systematically and identify the gaps in your understanding and skills. Spend extra time on the topics where improvement is required. Check your textbook and online resources for more information.
5. **Read actively.** Focus on understanding rather than memorizing. Recite key points and definitions using your own words. Try to solve every worked example and practice problem before looking at the answer. Make notes for future reference.
6. **Get ready for the exams.** Practice answering exam-style questions under a time constraint. Solve as many problems from past papers as you can. Take a trial exam using the papers at the end of this book.
7. **Optimize your exam approach.** Read all questions carefully, paying extra attention to command terms. Keep your answers as short and clear as possible. Double-check all numerical values and units. Label axes in graphs and annotate diagrams. Use exam tips from this book.
8. **Do not panic.** Take a positive attitude and concentrate on things you can improve. Set realistic goals and work systematically to achieve these goals. Be prepared to reflect on your performance and learn from your errors in order to improve your future results.

Key features of the book

Each chapter typically covers one core or option topic, and starts with "**You should know**" and "**You should be able to**" checklists. These outline the *understandings* and *applications and skills* sections of the IB diploma biology syllabus. Some assessment statements have been reworded or combined together to make them more accessible and simplify the navigation. These changes do not affect the coverage of key syllabus material, which is always explained within the chapter. **Chapters contain the features outlined on this page.**

Theoretical concepts and key definitions are discussed at a level sufficient for answering typical examination questions. Many concepts are illustrated by diagrams, tables or worked examples. Most definitions are given in a grey side box like this one, and explained in the text.

Example

Examples offer solutions to typical problems and demonstrate common problem-solving techniques. Many examples provide alternative answers and explain how the marks are awarded.

Nature of science relates a biology concept to the overarching principles of the scientific approach.

Assessment tip

This feature highlights the essential terms and statements that have appeared in past markschemes, warns against common errors and shows how to optimize your approach to particular questions.

Sample student answers show typical student responses to IB-style questions (most of which are taken from past examination papers). In each response, the correct points are often highlighted in green while incorrect or incomplete answers are highlighted in red. Positive or negative feedback on student's response is given in the green and red pull-out boxes. An example is given below.

Links provide a reference to relevant material, within another part of this book, that relates to the text in question.

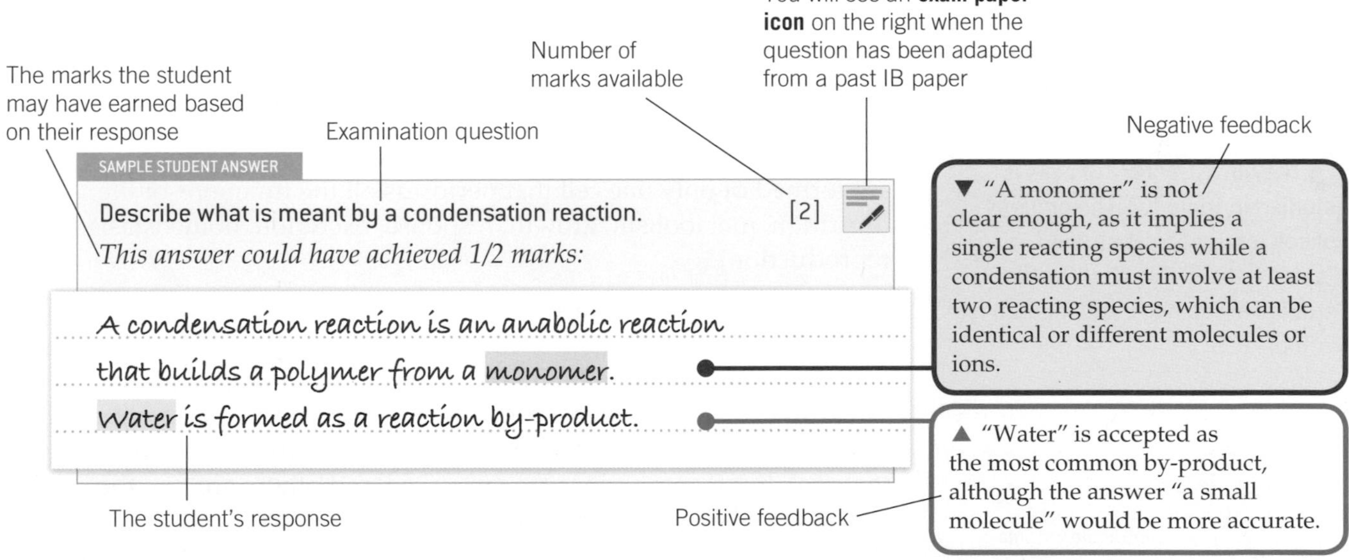

You will see an exam paper icon on the right when the question has been adapted from a past IB paper
Number of marks available
The marks the student may have earned based on their response
Examination question
Negative feedback
SAMPLE STUDENT ANSWER
Describe what is meant by a condensation reaction.
[2]
This answer could have achieved 1/2 marks:
A condensation reaction is an anabolic reaction that builds a polymer from a monomer. Water is formed as a reaction by-product.
▼ "A monomer" is not clear enough, as it implies a single reacting species while a condensation must involve at least two reacting species, which can be identical or different molecules or ions.
▲ "Water" is accepted as the most common by-product, although the answer "a small molecule" would be more accurate.
The student's response
Positive feedback

Questions not taken from past IB examinations will not have the exam paper icon.

Practice problems

Practice problems are given at the end of each chapter. These are IB-style questions that provide you with an opportunity to test yourself and improve your problem-solving skills. Some questions introduce factual or theoretical material from the syllabus that can be studied independently.

1 CELL BIOLOGY

1.1 INTRODUCTION TO CELLS

You should know:

- ✔ all living organisms are composed of cells.
- ✔ unicellular organisms consist of only one cell that carries out all functions of life in that organism.
- ✔ cell size is limited by the surface area to volume ratio of the cell.
- ✔ in multicellular organisms, specialized tissues can develop by cell differentiation.
- ✔ differentiation involves the expression of some genes and not others in a cell's genome.
- ✔ multicellular organisms have properties that emerge from the interaction of their cellular components.
- ✔ stem cell division and differentiation is necessary for embryonic development.

You should be able to:

- ✔ discuss exceptions to the cell theory, including striated muscle, giant algae and fungal hyphae.
- ✔ draw cells as seen under the light microscope.
- ✔ describe functions of life in *Paramecium* and a named photosynthetic unicellular organism.
- ✔ calculate the magnification and actual size of structures and ultrastructures shown in drawings or micrographs.
- ✔ explain the limitations of a cell having a large volume and small surface area.
- ✔ describe the therapeutic use of stem cells to treat Stargardt disease and one other example.
- ✔ discuss the ethics of using stem cells.

The ultrastructure of cells is studied in Topic 1.2. The process of cell respiration is studied in Topics 2.8 and 8.2. Nutrition is studied in Topic 6.1 and metabolism in Topic 8.1.

All living organisms are formed of cells. Unicellular organisms are formed of only one cell that performs all the functions of life (nutrition, metabolism, growth, response, excretion, homeostasis and reproduction).

Multicellular organisms are composed of many cells that become specialized by differentiation, forming different tissues. These tissues form organs which together make up organ systems. In order to differentiate, cells must express different genes and therefore produce different proteins. All cells in an organism have the same genetic material, but if some genes are expressed and others are not, the resulting cells will be different.

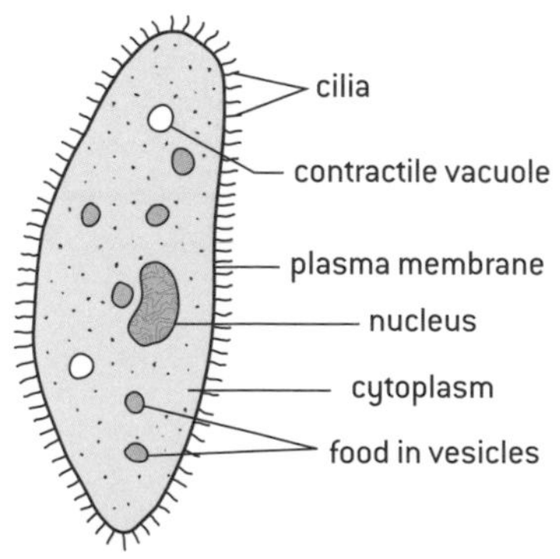

▲ Figure 1.1.1. *Paramecium* is a unicellular organism that obtains its food from the environment, digesting it in food vacuoles

- **Cells** are the basic units of life.
- **Emergent properties** are properties that appear in a complex system (or an organism) but do not appear in the individual units.
- **Differentiation** is the change in a cell to become more specialized.
- **Stem cells** are cells that are capable of differentiation.

Example 1.1.1.

The micrograph shows onion epidermal cells seen under the light microscope with a magnification of ×400.

a) (i) Label the nucleus of one cell.

(ii) Calculate the actual width of this cell. Show your working.

b) Suggest how the surface area to volume ratio of a cell can affect its function.

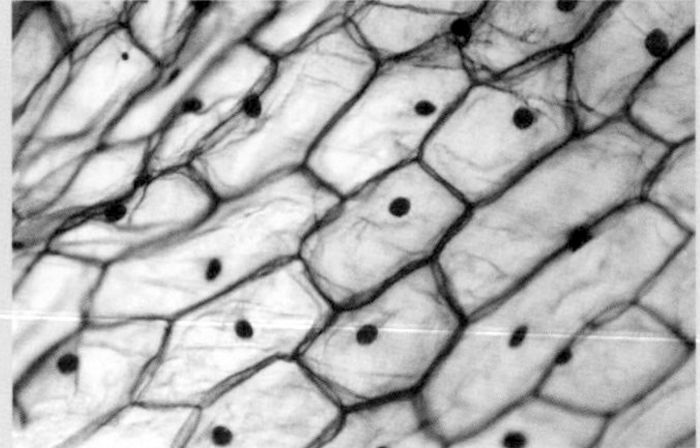

Solution

a) (i) Any dark circle labelled.

(ii) The magnification is the size of the image divided by the actual size of the cell. Therefore the formula for actual size is:

$$\text{Actual size of cell} = \frac{\text{size of image}}{\text{magnification}}$$

To calculate the width of the cell, you first use a ruler to measure the width of the chosen cell (size of image).

For example, measured cell width = 10 mm

$$\text{Actual width of cell} = \frac{10\text{ mm}}{400}$$

Actual width of cell = 0.025 mm

b) If the ratio is too small the exchange of substances will be too slow, waste substances will accumulate and heat will not be lost efficiently.

Because stem cells have the ability to differentiate into any type of cell, they are used in the development of medical treatments for a wide range of conditions. These include physical trauma, degenerative conditions and genetic diseases such as Stargardt disease. However, there are ethical issues regarding the use of stem cells in the treatment of diseases. Much of the debate surrounding stem cells concerns the use of human embryonic cells. The use of adult stem cells from sources such as blood from the umbilical cord is more convenient and less controversial.

Assessment tip

You have to check whether your result makes sense. 1 mm is equal to 1000 μm, therefore the cell width is 25 μm, which is the average cell width.

Example 1.1.2.

The pictures show drawings or micrographs from different cells with different magnifications not to scale. Which fully complies with the cell theory?

A Giant alga (*Acetabularia*)

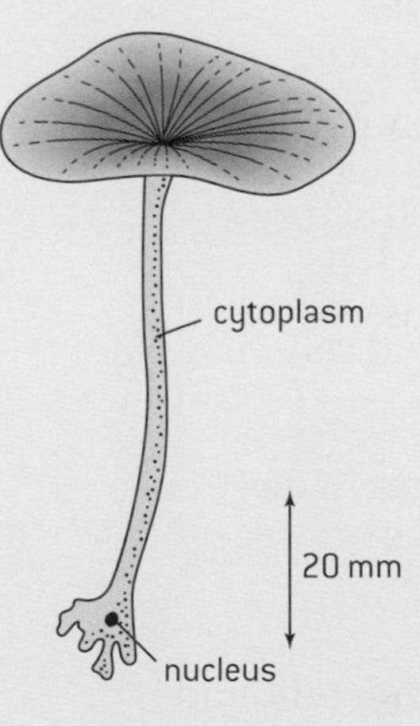

B White blood cell

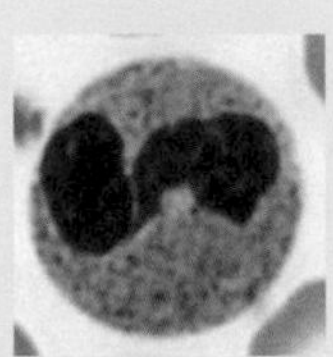

C Fungal aseptate hyphae

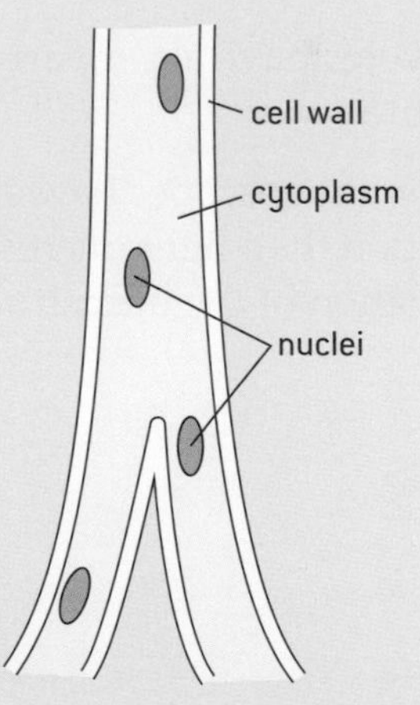

D Striated muscle

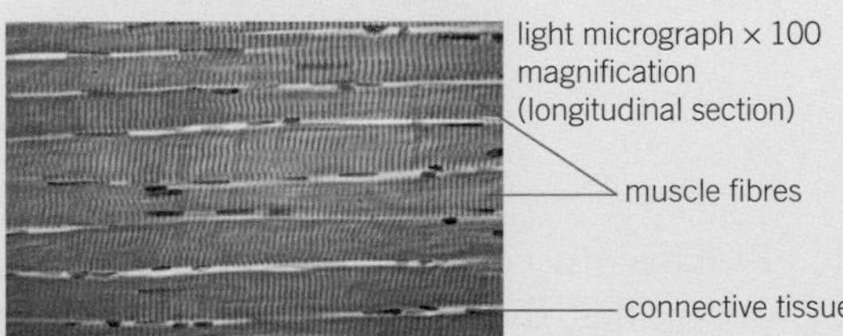

Solution

The answer is **B**, white blood cell, because it is surrounded by a cell membrane and has genetic material. The giant alga is a very large organism consisting of only one cell. The size does not correspond to the typical cell. The fungal aseptate hyphae do not have divisions between cells, therefore the cell contains many nuclei. The striated muscle cells are much larger than any average cell and also contain many nuclei. All these are exceptions to the cell theory.

SAMPLE STUDENT ANSWER

Outline the use of human embryonic stem cells (hESC) to treat Stargardt disease. [2]

This answer could have achieved 2/2 marks:

Human embryonic stem cells are unspecialized and can differentiate into almost any cell. These hESC are inserted into the retina of the eye and specialize to become healthy retinal cells which allow a person to regain their vision.

▲ The answer is correct as it outlines the use of stem cells specifically for Stargardt disease.

This answer could have achieved 0/2 marks:

As human embryonic stem cells are undifferentiated cells and are totipotent, they are able to be made into any cell that is needed. The Stargardt disease can be treated by stem cells as adequate cells can be created that the Stargardt disease kills.

▼ This answer only makes a vague reference of what the stem cells are used for, therefore scoring no mark. It is important to read the question carefully and answer what is being asked. In this case, the stem cells being used to replace retinal cells or photoreceptors was a key issue to include in the answer.

1.2 ULTRASTRUCTURE OF CELLS

You should know:

- ✔ eukaryotes have a much more complex cell structure than prokaryotes.
- ✔ prokaryotes do not have cell compartmentalization.
- ✔ eukaryotes have a compartmentalized cell structure.
- ✔ electron microscopes have a much higher resolution than light microscopes, allowing observation of the ultrastructure of cells.

You should be able to:

- ✔ describe the general structure and function of organelles within animal and plant cells.
- ✔ explain how prokaryotes divide by binary fission.
- ✔ draw the ultrastructure of prokaryotic cells.
- ✔ draw the ultrastructure of eukaryotic cells.
- ✔ compare and contrast the structure of prokaryotic and eukaryotic cells.
- ✔ compare and contrast animal cells and plant cells.
- ✔ interpret and label structures in electron micrographs.

An introduction to cells is given in Topic 1.1.

- **Magnification** is how much an image has been enlarged.
- **Resolution** is the minimal distance at which two points that are close together can be distinguished.

Prokaryotic cells do not have a nucleus or membrane-bound organelles, their nuclear material is found in the nucleoid or nuclear region and their DNA is naked, not bound to proteins. Prokaryotic cells have a cell wall, pili and flagella, and a plasma membrane enclosing cytoplasm that contains 70S ribosomes. Eukaryotic cells have a plasma membrane enclosing cytoplasm that contains larger (80S) ribosomes, a nucleus, mitochondria and other membrane-bound organelles. Plant cells are eukaryotic, but they also contain a cell wall and chloroplasts, which are not found in animal cells.

Example 1.2.1.

Complete the table using a tick (✓) for "possible presence" or a cross (✗) for "lack of" to distinguish prokaryotic and eukaryotic cells.

Solution

	Type of cell	
Characteristic	Prokaryotic	Eukaryotic
nuclear membrane	✗	✓
pili	✓	✗
flagellum	✓	✗
mitochondrion	✗	✓
70S ribosomes	✓	✗

The invention of electron microscopes led to greater understanding of cell structure. The maximum magnification of a light microscope is usually lower than × 2,000 and the maximum resolution is 0.2 μm. Beams of electrons have a much shorter wavelength compared with light waves, so electron microscopes have a much higher resolution. The maximum magnification of modern electron microscopes is around ×10,000,000 and the maximum resolution is less than 0.0001 μm.

SAMPLE STUDENT ANSWER

The electron micrograph shows the structures in a blood plasma cell.

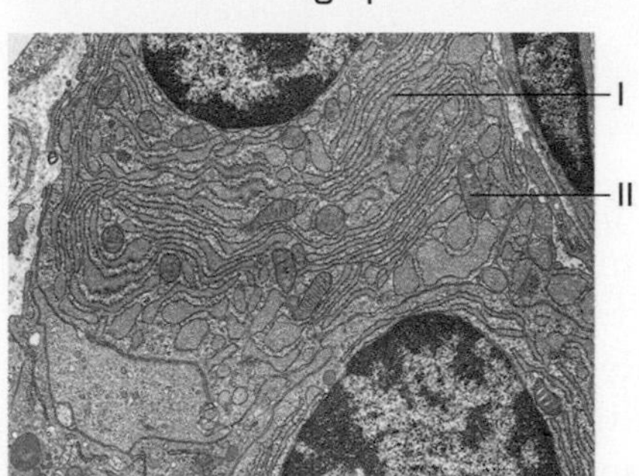

a) Using the table, identify the organelles labelled I and II on the electron micrograph with their principal role. [2]

This answer could have achieved 0/2 marks:

Organelle	Name	Principal role
I	rough endoplasmic reticulum (RER)	transport proteins across cell
II	mitochondria	secretes ATP

▼ Although the student did correctly identify the organelles, the functions are not correct. The RER does assist in transport across the cell, but in this case the principal role is to synthesize proteins. The mitochondria produce ATP, but they are not in charge of its secretion.

Assessment tip

You must be precise with the wording used.

b) Draw a labelled diagram of a eukaryotic plant cell as seen in an electron micrograph. [4]

This answer could have achieved 3/4 marks:

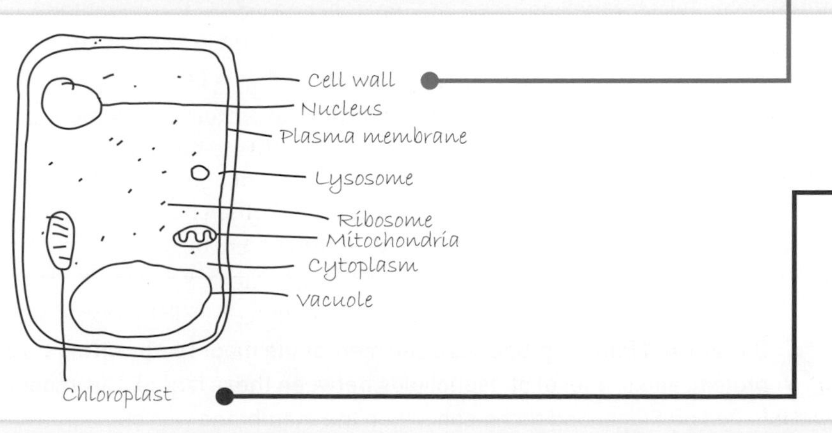

▲ This student scored the mark for correctly labelling the cell wall. The plasma membrane and the vacuole also scored a mark.

▼ The nucleus, ribosome, chloroplast and mitochondrion are not clear enough for a mark. Although the student correctly labelled the cytoplasm, the mark scheme did not include a mark for this.

Assessment tip

It is good practice to include multiple labels when answering this type of question because it increases the likelihood of identifying all the answers given in the mark scheme.

1.3 MEMBRANE STRUCTURE

You should know:

- membranes are formed by phospholipids, cholesterol, proteins, lipoproteins and glycoproteins.
- membrane proteins are diverse in terms of structure, position in the membrane and function.
- molecules that have hydrophilic and hydrophobic properties are said to be amphipathic.
- phospholipids form bilayers due to their amphipathic properties.
- cholesterol is a component of animal cell membranes.

You should be able to:

- draw the fluid mosaic model in two dimensions.
- explain the fluidity and permeability of the plasma membrane.
- analyse electron micrographs of plasma membranes.
- analyse information that led to the proposal of the Davson–Danielli model and its later falsification leading to the Singer–Nicolson model.

The ultrastructure of cells was studied in Topic 1.2.

The cell membrane is formed by a double layer of phospholipids. Phospholipids are amphipathic; this means they have a hydrophilic part and a hydrophobic part. The hydrophilic heads face both the outside and the inside of the cell while the hydrophobic part is in the middle of the bilayer. The low melting point of phospholipids in the bilayer is determined by the kinking of the long chain of fatty acids occurring at unsaturated bonds. This determines that some phospholipids are found in the liquid state while others are in the solid state, making the membrane fluid. In animal cells, cholesterol embedded in this double layer will control this fluidity and permeability to some solutes.

Proteins are embedded in the phospholipid bilayer. Some proteins are found crossing from side to side (integral transmembrane proteins), some partly inside (integral), whereas others are only on the outside (peripheral). Lipoproteins and glycoproteins can also be found on the outside of the cell membrane.

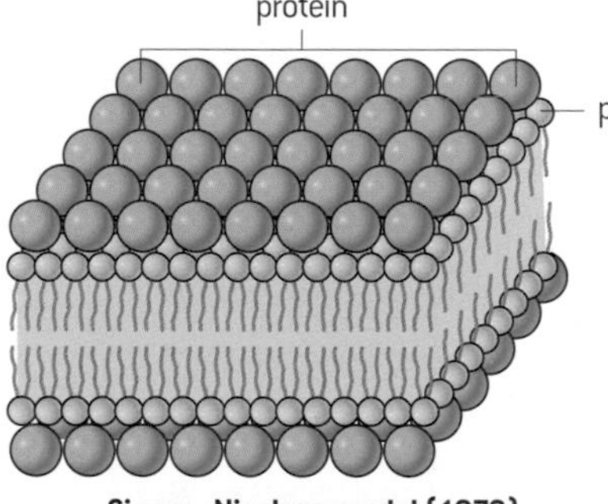

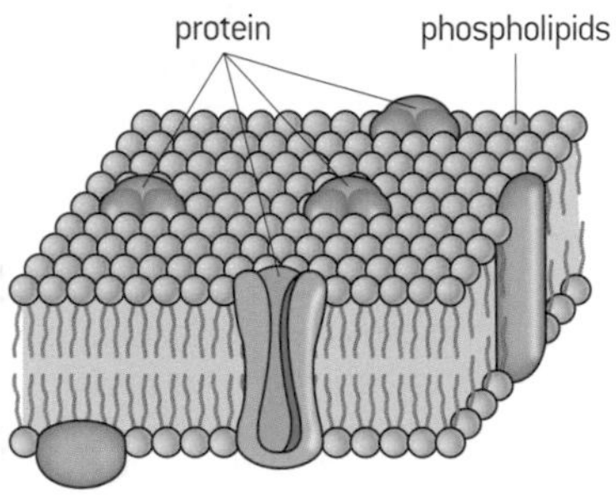

▲ Figure 1.3.1. Models of membrane structure

- **Amphipathic** molecules contain a **hydrophilic** (water-loving) and a **hydrophobic** (water-repelling) part.
- **Transmembrane proteins** are integral membrane proteins that span across the membrane. It is very hard to separate them from the membrane, but this can be done using detergents or solvents.
- **Integral proteins** are embedded in the phospholipid bilayer and protrude on only one side of the membrane. They are difficult to separate from the phospholipid bilayer.
- **Peripheral proteins** are temporarily attached either to the surface of the phospholipid bilayer or to integral proteins. They can be separated from the membrane using salts.

The structure of phospholipids is discussed in Topic 2.3, and transport across membranes is discussed in Topic 1.4.

Davson and Danielli proposed a cell membrane model with two layers of protein, and a layer of phospholipids between these layers. This model was falsified by Singer and Nicholson. When the membrane was split open it revealed irregular rough surfaces and therefore could not be a constant layer. Membrane proteins were shown to be mobile and not fixed in place, confirming the fluid mosaic structure.

SAMPLE STUDENT ANSWER

Draw a labelled diagram that shows the positions of proteins within the cell membrane. [3]

This answer could have achieved 2/3 marks:

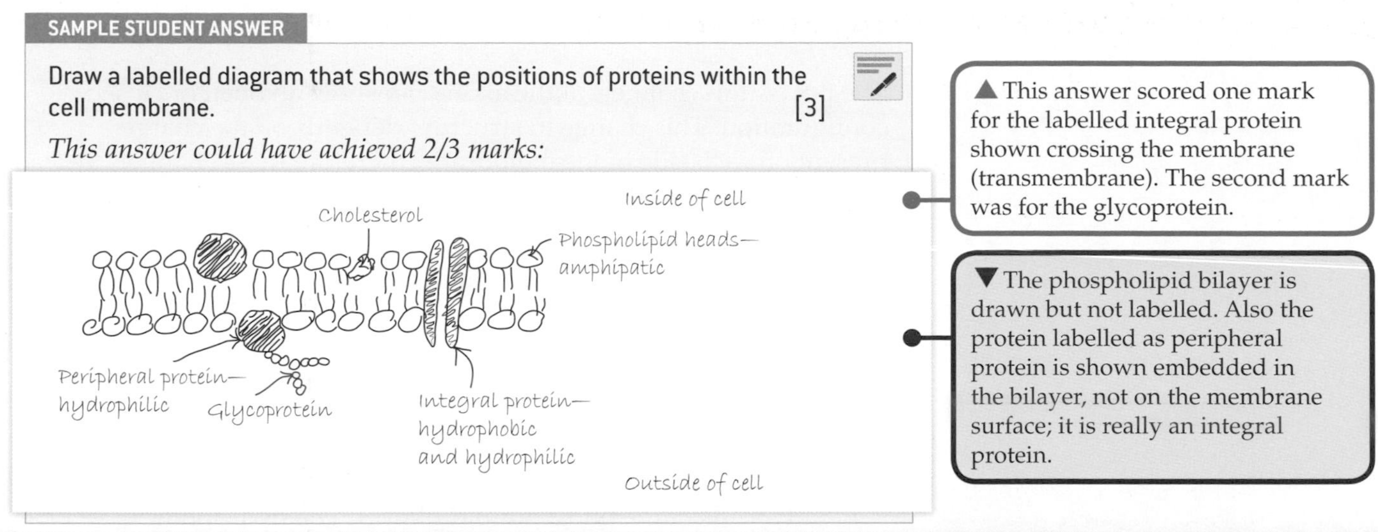

▲ This answer scored one mark for the labelled integral protein shown crossing the membrane (transmembrane). The second mark was for the glycoprotein.

▼ The phospholipid bilayer is drawn but not labelled. Also the protein labelled as peripheral protein is shown embedded in the bilayer, not on the membrane surface; it is really an integral protein.

1.4 MEMBRANE TRANSPORT

You should know:

- ✔ particles move across membranes by simple diffusion, facilitated diffusion, osmosis and active transport.
- ✔ fluidity of membranes allows materials to be taken into cells by endocytosis or released by exocytosis.
- ✔ vesicles move materials within cells.

You should be able to:

- ✔ describe the structure and explain the function of sodium–potassium pumps for active transport in axons.
- ✔ describe the structure and explain the function of potassium channels for facilitated diffusion in axons.
- ✔ provide reasons why tissues or organs used in medical procedures need to be bathed in a solution of the same osmolarity as the cytoplasm.
- ✔ estimate osmolarity in tissues by bathing samples in hypotonic and hypertonic solutions.

The cell membrane controls the entrance and exit of substances to and from the cell. Substances can pass in or out by active or passive transport. Active transport usually occurs against a concentration gradient; therefore it requires energy. Passive transport does not require energy and includes simple diffusion, facilitated diffusion and osmosis. Small molecules pass through by simple diffusion. However, charged molecules do not diffuse through the hydrophobic part of the membrane. Ions with positive or negative charges cannot easily diffuse through, while polar molecules, which have partial positive and negative charges over their surface, diffuse at very low rates. Slightly larger molecules pass by facilitated diffusion; channel proteins enable the diffusion of some molecules. Water molecules pass through by osmosis.

Substances that cannot enter through channel proteins because they are too large require bulk transport; this is transport in membrane-bound vesicles. Bulk transport into the cell is called endocytosis and bulk transport exiting the cell is called exocytosis. In endocytosis, the fluidity of the cell membrane allows the membrane to surround the particle to be ingested. In exocytosis, vesicles formed in the Golgi complex fuse with the membrane to transport the substances out of the cell.

The ultrastructure of cells was given in Topic 1.2 and the structure of cell membranes was discussed in Topic 1.3.

- **Passive transport** is the movement across the membrane without the use of energy.
- **Facilitated diffusion** is the passive transport of molecules or ions across the cell membrane through specific transmembrane proteins (channel proteins).
- **Active transport** is the movement across the membrane requiring energy in the form of ATP.
- **Osmosis** is the passage of water through a selectively permeable membrane, from a higher water potential (lower solute concentration) to a lower water potential (higher solute concentration).

Example 1.4.1.

The potassium channels in the axons can show an open or closed configuration. This change in structure depends on the charge present on each side of the membrane; therefore these channels are called voltage-gated.

a) Explain how the disposition of the proteins of the potassium channel in the membrane assists in the movement of ions.

b) Suggest the mode of transport of potassium through these channels.

Solution

a) The proteins of the channel are transmembrane proteins. The hydrophobic parts of the proteins are embedded in the tails of fatty acids of the phospholipid bilayer. The hydrophilic sections of the proteins are on the surface of the inner part of the membrane in contact with the cytoplasm, and on the surface of the part of the membrane in contact with the outside of the cell. The proteins make a tunnel, where the inside is also hydrophilic, allowing the passage of ions (and water) through the centre, acting therefore as a channel. These channels are very specific; they allow only potassium ions to pass through, not smaller sodium ions which have the same charge.

b) Facilitated diffusion, because it occurs through a protein channel and it does not require energy.

Assessment tip

You do not need to answer the question in a table, but it helps to make sure you are really comparing the two modes of transport. Remember you need at least one similarity and one difference. If the question is worth 3 marks you need to write at least three characteristics.

Example 1.4.2.

Compare and contrast osmosis and active transport.

Solution

The tick (✓) means it occurs, the cross (✗) that it does not occur.

	Osmosis	Active transport
Movement across cell membrane	✓	✓
Transmembrane protein required	✓	✓
Energy required	✗	✓
In the direction of a concentration gradient	✓	✗

The sodium–potassium pump allows a nervous impulse to occur along the axons of neurons. This involves the movement of sodium and potassium ions by facilitated diffusion through membrane proteins forming channels. The concentration gradient allowing for these movements is built up by the sodium–potassium pump protein, which carries out this process through active transport. Three sodium ions are transported across the protein to the outside of the cell against a gradient using 2 ATP molecules. Two potassium ions can then enter the cell by diffusion.

- **Osmolarity** is the measurement of the solute concentration of a solution, expressed as the total mass of solute (or osmoles) per litre of solution.
- An **isotonic solution** shares the same concentration as the tissues or cells it is bathing.
- A **hypertonic solution** has a higher solute concentration than the tissues (or cells) it bathes.
- A **hypotonic solution** has a lower solute concentration than the tissue it bathes.

SAMPLE STUDENT ANSWER

Red blood cell 1 is a normal red blood cell.

Red blood cell 2 and red blood cell 3 are two red blood cells that have been placed in solutions with different concentrations of solutes (osmolarities).

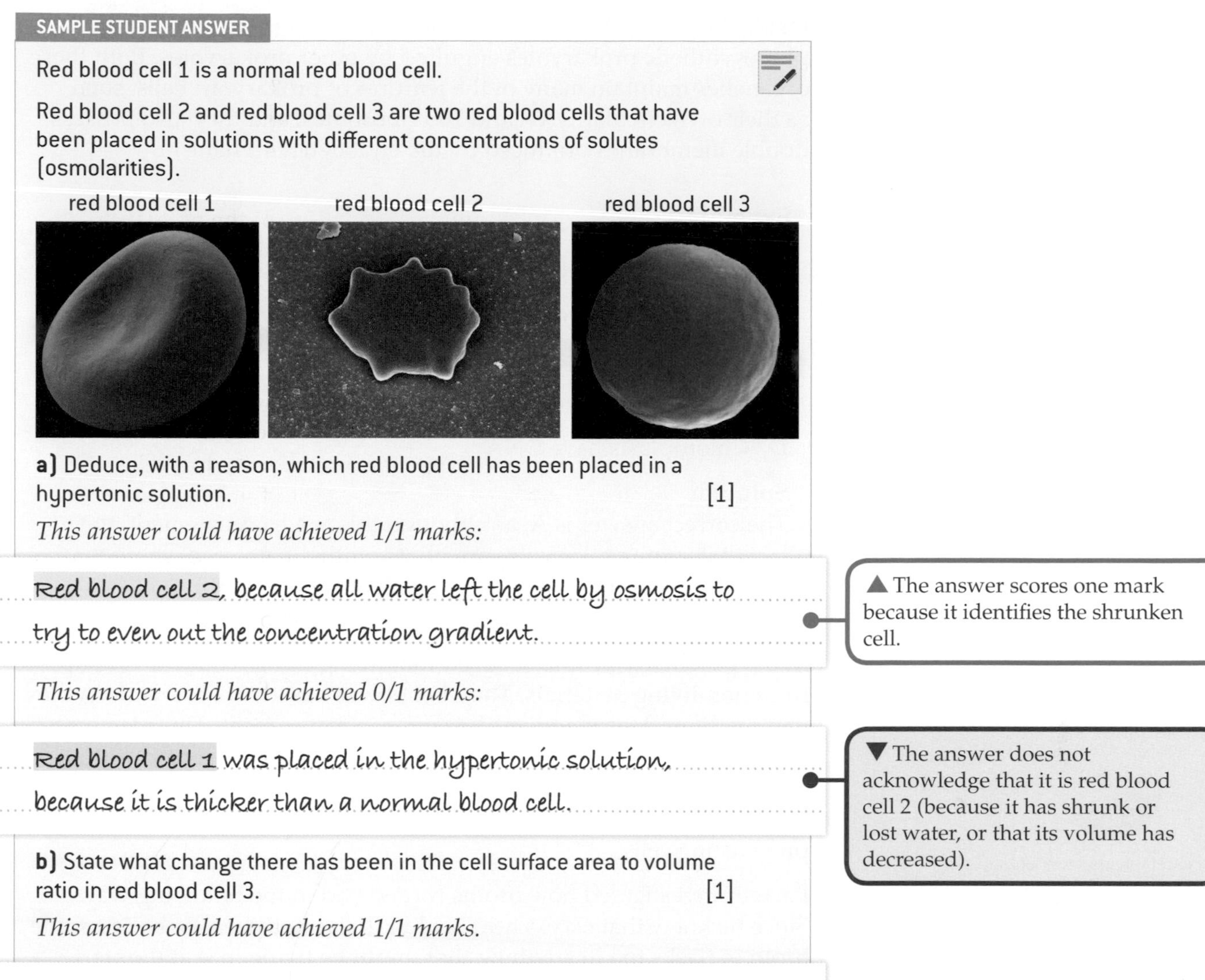

a) Deduce, with a reason, which red blood cell has been placed in a hypertonic solution. [1]

This answer could have achieved 1/1 marks:

Red blood cell 2, because all water left the cell by osmosis to try to even out the concentration gradient.

▲ The answer scores one mark because it identifies the shrunken cell.

This answer could have achieved 0/1 marks:

Red blood cell 1 was placed in the hypertonic solution, because it is thicker than a normal blood cell.

▼ The answer does not acknowledge that it is red blood cell 2 (because it has shrunk or lost water, or that its volume has decreased).

b) State what change there has been in the cell surface area to volume ratio in red blood cell 3. [1]

This answer could have achieved 1/1 marks.

It has decreased.

1.5 THE ORIGIN OF CELLS

You should know:

- ✔ cells can be formed only by division of pre-existing cells.
- ✔ the first cells must have arisen from non-living material.
- ✔ the endosymbiotic theory can explain the origin of eukaryotic cells.

You should be able to:

- ✔ analyse evidence from Pasteur's experiments that spontaneous generation of cells and organisms does not now occur on Earth.
- ✔ use modern apparatus to design an experiment that repeats Pasteur's experiment.

Stanley Miller and Harold Urey carried out experiments to show how the first cells might have arisen from non-living material. They passed steam through a mixture of methane, hydrogen and ammonia, representing the early Earth atmosphere. Electrical discharges were used to simulate lightning. They found that amino acids and other carbon compounds needed for life were produced.

In Topic 5.1 you will study the evidence for evolution.

The endosymbiotic theory explains the evolution of eukaryotic cells from prokaryotic cells. Mitochondria are thought to be aerobic prokaryotes that were engulfed by other prokaryotes and remained inside the cells. Likewise, chloroplasts are thought to have been photosynthetic prokaryotes engulfed by other prokaryotes. Both these organelles maintain many of the features of prokaryotic cells, such as their own circular DNA and 70S ribosomes, and they also possess double membranes produced by the endocytic mechanism.

Assessment tip

Negative questions are not very frequent, but you must look out for them, as they can lead to confusion.

Example 1.5.1.

Which is **not** evidence for the endosymbiotic theory?

A. Prokaryotes can carry out photosynthesis

B. Mitochondria have a double membrane

C. Chloroplasts have 70S ribosomes

D. Chloroplasts have DNA

Solution

The correct answer is **A**, as although this is a true statement, the fact that some prokaryotes are photosynthetic does not show that these were engulfed by other prokaryotes.

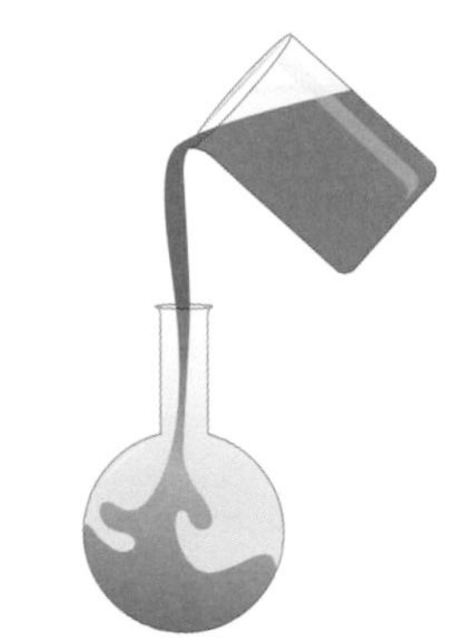

1. Pasteur poured nutrients into two flasks like this.

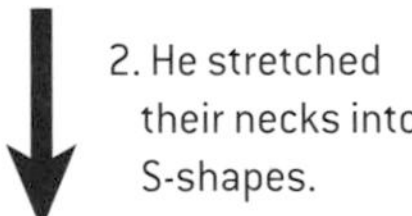

2. He stretched their necks into S-shapes.

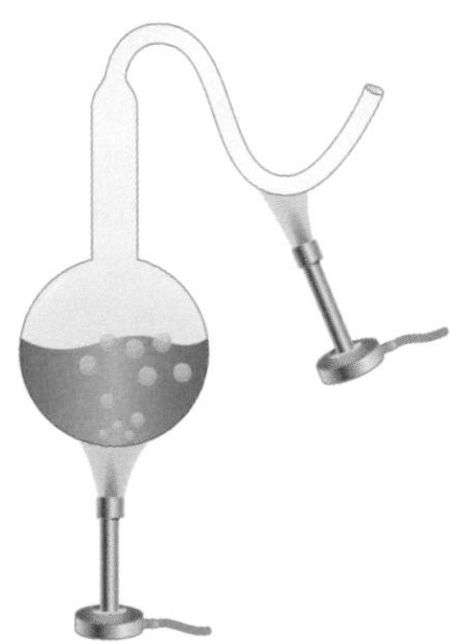

3. He boiled the nutrients.

▲ Figure 1.5.1. Pasteur's experiments

In the past many people believed that living things could spring up from non-living materials. Their belief was based on observations they had made, such as maggots appearing in rotten meat. This idea was called spontaneous generation. Pasteur's experiments with broth in swan-necked flasks were carried out to prove whether microbes could be spontaneously generated or whether they could come only from pre-existing cells.

Pasteur investigated how broths turned bad in the following way. Since he knew that excess heat could kill living things, he boiled some broth in flasks to kill anything that might be living in it at the start. He then heated the necks of the glass flasks until they were soft, and pulled them out into a long, thin, curving tube called a swan-neck. The broths in the flasks did not go bad. Then Pasteur broke open one of the flasks and exposed the broth to the open air and this time he noticed that the broth did go bad. This made Pasteur conclude that there was something in the air that was affecting broth. In the swan-necked flasks where the broth was clear, whatever was affecting the broth might have settled in the bend of the neck and therefore not reached the broth. To test his idea he tipped a swan-necked flask so that some of the broth went into the bend where dust and other particles may have collected, and then he tipped it back again. The broth in this second flask then went bad. Pasteur concluded that whatever was causing the change could be carried by air currents, but it must be heavier than air as it settled in the bend in the swan-neck.

Example 1.5.2.

What theory did Pasteur falsify with his experiments?

A. Independent assortment

B. Spontaneous generation

C. Endosymbiosis

D. Evolution

Solution

The correct answer is **B**, as with his experiments Pasteur showed that microorganisms could not grow in a broth unless dust particles (covered in microbes) were allowed into the flask.

1.6 CELL DIVISION

You should know:

- two genetically identical daughter nuclei are produced from the division of the nucleus during mitosis.
- chromosomes condense by supercoiling.
- interphase is a very active phase of the cell cycle with many processes occurring in the nucleus and cytoplasm.
- the stages of mitosis are prophase, metaphase, anaphase and telophase.
- cytokinesis is cell division occurring after mitosis.
- cyclins are involved in the control of the cell cycle.
- mutagens, oncogenes and metastasis are involved in the development of primary and secondary tumours.

You should be able to:

- analyse data to see the correlation between smoking and the incidence of cancers.
- identify the phases of mitosis in cells viewed with a microscope or in a micrograph.
- determine the mitotic index from a micrograph.

Sister chromatids are two parts of a chromosome attached to each other by a centromere in the early stages of mitosis. When sister chromatids have separated to form individual structures they are referred to as chromosomes.

- **Interphase** is the stage of the cell division before mitosis. Cells grow, forming organelles (G1 stage), DNA is duplicated (S stage) and synthesis of proteins that are involved in nuclear division occurs (G2 stage).
- **Mitosis** is nuclear division consisting of four stages: prophase, metaphase, anaphase and telophase.
- **Prophase** is the stage where chromatin condenses and associates with histones forming chromosomes, the nuclear membrane disappears and spindle fibres are formed. Chromosomes attach to spindle fibres.
- **Metaphase** is the stage where chromosomes are aligned in the equator of the cell.
- **Anaphase** is the stage where sister chromatids (V shape, with the vertex pointing to the poles) are separated to the opposite poles of the cell.
- **Telophase** is the last stage of mitosis, where a nuclear membrane forms around each set of chromosomes that begin to uncoil.
- **Cytokinesis** is a process that occurs along with telophase. The cytoplasm of the parental cell divides into two daughter cells. Cytokinesis in animals is produced by cell strangling while in plants it is by formation of a plate.

You will study how DNA replicates before cell division in Topic 2.7. You will also study how cells divide to produce gametes in Topic 3.3.

Assessment tip

When answering a question relating to mitosis, you need to be clear that mitosis is the nuclear division while cytokinesis is cell division.

Example 1.6.1.

a) State two processes occurring during prophase.

b) (i) Define 'centromere'.

(ii) Explain the reason centromeres are facing either opposite pole.

c) Define 'mitotic index'.

Sir Richard Timothy Hunt discovered a new protein in fertilized sea urchin eggs (*Arbacia punctulata*). Cyclin was synthesized soon after the eggs were fertilized and increased in levels during interphase, but the levels decreased quickly in the middle of mitosis. The fact that the amount of cyclins drops periodically at different cell division stages proved to be of importance for cell cycle control.

Solution

a) During prophase chromatin condenses and associates with histones forming chromosomes, the nuclear membrane disappears and spindle fibres are formed.

b) (i) The centromere is the structure of the chromosome (DNA) that holds together both chromatids. It is also the point of attachment to the spindle fibres.

(ii) During anaphase, the sister chromatids are pulled by the spindle fibres to opposite poles. As the centromere is attached to the spindle fibre, the centromere always goes first towards the pole. One sister chromatid migrates to one pole and the other sister chromatid migrates to the opposite pole.

c) Number of cells in mitosis divided by total number of cells seen under the microscope.

SAMPLE STUDENT ANSWER

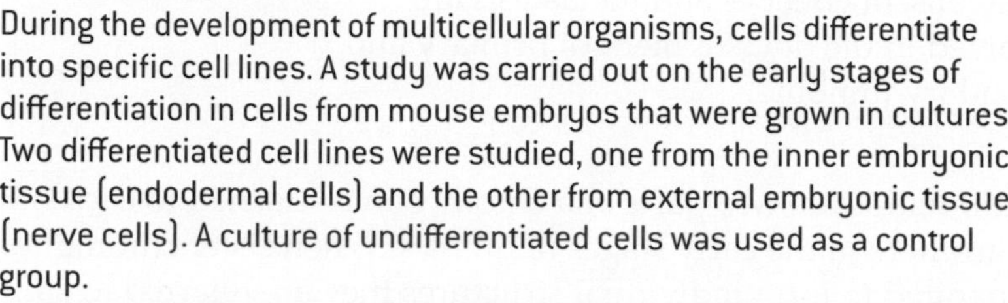

During the development of multicellular organisms, cells differentiate into specific cell lines. A study was carried out on the early stages of differentiation in cells from mouse embryos that were grown in cultures. Two differentiated cell lines were studied, one from the inner embryonic tissue (endodermal cells) and the other from external embryonic tissue (nerve cells). A culture of undifferentiated cells was used as a control group.

The role of regulators during cell differentiation was studied. After 96 hours of incubation, a sample was taken of each cell line and the cyclins separated by gel electrophoresis. The presence of different cyclins D1, D2 and D3 was analysed in the three cell lines. The image shows the results. The size and intensity of the bands is an indicator of the quantity of cyclins.

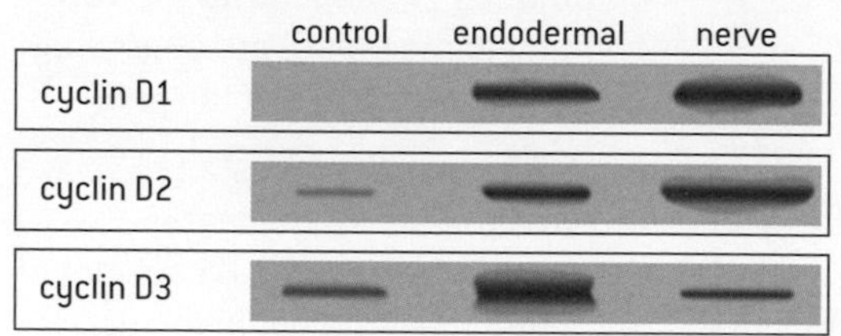

a) Compare and contrast the amounts of the different cyclins in nerve cells and control cells. [2]

This answer could have achieved 2/2 marks:

There is no cyclin D1 in control cells, while there is a very high amount in nerve cells. There is about three times as much cyclin D2 in nerve cells than control cells. Cyclin D3 is the most prominent of the cyclins in the control cells and the least prominent in the nerve cells. However, there is almost the same amount of cyclin D3 in nerve cells than control cells.

▲ This answer scores full marks because it compares all cyclins in both types of tissues.

b) Using the data, discuss the possible role of the three cyclins in the differentiation of nerve and endodermal cell lines. [3]

This answer could have achieved 2/3 marks:

As cyclin D3 is in greater quantity in the endodermal cells, it could indicate that this leads to an increase in differentiation. As both the endodermal and control cells have the same amount of cyclin D1, it may be responsible for the differentiation of undifferentiated cells, but not of nerve or endodermal cell lines. Cyclin D2 is more present in nerve cells than endodermal cells. This suggests that the higher presence of cyclin D2 lowers the rate of differentiation.

▲ The answer correctly mentions cyclin D3 in relation to endodermal differentiation, and D2 specific for nerve differentiation.

▼ The answer does not mention that cyclin D1 is most likely what causes differentiation as the control group contains none of it.

Assessment tip

As this is a discussion, the answer could have mentioned that there is limited data to determine roles of cyclins as there are very complex processes involved.

Practice problems for Topic 1

Problem 1

The electron micrograph shows a section of the epithelium of the small intestine.

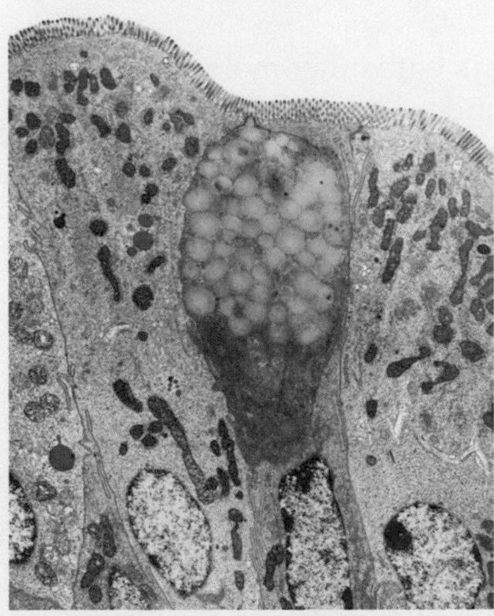

Identify **two** structures present in these cells that show they are involved in the uptake of food.

Problem 2

Glucose is a six-carbon sugar that provides energy needed by cells. Because glucose is a large molecule, it is difficult for it to be transported across the membrane through simple diffusion. Explain how glucose is transported into a cell such as a red blood cell down a concentration gradient.

Problem 3

a) Describe the experimentation that led to the proposal of the Davson–Danielli model of the cell membrane and its later falsification leading to the Singer–Nicolson model.

b) Outline the process of endocytosis.

Problem 4

Describe the experiment of Miller and Urey into the origin of organic compounds.

Problem 5

Cells go through a repeating cycle of events in growth regions such as plant root tips and animal embryos. Outline this cell cycle.

2 MOLECULAR BIOLOGY

2.1 MOLECULES TO METABOLISM

You should know:

- ✔ molecular biology describes living processes via the interactions of the chemical substances involved.
- ✔ carbon atoms form four covalent bonds with other atoms.
- ✔ living organisms are created from carbon compounds including carbohydrates, lipids, proteins and nucleic acids.
- ✔ metabolism refers to the enzyme-catalysed reactions in a cell or organism.
- ✔ anabolism is the synthesis of complex molecules from simpler molecules.
- ✔ catabolism is the breakdown of complex molecules into simpler molecules.

You should be able to:

- ✔ explain that some carbon compounds, such as urea, are produced by living organisms and are also synthesized artificially.
- ✔ draw molecular diagrams of glucose, ribose, a saturated fatty acid and a generalized amino acid.
- ✔ use molecular diagrams to identify sugars, lipids and amino acids.

In 1828, Friedrich Wöler synthesized urea, an organic carbon compound, from inorganic salts. Carbon compounds were called "organic" from the earlier belief that they had an inner vital force that could only be found in living organisms. This theory was called "vitalism". Because Wöler did not require any living organism to obtain urea, the theory of vitalism was falsified.

In Topic 1.3 you studied the molecules that form membranes. In Topics 8.1 and D.5 you will study how these molecules are used in metabolism.

Carbon atoms form four covalent bonds with other molecules, allowing the production of a variety of compounds. All living organisms are created from carbon compounds, including carbohydrates, proteins, lipids and nucleic acids. These compounds are used by living organisms in a complex web of chemical reactions called metabolism.

- **Carbohydrates** are molecules composed of carbon, hydrogen and oxygen which serve as immediate energy and for structural purposes. They have the general formula $C_n(H_2O)_n$.
- **Lipids** are molecules composed of carbon, hydrogen and, in a lesser amount, oxygen. They serve as an energy storage molecule and form the cell membrane. They can also act as hormones.
- **Proteins** are macromolecules formed from chains of amino acids. They contain carbon, hydrogen, oxygen and nitrogen. They can also contain sulfur. They have many different functions in living organisms.
- **Nucleic acids**, such as DNA and RNA, are molecules in charge of the genetic information in cells. They are composed of a nitrogenous base joined to a sugar and a phosphate group.

Example 2.1.1.

The diagram shows the structure of different biological molecules. Which diagram shows the structure of D-ribose?

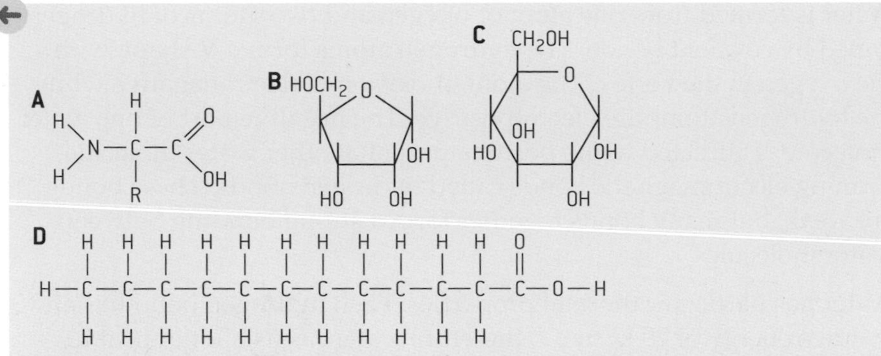

Solution

The correct answer is diagram **B**. Diagram **A** is a generalized formula for an amino acid; you can clearly see the acid group on the right, the amine group on the left containing nitrogen, and in the centre the α-carbon joined to the R group (side chain) that could be a single hydrogen atom or more complex groups of atoms. Diagram **C** is α-glucose, as can be seen from the six-membered ring and the hydroxyl group of carbon 1 facing downwards. Diagram **D** is a fatty acid, a long chain of carbons and hydrogens with an acid group at carbon 1.

>> **Assessment tip**

You should be able to identify biochemicals such as sugars, lipids or amino acids from molecular diagrams. You will only be asked to identify a molecule type from a series of molecular structures you already know.

- In **anabolism**, two or more molecules join to form a larger molecule.
- In **catabolism**, molecules are broken down into smaller molecules.

All living organisms perform metabolic reactions. These are a series of chemical and biological changes that require enzymes that are continually being produced in the cell. Anabolic and catabolic reactions allow organisms to obtain the chemicals and energy needed to perform the processes that allow them to live.

An example of anabolism is the formation of sucrose, common sugar. A six-carbon monosaccharide glucose molecule joins a six-carbon monosaccharide fructose molecule to form the disaccharide sucrose, with the loss of one molecule of water.

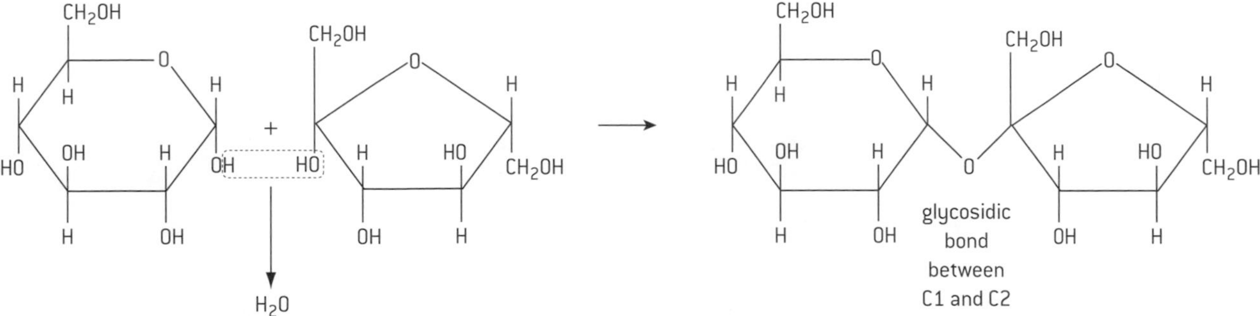

▲ **Figure 2.1.1.** The formation of sucrose (a disaccharide) from the condensation of glucose and fructose (two monosaccharides)

2.2 WATER

You should know:

- ✔ water molecules are polar, they have a positive end and a negative end.
- ✔ water's cohesive, adhesive, thermal and solvent properties can be explained by hydrogen bonding and dipolarity.
- ✔ hydrophilic substances are attracted to water (water-loving) and hydrophobic substances are repelled by water (water-fearing).

You should be able to:

- ✔ compare the thermal properties of water with those of methane.
- ✔ explain how water acts as a coolant in sweat.

Carbon compounds and carbon cycling will be studied in Topic 4.3. Topic 4.4 describes how carbon dioxide released from metabolic reactions causes climate change.

Water is formed from one atom of oxygen and two atoms of hydrogen joined by covalent bonds. The hydrogen atoms form a V shape with the oxygen in the vertex. The atom of oxygen is electronegative while the hydrogen atoms are electropositive. The negative part of one water molecule is attracted to the positive part of another water molecule, forming electromagnetic bonds called hydrogen bonds. These bonds are weak, but many bonds together form a strong bonding between water molecules.

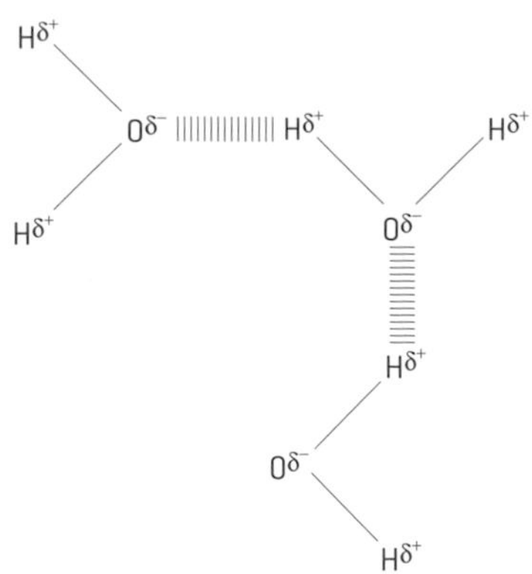

▲ Figure 2.2.1. Hydrogen bonding between two water molecules

Water has particular thermal properties. Each hydrogen bond has an average energy of 20 kJ mol^{-1}, therefore water needs a lot of heat to change state, causing the boiling and melting points to be high. Water has a high specific heat capacity, a high heat of vaporization and a high heat of fusion.

- **Heat capacity** is the energy required to raise the temperature of 1 g of substance by 1°C.
- **Heat of vaporization** is the energy required to change 1 g of liquid to vapour.
- **Heat of fusion** is the energy required to change 1 g of solid to liquid.

Example 2.2.1.

a) Complete the table to distinguish the solubility of molecules. Place a tick (✓) in the cases where the molecules are soluble in water.

Solution

a)

Molecule	glucose	amino acids	fats	cholesterol	oxygen	sodium chloride
Solubility in water	✓	✓			✓	✓

b) Explain the relationship between the solubility in water and the possibility of transport in blood plasma of these molecules.

Solution

b) Water-soluble substances can be transported in blood plasma which is composed mainly of water. Substances that are not soluble in water, such as lipids, need to be transported through vesicles or attached to soluble substances such as proteins. Cholesterol is an example of an insoluble molecule that has to be attached to proteins, forming high-density lipoprotein cholesterol, for transport.

2.3 CARBOHYDRATES AND LIPIDS

You should know:

- ✔ monosaccharide monomers are joined by condensation reactions to form disaccharides and polysaccharides.
- ✔ fatty acids can be saturated, monounsaturated or polyunsaturated.
- ✔ unsaturated fatty acids can be *cis* or *trans* isomers.
- ✔ triglycerides are formed by condensation from three fatty acids and one glycerol.

You should be able to:

- ✔ identify and analyse the structure and function of cellulose, starch and glycogen.
- ✔ compare and contrast cellulose, starch and glycogen with respect to their structure and function, using molecular visualization software.
- ✔ explain the health risks of *trans* fats and saturated fatty acids in the diet.
- ✔ compare lipids and carbohydrates in terms of energy storage.
- ✔ evaluate evidence and the methods used to obtain the evidence for health claims made about lipids.
- ✔ determine body mass index by calculation or use of a nomogram.

Monosaccharide monomers, such as glucose and fructose, can be linked together with glycosidic bonds to form disaccharides, such as sucrose. Because this reaction liberates water it is called a condensation reaction. Sucrose is found in plants. Lactose, the sugar found in milk, and maltose, the sugar found in seeds, are also disaccharides. Lactose is formed from glucose and galactose. Maltose is composed of two glucose molecules and is produced by the hydrolysis of starch.

Polysaccharide polymers are composed of long chains of monosaccharides. The most abundant polysaccharides in nature are starch and cellulose, which are found in plants, and glycogen, which is mainly found in animals. Starch is composed of two different polymers of α-glucose, amylose and amylopectin. Amylose is a linear molecule, while amylopectin is branched. In amylose, the glucose molecules are joined through α-1,4-glycosidic bonds. In amylopectin, these same bonds are formed in the main branch, but other α-glucose molecules are joined at carbon 6 forming α-1,6-glycosidic bonds.

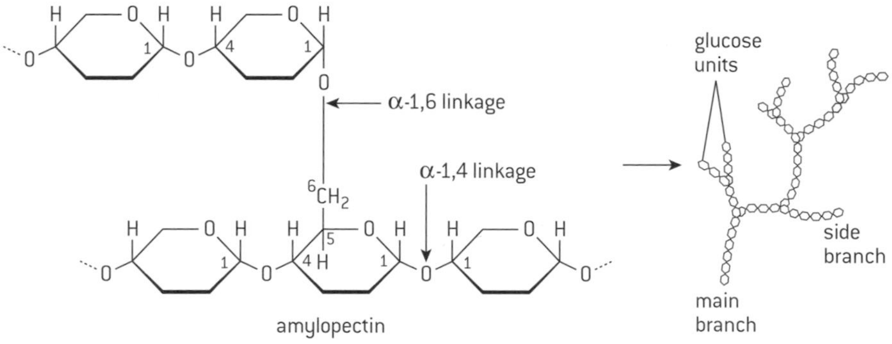

▲ **Figure 2.3.1.** Structure of starch

Cellulose is a polymer of β-glucose molecules joined by 1,4-glycosidic bonds. Because the hydroxyl group on carbon 1 is on the opposite side of the chain to the hydroxyl group on carbon 4, every second glucose is upside down in order to position these two hydroxyl groups together. This makes cellulose a strong molecule, difficult to digest in the human body. Many cellulose molecules join together to form microfibres. The fibres found in the cell walls of plants are made from these cellulose microfibres.

Example 2.3.2.

a) Compare and contrast the structure of glycogen and starch.

b) Explain how the bonding in starch and cellulose molecules affects their structure.

Solution

a) Glycogen, found mostly in animals, and starch, found in plants, are polymers of glucose molecules. Starch is formed by two molecules, amylose and amylopectin. Amylose is a linear molecule while amylopectin is branched. In both starch and glycogen, the bond joining the main chain of glucose molecules is an α-1,4-glycosidic bond. The branching occurs at carbon 6 with an α-1,6-glycosidic bond. Glycogen is also formed from branched molecules, but in glycogen the branching is greater than in starch, therefore the molecule is more compact.

b) Starch is formed by α-glucose while cellulose is formed by β-glucose. In α-glucose the hydroxyl group on carbon 1 is on the same side of the chain as the one on carbon 4, allowing the

- **Carbohydrates** are carbon compounds containing hydrogen and oxygen. They serve as energy storage and structural components of organisms.
- **Monosaccharides** are simple sugars and are the basic building blocks of carbohydrates.
- **Disaccharides** are two monosaccharides joined together.
- **Polysaccharides** are polymers of monosaccharides.

Example 2.3.1.

A disaccharide is broken down into glucose and fructose.

a) State the name of the disaccharide.

b) Identify the type of bond present in the disaccharide.

Solution

a) Sucrose

b) Glycosidic bond (formed by condensation)

Lipids in the diet have been said to be bad for human health. Nevertheless, they are necessary for many metabolic processes. Health problems linked to dietary lipids include obesity and coronary heart disease.

- **Lipids** are carbon compounds containing mainly hydrogen and some oxygen. They form a diverse group containing oils, fats, membrane components and hormones.
- **Fatty acids** are formed by long chains of carbons (4 to 34) with hydrogens attached and an acid group at one end.
- **Unsaturated** fatty acids have double or triple bonds between two or more carbons.
- **Saturated** fatty acids do not have any double or triple bonds.
- ***Cis* fatty acids** have the two hydrogen atoms adjacent to the double bond located on the same side of the chain.
- ***Trans* fatty acids** have the two hydrogen atoms adjacent to the double bond located on opposite sides of the chain.

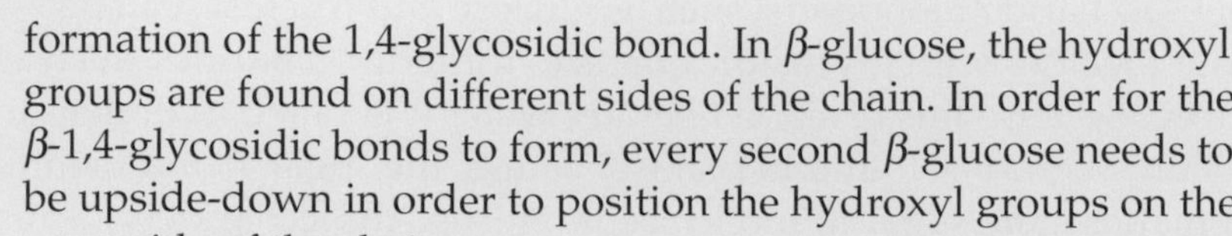
formation of the 1,4-glycosidic bond. In β-glucose, the hydroxyl groups are found on different sides of the chain. In order for the β-1,4-glycosidic bonds to form, every second β-glucose needs to be upside-down in order to position the hydroxyl groups on the same side of the chain.

Lipids are carbon compounds such as glycerides and steroids that are insoluble in water. They contain double the energy of carbohydrates. Fatty acids are the main components of simple lipids. They are formed by a long chain of carbons and hydrogens with an acid group at carbon 1. The carbons can be joined by single or double bonds. When the fatty acid has no double bonds it is said to be saturated; if it has one or more double bonds it is said to be unsaturated. Monounsaturated fatty acids have only one double bond while polyunsaturated fatty acids have two or more double bonds.

Unsaturated fatty acids can have the hydrogen atoms next to the double bond on the same side or on opposite sides of the chain. If the hydrogen atoms are on the same side of the chain, the unsaturated fatty acid is said to be *cis*; if they are on opposite sides of the chain it is *trans*. The tail in *cis* fatty acids kinks, separating the molecules. This gives *cis* fatty acids a lower melting point than *trans fatty acids*.

There are many types of lipids. Some have a simple structure, such as the monoglycerides, formed from glycerol and one fatty acid, while others have very complex molecules, such as steroids. You need to know only the structure of triglycerides. Triglycerides are formed by condensation from three fatty acids and one glycerol molecule. These fatty acids do not need to be the same. In some cases, one carbon joins to a phosphate group instead of a fatty acid, forming a phospholipid. This phosphate group can at the same time join to another group, for example choline, forming phosphatidylcholine, a phospholipid that forms an important part of the cell membrane.

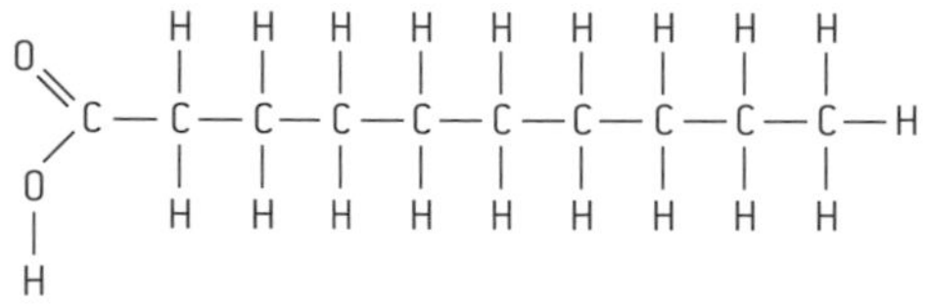

saturated fatty acid

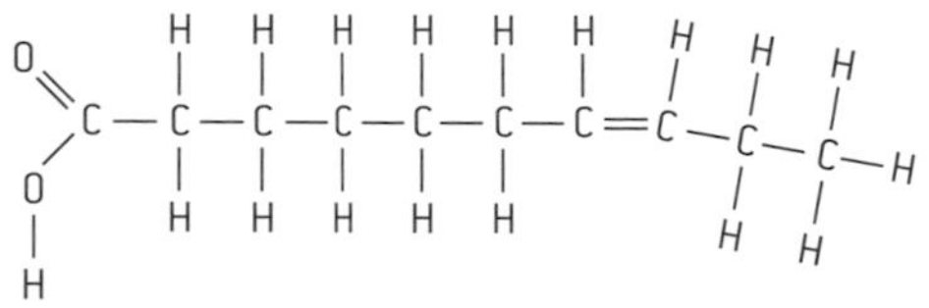

unsaturated fatty acid

▲ **Figure 2.3.2.** Saturated fatty acid with no carbon–carbon double bonds and unsaturated fatty acid showing one carbon–carbon double bond

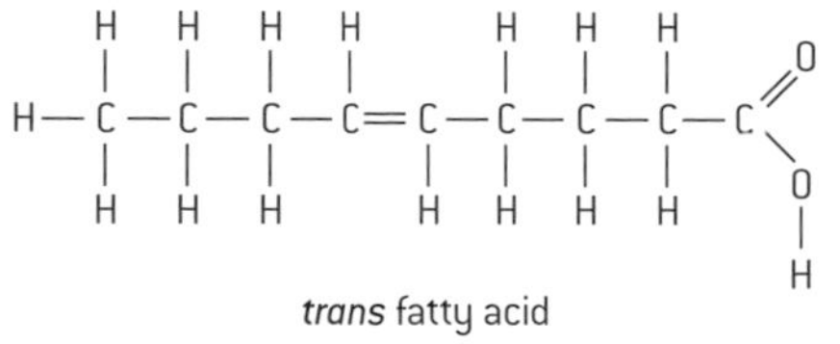

trans fatty acid

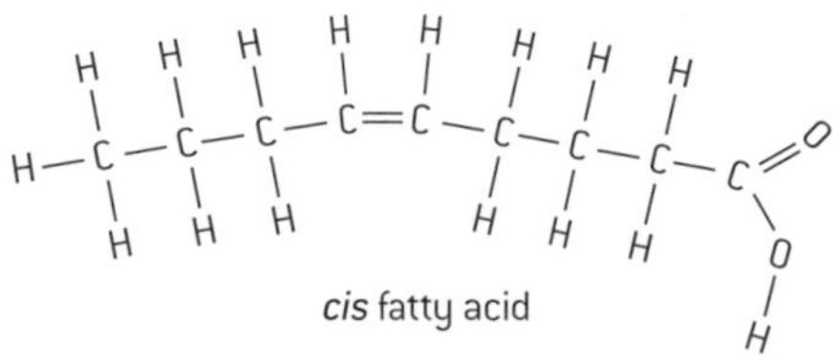

cis fatty acid

▲ **Figure 2.3.3.** *Cis* and *trans* fatty acids

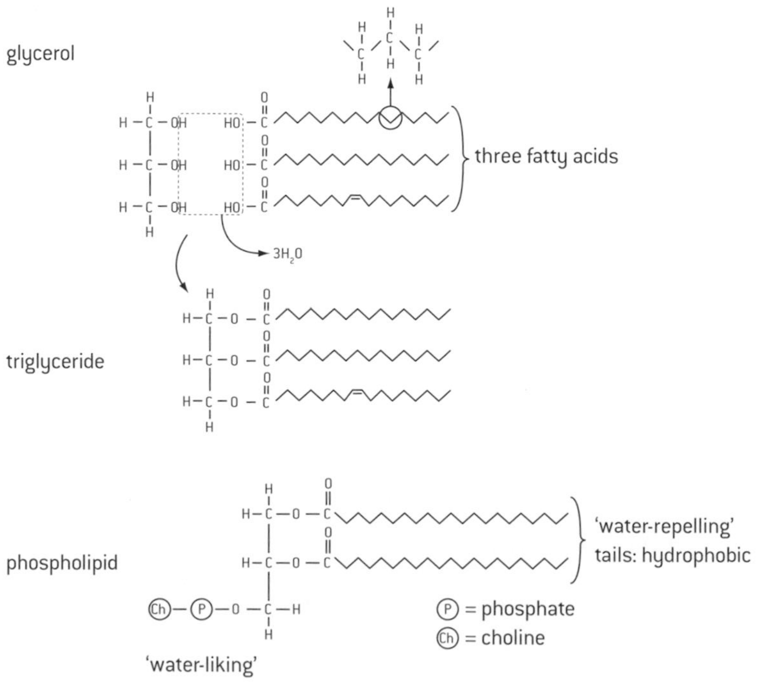

glycerol

three fatty acids

$3H_2O$

triglyceride

phospholipid

'water-repelling' tails: hydrophobic

Ⓟ = phosphate

Ⓒh = choline

'water-liking'

▲ **Figure 2.3.4.** Formation of triglycerides and structure of a phospholipid

Example 2.3.3.

Explain the health risks of *trans* fats and saturated fatty acids in the diet.

Solution

Triglycerides formed from saturated fatty acids or from *trans* unsaturated fatty acids (fats made with these fatty acids are called *trans* fats) form rods that pack tightly together, forming a high density of intermolecular contacts. Hydrocarbons with *cis* double bonds or branches are irregular in shape and cannot pack so tightly. High levels of saturated or *trans* fats may contribute to hardening of the arteries (arteriosclerosis) or the formation of plaques or atheroma on the artery walls (atherosclerosis). An atheroma consists mainly of macrophages, lipids and connective tissue. It forms a swelling in the artery wall which slows down or even blocks the flow of blood. The loss of blood flow to the heart can result in coronary heart disease, a heart attack or myocardial infarction, or other cardiovascular diseases.

SAMPLE STUDENT ANSWER

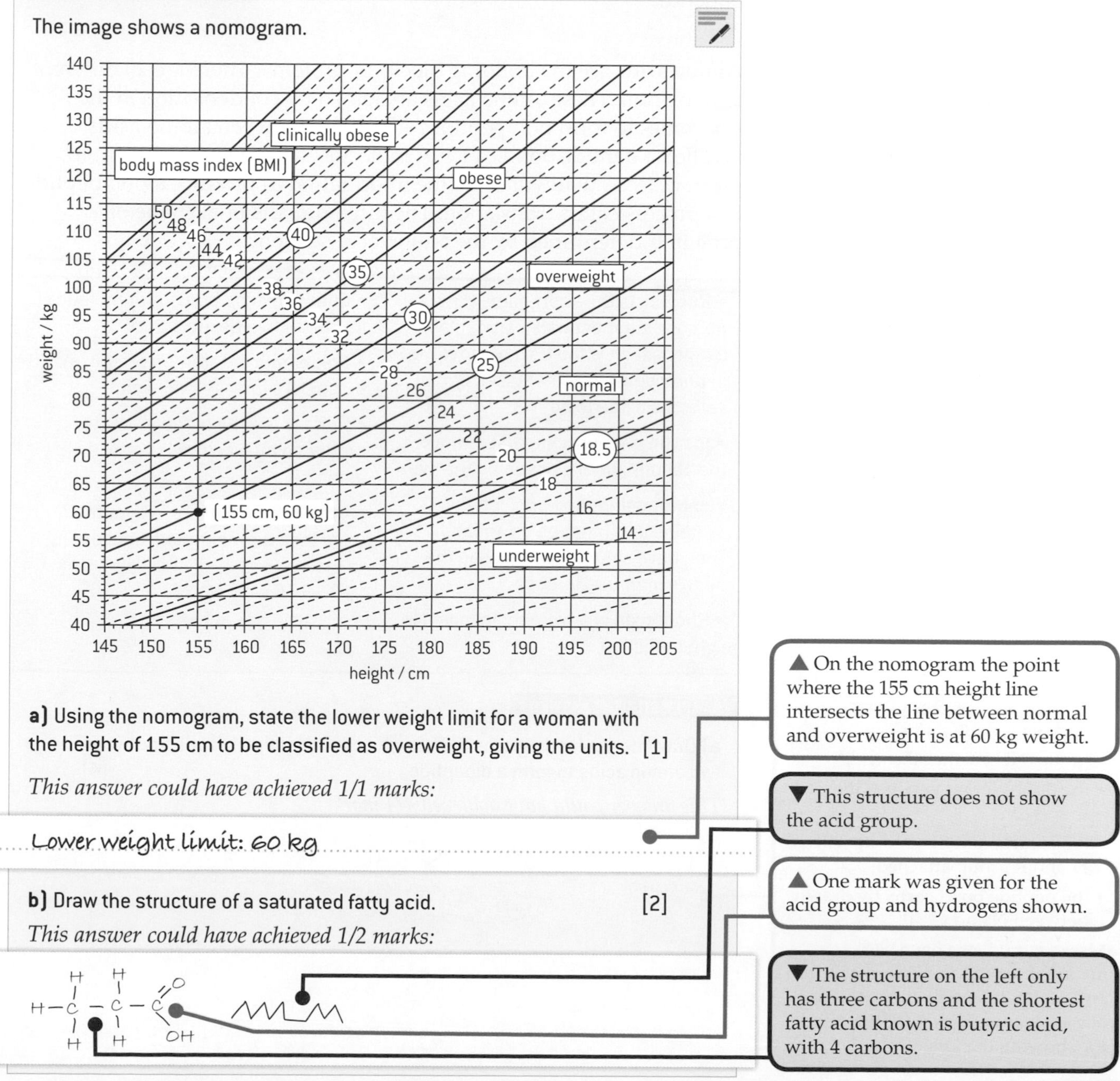

The image shows a nomogram.

a) Using the nomogram, state the lower weight limit for a woman with the height of 155 cm to be classified as overweight, giving the units. [1]

This answer could have achieved 1/1 marks:

b) Draw the structure of a saturated fatty acid. [2]

This answer could have achieved 1/2 marks:

▲ On the nomogram the point where the 155 cm height line intersects the line between normal and overweight is at 60 kg weight.

▼ This structure does not show the acid group.

▲ One mark was given for the acid group and hydrogens shown.

▼ The structure on the left only has three carbons and the shortest fatty acid known is butyric acid, with 4 carbons.

2.4 PROTEINS

You should know:

- ✔ polypeptides are formed from amino acids linked together by condensation.
- ✔ polypeptides are synthesized in ribosomes from 20 different amino acids.
- ✔ a huge range of polypeptides is possible because the amino acids can be linked together in any sequence.
- ✔ genes code for the amino acid sequence of polypeptides.
- ✔ a protein may consist of one polypeptide or of multiple polypeptides linked together.
- ✔ the three-dimensional conformation of a protein is determined by the amino acid sequence.
- ✔ proteins have a very wide range of functions in living organisms.
- ✔ a proteome is the full set of proteins in an individual; it is unique to each individual.

You should be able to:

- ✔ describe the functions of rubisco, insulin, immunoglobulins, rhodopsin, collagen and spider silk.
- ✔ explain how heat and changes to pH can denature a protein.
- ✔ draw molecular diagrams to show the formation of a peptide bond.

The synthesis of proteins through translation will be studied in Topic 2.7. The industrial use of organisms to produce proteins will be discussed in Topic B.1.

Amino acids are the building blocks of proteins. There are 20 different amino acids. These amino acids are joined by condensation in the ribosomes to form polypeptides. The bond between amino acids is called a peptide bond. A DNA molecule is used as a blueprint to produce RNA that will carry the information to produce a polypeptide in a process called translation. The three-dimensional shape of a protein is determined by the amino acid sequence.

Most organisms assemble proteins from the same 20 amino acids. However, pyrrolysine and selenocysteine, which were discovered fairly recently, are two additional amino acids that are not present in all organisms, and they are genetically encoded in a different way from the other 20.

- **Rubisco** is the short name for ribulose bisphosphate carboxylase/oxygenase. It is the enzyme involved in photosynthesis that takes part in carbon dioxide fixation.
- **Insulin** is a hormone that promotes the absorption of glucose by the liver.
- **Immunoglobulins** are globular proteins that function as antibodies. They have a number of roles in the body's immune defence.
- **Rhodopsin** is a pigment involved in light detection.
- **Collagen** is a fibrous protein found in connective tissues.
- **Spider silk** is a fibrous protein made by spiders to spin their web.
- **Proteome** is all of the proteins produced by a cell, a tissue or an organism.
- **Gel electrophoresis** is a method used to separate proteins according to their size.

SAMPLE STUDENT ANSWER

a) Draw molecular diagrams to show the condensation reaction between two amino acids to form a dipeptide. [4]

This answer could have achieved 4/4 marks:

```
H         H    O        H       H    O
 \        |   //         \      |   //
  N  -  C  - C     +      N  -  C - C
 /        |    \ OH  H /        |    \
H         R                     R1    OH

                   ↓

H        H   O    H    H    O
 \       |   ||   |    |   //
H - N  - C - C  - N  - C - C         + H2O
         |             |    \
         R             R1    OH
```

▲ The answer scored a mark for showing:

- each amino acid with an acid group (COOH) at one end and an amine group (NH_2) at the other end;
- the double bond between the C and O;
- the α-carbon in the middle with H and R group attached;
- the peptide bond correctly drawn between N and C = 0.

Marks would also have been awarded for the acid group at one end of the dipeptide and the amine at the other end, plus another mark for showing the loss of water.

Example 2.4.1.

Explain how heat can denature a protein.

Solution

Heat can modify the three-dimensional structure of a protein, thereby affecting its function. In the case of an enzyme such as rubisco, it would cease to catalyse the reaction. When temperature is increased, the increased vibrations within the molecule can cause the interactions between the R groups of different amino acids to be broken, changing the structure of the protein.

Assessment tip

It is possible to provide more correct answers than there are marks available in the question. If you write a good answer, it won't matter if you missed something on the mark scheme because you can still get full marks for another correct point.

To detect the proteome of a tissue, proteins are extracted then separated in a gel electrophoresis run.

2.5 ENZYMES

You should know:

- ✔ enzymes have an active site to which specific substrates bind.
- ✔ molecular movement and collision of the enzyme's active site with the substrate are required for catalysis.
- ✔ the rate of activity of enzymes depends on temperature, pH and substrate concentration.
- ✔ enzymes can be denatured.
- ✔ enzymes can be immobilized to be used in industry.

You should be able to:

- ✔ design experiments to test the effect of temperature, pH and substrate concentration on the activity of enzymes.
- ✔ sketch graphs to show the expected effects of temperature, pH and substrate concentration on the activity of enzymes.
- ✔ describe methods to produce lactose-free milk and its advantages.

Enzymes control cell metabolism. They are globular proteins that have a catalytic zone called the active site. This site can consist of around three to twelve amino acids. Changes in temperature and pH can alter the spatial disposition of the protein, affecting the rate of reaction of the enzyme. If the change is extreme, the protein can denature, leaving the enzyme unable to work.

How enzymes are used in metabolism, cell respiration and photosynthesis will be discussed in Topic 8.

- An **enzyme** is a globular protein that acts as a biological catalyst.
- A **substrate** is the molecule changed by the enzyme.
- The **active site** is the part of the enzyme that binds to the substrate.
- **Denaturation** is the loss of the tertiary structure of the enzyme.
- **Optimum** describes the ideal conditions required for an enzyme to work.

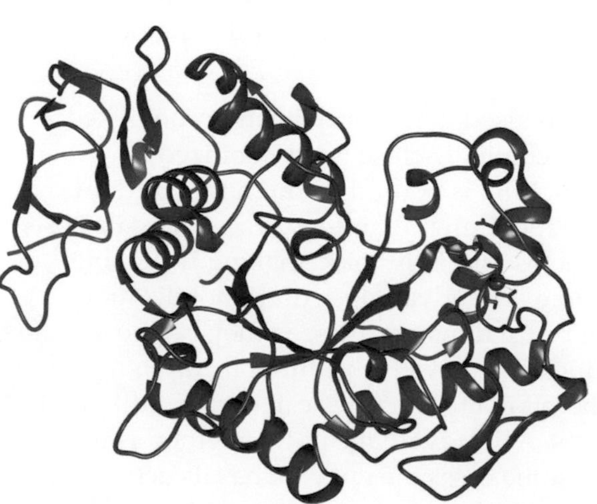

▲ Figure 2.5.1. Structure of an enzyme showing its active site in red

SAMPLE STUDENT ANSWER

Keratin is a protein found in hair, nails, wool, horns and feathers. The graphs show the relative keratinase activity obtained in experiments of keratin digestion at different pH values and at different temperatures.

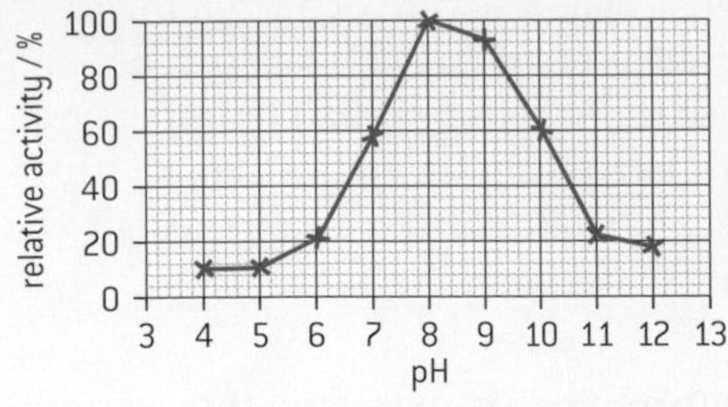

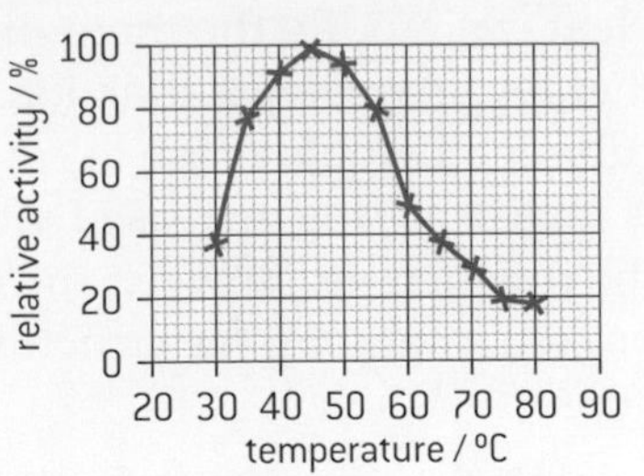

a) Determine the optimum pH and temperature of keratinase. [1]

This answer could have achieved 1/1 marks:

Optimum pH = 8.

Optimum temperature: 45°C.

▲ The student identified the optimum temperature in the range 44 to 48°C and the pH between 7.8 and 8.5.

b) Suggest **two** changes occurring in the reaction vessel that could be used to indicate keratinase activity. [2]

This answer could have achieved 2/2 marks:

The change in mass of keratin.

The change in mass of the newly produced amino acids.

▲ Marks could be scored for mentioning changes in colour or absorbance or any chemical changes indicating the presence of amino acids.

c) State two conditions that should be kept constant in both experiments. [2]

This answer could have achieved 2/2 marks:

The mass of keratinase used.

The time of keratinase-keratin reaction.

▲ The answers are correct. The amount of buffer would have also scored a mark, but not temperature or pH, as these were the variables of the experiment.

Increasing substrate concentration will increase the rate of reaction of enzymes, but only up to a point, as all the active sites will become occupied and the enzyme will not be able to react any faster.

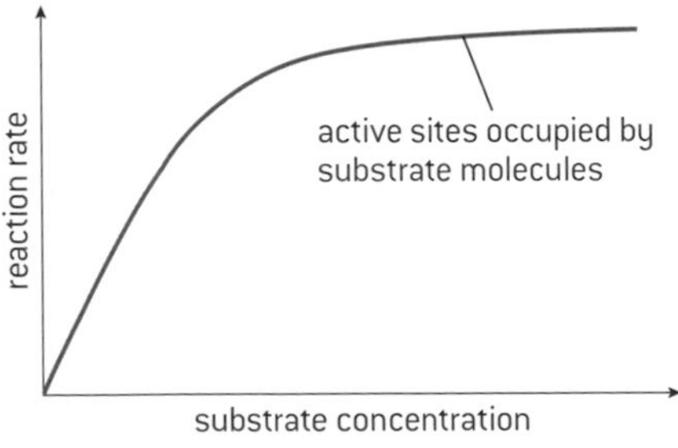

▲ **Figure 2.5.2.** Effect of the substrate concentration on the rate of reaction of an enzyme

Example 2.5.1.

Describe a method to obtain a **named** immobilized enzyme, pointing out the advantages and disadvantages of the method.

Solution

Lactase is an enzyme widely used in industry in an immobilized form to produce lactose-free milk. Lactase is immobilized in alginate beads in a column. Milk is poured through this column. Lactase digests the lactose found in milk into glucose and galactose, reducing the amount of lactose sugar in milk, making it suitable for lactose-intolerant people. The advantages of immobilization are that the enzymes can be used many times, as they can be separated from the product. In many cases, immobilization stabilizes the enzyme, therefore the reactions can occur at different temperatures or pH values. A disadvantage of the method is that it is more expensive and in some cases the enzyme has its active site sequestered in the matrix, reducing its reactivity.

2.6 STRUCTURE OF DNA AND RNA

You should know:

✔ the nucleic acids DNA and RNA are polymers of nucleotides.

✔ DNA differs from RNA in the number of strands present, the base composition and the type of pentose.

✔ DNA is a double helix composed of two antiparallel strands of nucleotides linked by hydrogen bonding between complementary base pairs.

You should be able to:

✔ explain how Crick and Watson elucidated the structure of DNA using model making.

✔ draw simple diagrams of the structure of single nucleotides of DNA and RNA, pentoses and bases.

DNA is a chain of nucleotides forming a double helix. The strands are aligned in an antiparallel manner and are held together by hydrogen bonds between complementary nitrogenous bases, C with G and A with T.

The genetic modification of DNA and its use in biotechnology will be described in Topic 3.5. In Topic 7 the structure of nucleic acids will be studied in more detail.

SAMPLE STUDENT ANSWER

Label the parts of two paired nucleotides in the polynucleotide of DNA. [3]

This answer could have achieved 2/3 marks:

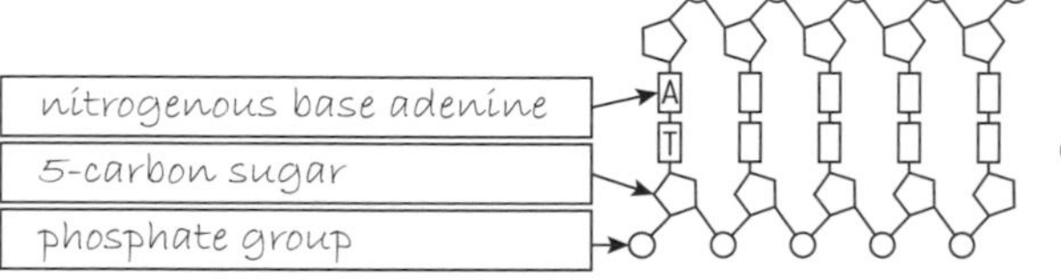

▼ "5-carbon sugar" is too vague for a mark. The correct answer is deoxyribose.

- **DNA (deoxyribonucleic acid)** is a polymer of nucleotides containing the genetic instructions responsible for inheritance. The segments carrying the information are called genes.
- **RNA (ribonucleic acid)** is a polymer of nucleotides containing the genetic instructions for protein synthesis.
- A **nucleotide** is a molecule containing a phosphate, a sugar and a nitrogenous base.
- The **nitrogenous bases** adenine (A), guanine (G), cytosine **(C)** and thymine (T) are found in DNA; uracil (U) replaces thymine in RNA.
- **Complementary** nitrogenous bases are linked by hydrogen bonds on opposite strands of DNA or double-stranded RNA.

Assessment tip

Use circles, pentagons and rectangles to represent phosphates, sugars and bases respectively. The two DNA strands should be shown antiparallel.

Crick and Watson created models to discover the structure of DNA. They used information from Rosalind Franklin's X-ray diffraction images of DNA and Erwin Chargaff's measurements of the proportions of each of the four bases.

Example 2.6.1.

The table shows the ratio of nucleotides in the DNA of different organisms.

Source	A to G	A to T	C to T	C to G
human	1.29	1.04	0.57	1.00
salmon	1.43	1.02	0.70	0.98
wheat	1.22	1.00	0.85	1.03
E. coli	1.05	1.09	1.05	1.01

a) State the two nucleotide ratios that are closest to 1 for most organisms.

b) Discuss whether this data supports the hypothesis that adenine (A) always pairs with thymine (T) and cytosine (C) pairs with guanine (G) by complementary base pairing in the DNA structure.

Solution

a) A to T and C to G.

b) DNA is a double-stranded helix. Each base in one strand is paired with its complementary base in the second strand; there are no unpaired nucleotides. The ratios A to G and C to T are not close to 1, while A to T and C to G are. This means that the total amount of A in the DNA of these organisms is similar to the total amount of T. The same happens with C and G. This supports the idea that they are joining by complementary base pairing.

2.7 DNA REPLICATION, TRANSCRIPTION AND TRANSLATION

You should know:

- ✔ DNA replication is semi-conservative and depends on complementary base pairing.
- ✔ helicase unwinds the double helix and separates the two strands by breaking hydrogen bonds.
- ✔ DNA polymerase uses the DNA strand as a template in replication.
- ✔ mRNA is copied from the DNA base sequences by RNA polymerase in transcription.
- ✔ synthesis of polypeptides occurs on ribosomes in translation.
- ✔ mRNA determines the amino acid sequence of polypeptides according to the genetic code.
- ✔ codons of three bases on mRNA correspond to one amino acid in a polypeptide.
- ✔ translation depends on complementary base pairing between codons on mRNA and anticodons on tRNA.

You should be able to:

- ✔ explain the processes of replication, transcription and translation.
- ✔ explain how Taq DNA polymerase produces multiple copies of DNA by the polymerase chain reaction (PCR).
- ✔ explain the universality of the genetic code using the production of human insulin in bacteria as an example of gene transfer between species.
- ✔ deduce the DNA base sequence for the mRNA strand.
- ✔ deduce which codon(s) corresponds to which amino acid in the genetic code.
- ✔ analyse Meselson and Stahl's results to obtain support for the theory of semi-conservative replication of DNA.

In Topic 3.5 the importance of DNA replication, transcription and translation in genetic modification and biotechnology will be studied. Transcription and gene expression will be studied in more detail in Topic 7.2; translation will be covered in Topic 7.3.

The central principle of genetics is that DNA forms RNA which then makes proteins. Therefore DNA carries all the information for inheritance. In replication, DNA polymerase uses the DNA strand as a template to link nucleotides together to form a new strand. Replication is semi-conservative because one strand is an original strand and the other is a new strand.

Example 2.7.1.

The first, second and third positions of the three bases of the codons on the messenger RNA (mRNA) are shown in this genetic code.

		Second letter									
		U		C		A		G			
	U	UUU	Phe	UCU		UAU	Tyr	UGU	Cys	U	
		UUC		UCC	Ser	UAC		UGC		C	
		UUA	Leu	UCA		UAA	Stop	UGA	Stop	A	
		UUG		UCG		UAG	Stop	UGG	Trp	G	
	C	CUU		CCU		CAU	His	CGU		U	
		CUC	Leu	CCC	Pro	CAC		CGC	Arg	C	
		CUA		CCA		CAA	Gln	CGA		A	
First letter		CUG		CCG		CAG		CGG		G	Third letter
	A	AUU		ACU		AAU	Asn	AGU	Ser	U	
		AUC	Ile	ACC	Thr	AAC		AGC		C	
		AUA		ACA		AAA	Lys	AGA	Arg	A	
		AUG	Met	ACG		AAG		AGG		G	
	G	GUU		GCU		GAU	Asp	GGU		U	
		GUC	Val	GCC	Ala	GAC		GGC	Gly	C	
		GUA		GCA		GAA	Glu	GGA		A	
		GUG		GCG		GAG		GGG		G	

A sequence of nucleotides of mRNA is shown:

AUGUUUCCACUAAAUGUAAGCAGA

a) **(i)** Identify the DNA sequence that has been transcribed to produce this mRNA.

(ii) Deduce the peptide sequence translated from this mRNA sequence.

b) Discuss whether all proteins contain Met as a first amino acid when leaving the ribosomes.

c) **(i)** Explain what would have happened if instead of AGA you had UGA in the last codon of this portion of this mRNA.

(ii) Suggest a reason how the change in c) (i) could have happened.

Solution

a) **(i)** TAC AAA GGT GAT TTA CAT TCG TCT

(ii) Met Phe Pro Leu Asn Val Ser Arg

b) The start codon of most open reading frames (ORF) is AUG. Transcription starts at TAC (or ATG in the DNA coding sequence) as this is where the polymerase starts transcription. There are some cases of alternative start codons, but these are very rare in eukaryotes.

c) **(i)** UGA is a stop codon, so the translation would be stopped here. There is no tRNA capable of joining this codon, so the peptide leaves the ribosome. The peptide produced would be short.

(ii) A point mutation replacing A with U could have caused the change.

Example 2.7.2.

In what way are enzymes and DNA related?

A. enzymes produce RNA during replication

B. the structure of enzymes is determined by DNA

C. both enzymes and DNA have globular shapes

D. enzymes contain the code for DNA

Solution

The answer is **B**, as DNA contains the blueprint for the synthesis of mRNA, which will determine the primary structure of the protein, which will determine the structure of the enzyme.

- **Replication** is the semi-conservative synthesis of DNA.
- **Transcription** is the synthesis of mRNA copied from the DNA base sequences by RNA polymerase.
- **Translation** is the synthesis of polypeptides on ribosomes.

Assessment tip

To identify the DNA sequence, you have to find the complementary base: for A it is T and for C it is G. Remember that in DNA you have T instead of U.

Assessment tip

To deduce the peptide sequence, you first separate the sequence of mRNA every three nucleotides. You then look in the table for A as the first letter, U for the second letter and G for the third. This is done for each codon.

2.8 CELL RESPIRATION

You should know:

- ✔ cell respiration is the controlled release of energy from organic compounds (such as glucose) to produce ATP.
- ✔ ATP is an immediate source of energy in the cell.
- ✔ anaerobic cell respiration yields a small amount of ATP.
- ✔ aerobic cell respiration requires oxygen and gives a large yield of ATP.

You should be able to:

- ✔ describe the use of yeasts in baking to produce ethanol and carbon dioxide through anaerobic respiration.
- ✔ explain lactate production in humans when anaerobic respiration is used to maximize the power of muscle contractions.
- ✔ analyse results from experiments using a respirometer.

Assessment tip

You do not need to know all the metabolic pathways of cell respiration. You need to recall only substrates and final waste products.

Most living processes require energy, which is obtained from cell respiration through the breakdown of glucose molecules. In anaerobic respiration, glucose is broken down to produce a small amount of ATP. The end products in anaerobic respiration can be lactic acid (in muscles or bacteria) or carbon dioxide and ethanol (in yeast or bacteria). In aerobic respiration, large amounts of ATP are produced from the breakdown of glucose; carbon dioxide and water are also produced.

- **Anaerobic respiration** is the catabolic process through which small amounts of energy in the form of ATP are produced from glucose without the use of oxygen.
- **Aerobic respiration** is the catabolic process involving the use of oxygen through which glucose is broken down to carbon dioxide, water and large amounts of energy in the form of ATP.
- **ATP** or **adenosine triphosphate** is a molecule that is used as a fast source of energy. It is formed by a ribose that has a nitrogenous base joined to carbon 1 and three phosphate groups joined to carbon 5.

In Topic 1.2, the structure of mitochondria were studied. How respiration is involved in carbon cycling will be studied in Topic 4.3. In Topic 8.2, cell respiration will be explained in more detail.

Example 2.8.1.

Describe the use of a respirometer to measure cell respiration in germinating seeds.

Solution

In a simple respirometer, germinating seeds are placed in a tube on a mesh. Underneath, an alkali such as sodium hydroxide is used to absorb CO_2. This ensures that reductions in volume are due to oxygen use. Temperature is kept constant using a water bath. This prevents volume changes due to temperature fluctuations.

Assessment tip

You can illustrate your answer with a drawing, but you must make sure to annotate it clearly.

SAMPLE STUDENT ANSWER

The oxygen consumption rate of the fish *Oplegnathus insignis* was examined in a respirometer at three different water temperatures and at four different body masses.

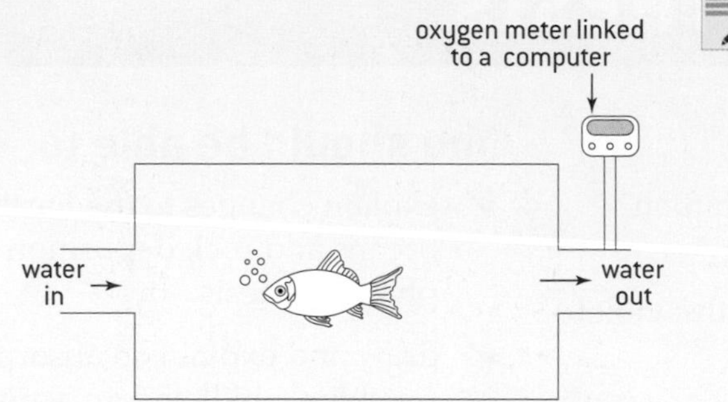

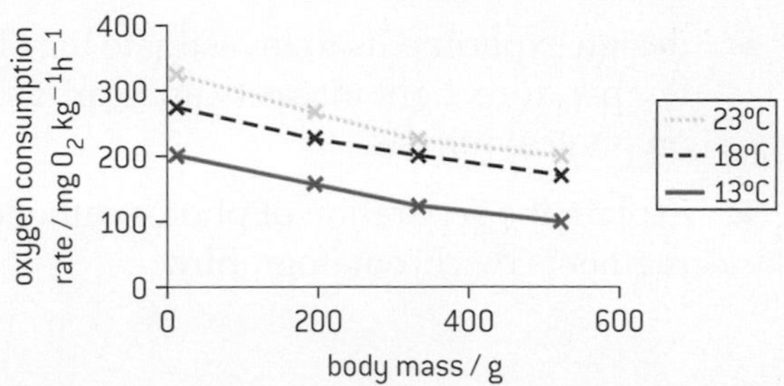

a) Suggest how the oxygen consumption rate is determined using this apparatus. [2]

This answer could have achieved 1/2 marks:

The "water out" part of the apparatus contains water from which the fish has consumed oxygen. The water is passed to the oxygen meter, which detects oxygen levels in the water. The result is compared with the oxygen level of the supplied water to calculate the rate using the computer.

▲ The answer gets one mark because it mentions that the data logger measures the difference in oxygen concentration.

▼ The answer is lacking the fact that the measurement is done over time and that the mass of the fish needs to be measured to calculate the rate.

b) State the relationship between body mass and the oxygen consumption of fish. [1]

This answer could have achieved 1/1 marks:

The greater the body mass of the fish, the lower the oxygen consumption rate it has. This means it consumes less oxygen per kilogram.

c) Predict the effects of global warming on aerobic respiration in fish. [2]

This answer could have achieved 2/2 marks:

A higher temperature causes a higher oxygen consumption rate. Therefore global warming will cause fish to have more metabolism and therefore consume more oxygen.

Assessment tip

A mark would also have been given for saying that less oxygen can dissolve in warmer water so less aerobic respiration can occur. Another mark could be given for saying that larger fish are favoured as they have a lower oxygen consumption rate.

2.9 PHOTOSYNTHESIS

You should know:

- ✔ photosynthesis is the production of carbon compounds in cells using light energy.
- ✔ visible light has a range of wavelengths: violet is the shortest and red the longest.
- ✔ chlorophyll absorbs red and blue light most effectively and reflects green light more than other colours.
- ✔ oxygen is produced in photosynthesis from the photolysis of water.
- ✔ energy is needed to produce carbohydrates and other carbon compounds from carbon dioxide.
- ✔ temperature, light intensity and carbon dioxide concentration are possible limiting factors on the rate of photosynthesis.

You should be able to:

- ✔ explain changes to the Earth's atmosphere, oceans and rock deposition due to photosynthesis.
- ✔ draw and explain an absorption spectrum for chlorophyll and an action spectrum for photosynthesis.
- ✔ design experiments to investigate the effect of temperature, light intensity and carbon dioxide on photosynthesis.
- ✔ explain the separation of photosynthetic pigments by chromatography.

In Topic 2.4, an enzyme involved in photosynthesis was discussed In Topic 8.3, knowledge of photosynthesis will be deepened.

- **Photosynthesis** is the process by which green plants (and some microorganisms) use sunlight to synthesize nutrients from carbon dioxide and water. Light is absorbed by pigments and generates oxygen as a by-product.
- **Chlorophyll** is a green pigment that absorbs light in most plants.
- An **action spectrum** is a graph of the rate of photosynthesis performed by an organism at different light wavelengths.
- An **absorption spectrum** is a graph of the amount of light absorbed by each pigment at different light wavelengths.

Visible light has a range of wavelengths with violet the shortest wavelength (400 nm) and red the longest (700 nm). Chlorophyll absorbs red and blue light most effectively and reflects green light more than other colours. Photosynthesis occurs in plants, algae and some microorganisms that contain chlorophyll. Photosynthesis uses light energy to split water molecules (photolysis). The purpose of photolysis is to release electrons that can eventually be used to reduce carbon dioxide and hydrogen ions to produce energy. Oxygen is a by-product. Energy from ATP and the hydrogen from photolysis are used to produce carbohydrates and other carbon compounds from carbon dioxide.

Although chlorophyll is the main pigment found in plants, there are other pigments that absorb light at different wavelengths. Examples of these pigments are xanthophylls and carotenes. It is possible to separate photosynthetic pigments using chromatography.

SAMPLE STUDENT ANSWER

Outline the production of carbohydrates in photosynthesis. [4] 

This answer could have achieved 3/4 marks:

Glucose is a type of carbohydrate formed through photosynthesis.

The photosynthesis equation allows us to see how and when in the process glucose is produced:

carbon dioxide + water —LIGHT→ oxygen + glucose

There are two stages to the production of glucose or carbohydrates.

Stage 1: light-dependent reaction. Sunlight is used to bring about photolysis, which is the splitting of water molecules into hydrogen and oxygen ions.

Stage 2: light-independent reaction. Carbon fixation occurs.

Through these two stages of getting hydrogen and oxygen ions and fixating carbon we get the carbohydrate molecule glucose.

▲ The answer scores a mark for photolysis and for mentioning oxygen and hydrogen as products of photolysis. The second mark is for carbon fixation.

▼ The answer does not mention that light is absorbed by chlorophyll and is converted to chemical energy. Some of the energy is used for the production of ATP which is needed to produce carbohydrates or starch.

Example 2.9.1.

How is the oxygen released by photosynthesis produced?

A. Water molecules split with the energy from ATP to liberate oxygen.

B. Water molecules split with the energy from light to liberate oxygen.

C. Water molecules combine with carbon dioxide molecules to form oxygen.

D. Carbon dioxide molecules split with the energy from light producing oxygen.

Solution

The answer is **B.** Light energy is used to split the water molecule in a process called photolysis. The hydrogen is used in photosynthesis and the oxygen is liberated as a by-product.

Practice problems for Topic 2

Problem 1

Describe the properties of water that allow cooling by sweating.

Problem 2

Outline the action of enzymes.

Problem 3

Design an experiment using egg white to show the effect of heat on proteins.

Problem 4

Outline the production of carbohydrates in photosynthesis.

3 GENETICS

3.1 GENES

You should know:

- ✔ every living organism inherits a blueprint for life from its parents.
- ✔ a gene consists of a length of DNA that carries information for a specific characteristic.
- ✔ each gene occupies a specific place on a chromosome.
- ✔ alleles are different forms of a gene that vary by only a few bases.
- ✔ new alleles are formed by mutation.
- ✔ the genome is the whole genetic information of an organism.
- ✔ the Human Genome Project sequenced the whole human genome.

You should be able to:

- ✔ explain how sickle cell anemia arises if there is a mutation in the gene for hemoglobin.
- ✔ explain the link between malaria and sickle cell anemia.
- ✔ compare the numbers of genes in humans and other organisms.
- ✔ explain how human diversity is due to single nucleotide polymorphisms.
- ✔ describe the use of a database to determine differences in the base sequence of a gene in two species.

- A **gene** is a heritable factor that consists of a length of DNA and influences a specific characteristic.
- The **locus** is the place on the chromosome where a gene is located.
- **Alleles** are various specific forms of a gene.

The development of gene sequencers has allowed for the sequencing of genes in many organisms. Frederick Sanger was a pioneer in DNA sequencing, as in 1977, together with his co-workers, he developed the chain-termination method. This method consisted of the selective incorporation of chain-terminating dideoxynucleotides by DNA polymerase during replication. Other faster methods have been developed since, allowing for larger sections of DNA to be sequenced.

Each gene is located in a specific place or locus on the chromosome. There can be several variants of the same gene; these are called alleles. Humans have around 20,000 genes, while some bacteria have 5,000. If an organism has two alleles for a particular gene that are the same, it is called a homozygote. If an organism has two alleles for a particular gene that are different (for example, alleles for both blue eyes and brown eyes), it is called a heterozygote.

In sickle cell anemia, one specific base substitution causes glutamic acid to be substituted by valine as the sixth amino acid in the hemoglobin polypeptide. As a result the structure of the hemoglobin protein is altered and it does not carry oxygen efficiently in erythrocytes; the shape of affected erythrocytes is also altered (they are 'sickle-shaped').There appears to be a correlation between malaria and the incidence of sickle cell anemia. Regions where malaria is endemic also show a high incidence of sickle cell anemia in the local population; sickle cell anemia has a low rate of incidence in regions that do not have malaria. This is probably due to the fact that the malarial parasite cannot enter into 'sickle-shaped' erythrocytes, so heterozygotes for the sickle cell gene are relatively protected against malaria. Therefore, the gene can survive and be inherited more often. On the other hand, patients who are homozygous for the sickle cell gene, and therefore suffer from sickle cell anemia, are highly susceptible to the lethal effects of malaria.

Example 3.1.1.

The ABO blood group gene locus is located on the long arm of human chromosome 9, and encodes for an enzyme that mediates the expression of A and B antigens on erythrocytes. The A and B genes differ in a few single-base substitutions. The protein sequence of the allele for blood group O differs in more amino acids from the sequences for blood groups A and B. The table shows part of a protein sequence alignment of the alleles of blood groups (an asterisk means the sequence is the same and a dash means the amino acid is not present).

Allele	Amino acid
Blood group A	P K V L T P W K – D V L V V T P W L A P I V W E G T F
Blood group B	* * * * * * * C R * * * * * * * * * * * * * * * * * *
Blood group O	* * * * * * * C R * * * * * P L G W L P L S G R A H S

a) (i) Identify the number of amino acids that are different in this sequence between blood group A and blood group B.

(ii) Suggest whether this data shows that human diversity is due to single nucleotide polymorphisms (SNPs).

b) Explain how a base substitution in the DNA molecule of one gene can cause a difference in the protein that determines the blood group of a person.

Solution

a) (i) 2

(ii) There are more SNPs between the sequence for blood group O and both A and B. There are few differences between A and B. Therefore, the SNPs make a difference in the blood group, confirming that SNPs can contribute to human diversity.

b) The DNA molecule is the blueprint for the formation of mRNA through transcription. A change in the DNA could determine a change in the mRNA codon. When the mRNA enters the ribosome, translation will occur. The change in mRNA could cause a different tRNA to join by its anticodon, determining a different primary sequence of the protein. Because the genetic code is degenerate, if the change in nucleotide determines for a codon that codes for the same amino acid, no change will occur to the protein (neutral mutation). If the new codon is a stop codon, the protein will be much shorter.

- A **mutation** is a change to the structure of a gene caused by the alteration of single base units in the DNA. A mutation may be inherited by subsequent generations, or may affect only the individual in which it occurs. Mutations can be caused by exposure to radiation or carcinogens, or by errors in replication or transcription.
- **Single nucleotide polymorphisms** are caused by point mutations that give rise to different alleles containing alternative bases at a given nucleotide position within a locus.

Assessment tip

You do not need to study deletions, insertions and frame shift mutations.

Topic 3.4 shows how genetic information is inherited. Topic 5.4 explains how cladistics is used to describe the way organisms are related phylogenetically.

SAMPLE STUDENT ANSWER

Mutations are the ultimate source of genetic variation and are essential to evolution.

a) State one type of environmental factor that may increase the mutation rate of a gene. [1]

This answer could have achieved 1/1 marks:

Exposure to radiation.

▲ Correct answer: mutations can arise from exposure to radiation such as UV rays, presence of chemical mutagens, carcinogens, papilloma virus or cigarette smoke. Although it is less common, mutations could occur from errors by polymerases in transcription or translation.

b) Identify one type of gene mutation. [1]

This answer could have achieved 1/1 marks:

Base substitution mutation.

▲ Correct answer: mutations can occur from base substitution. They could also occur from insertion, deletion or frameshift.

3.2 CHROMOSOMES

You should know:

- ✔ prokaryotes have one chromosome consisting of a circular DNA molecule.
- ✔ plasmids are small circular DNA molecules present only in some prokaryotes.
- ✔ eukaryote chromosomes are linear DNA molecules associated with histone proteins.
- ✔ eukaryotes have different chromosomes that carry different genes.
- ✔ homologous chromosomes carry the same sequence of genes but not necessarily the same alleles of those genes.
- ✔ diploid nuclei have pairs of homologous chromosomes.
- ✔ haploid nuclei have one chromosome of each pair.
- ✔ each species has a specific number of chromosomes.
- ✔ a karyogram shows the chromosomes of an organism in homologous pairs of decreasing length.
- ✔ sex chromosomes determine sex and autosomes are chromosomes that do not determine sex.

You should be able to:

- ✔ explain how Cairns' technique is used for measuring the length of DNA molecules.
- ✔ compare genome size in T2 phage, *Escherichia coli, Drosophila melanogaster, Homo sapiens* and *Paris japonica.*
- ✔ compare the diploid chromosome numbers of *Homo sapiens, Pan troglodytes, Canis familiaris, Oryza sativa* and *Parascaris equorum.*
- ✔ analyse karyograms to deduce sex and diagnose Down syndrome in humans.
- ✔ describe the use of databases to identify gene loci and the protein products of genes.

Chromosome separation during cell division is studied in Topic 1.6.

In 1963, John Foster Cairns used autoradiography to establish the length of DNA molecules in chromosomes. *E. coli* cells were grown with radioactive thymidine, which was incorporated into the DNA during replication. Radioactively labelled DNA from broken cells was spread on a glass slide and all the molecules covered with a light-sensitive emulsion. The tracks of silver grains resulting from exposure of DNA to radioactive thymidine were used to measure the DNA fibres.

- The **karyotype** is the number and type of chromosomes present in the nucleus.
- A **karyogram** is a photograph of chromosomes ordered by size.
- The **genome size** is the total length of DNA in an organism.
- The **centromere** is the area of the chromosome where the sister chromatids are linked. During cell division, it is where the spindle fibres attach.
- A **chromatid** is a DNA molecule that is the copy of a newly formed chromosome.
- **Sister chromatids** are two DNA molecules formed by DNA replication prior to cell division until the splitting of the centromere at the start of anaphase.

Chromosomes are linear DNA molecules associated with histone proteins. All members of a species have the same number of chromosomes (with a few exceptions, for example in diseases such as Down syndrome). Prokaryotes have one chromosome consisting of a circular DNA molecule. Because prokaryotes do not have a nuclear membrane, the naked DNA is found in the cytoplasm in an area called the nucleoid. Prokaryotes may also contain plasmids, which are small circular DNA molecules. Eukaryote chromosomes are found inside the nucleus of the cell. Diploid nuclei have pairs of homologous chromosomes. The pairs of chromosomes carry the same sequence of genes but not necessarily the same alleles of those genes. Haploid nuclei have half the number of chromosomes as diploid nuclei—they have one chromosome of each pair.

Example 3.2.1.

The scanning electron micrograph shows part of a cell in mitosis (×1,600).

a) On the micrograph, label a centromere.

b) Identify on the micrograph two sister chromatids.

c) Identify, with a reason, the stage of mitosis this cell is in.

d) Describe how you could make a karyogram from this picture.

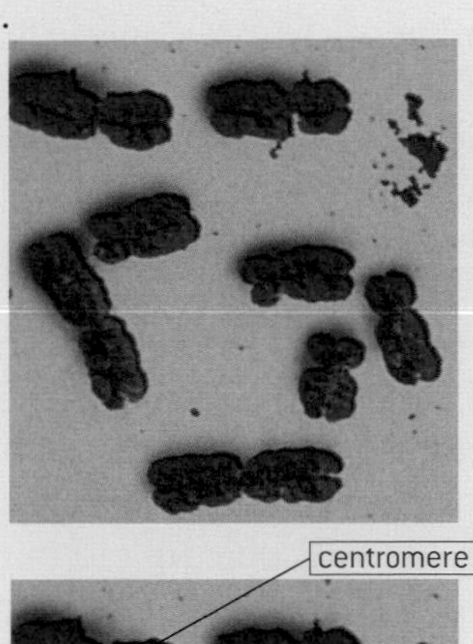

Solution

a) and b) See diagram on the right

c) Metaphase, because chromosomes are highly condensed, and chromatids become apparent. This is the stage where chromosomes are more clearly seen.

d) Each chromosome is cut from the picture (using scissors) and ordered by homologous pairs according to their size and the length of each arm of chromatid.

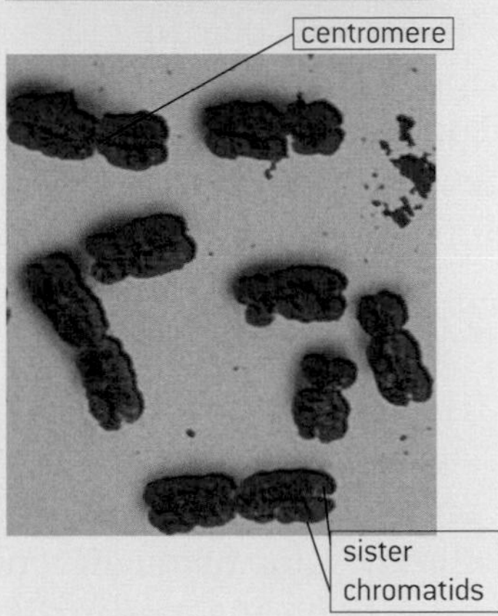

The genome size does not give an idea of how complex in form the organism is. Humans have a genome of around 3,000 million base pairs, while *Paris japonica*, a woodland plant, has 150,000 million base pairs. The number of chromosomes is not an indication of evolution either. Humans have 46 chromosomes in diploid cells (23 in haploid) while dogs have 78 chromosomes (36 in haploid).

- **Gametes** have a haploid number of chromosomes in their nucleus.
- **Autosomal cells** have a diploid number of chromosomes in their nucleus.

3.3 MEIOSIS

You should know:

- ✔ one diploid nucleus divides by meiosis to produce four haploid nuclei.
- ✔ the chromosome number is halved to allow a sexual life cycle with fusion of gametes.
- ✔ DNA is replicated before meiosis so that all chromosomes consist of two sister chromatids.
- ✔ pairing of homologous chromosomes and crossing over is followed by condensation during early meiosis.
- ✔ orientation of pairs of homologous chromosomes prior to separation is random.
- ✔ separation of pairs of homologous chromosomes in the first division of meiosis halves the chromosome number.
- ✔ alleles segregate during meiosis allowing new combinations to be formed by the fusion of gametes.
- ✔ genetic variation is promoted by crossing over, random orientation of chromosomes and fusion of gametes from different parents.

You should be able to:

- ✔ explain how non-disjunction can cause Down syndrome and other chromosome abnormalities.
- ✔ describe methods used to obtain cells for karyotype analysis (for example, chorionic villus sampling and amniocentesis) and the associated risks.
- ✔ draw the stages of meiosis resulting in the formation of four haploid cells.
- ✔ analyse microscope slides showing meiosis.
- ✔ interpret studies investigating whether the age of parents influences the chances of non-disjunction.

Cell division is studied in Topic 1.6. Topic 10.1 covers meiosis in greater depth and Topic 11.4 discusses the formation of gametes in sexual reproduction.

Meiosis was discovered by careful microscope examination of dividing germ-line cells. In 1876 Oscar Hertwig first described meiosis in sea urchin eggs, but it was not until 1890 that August Weismann described the change from a diploid to four haploid cells.

Meiosis produces gametes. Gametes are cells that contain the haploid number of chromosomes, allowing two gametes (one from each parent) to fuse and create an embryo with the diploid number of chromosomes. Meiosis is divided into two processes, meiosis I and meiosis II. The halving of chromosome number occurs in meiosis I, where each chromosome of a homologous pair migrates to the opposite pole. DNA is replicated before meiosis so that all chromosomes consist of two sister chromatids. In metaphase I the homologous chromosomes pair up, and crossing over occurs. The homologous chromosomes are randomly orientated prior to separation in anaphase I. Meiosis II is similar to a mitotic cell division, as each sister chromatid migrates to an opposite pole. Overall, one diploid nucleus divides by meiosis to produce four haploid nuclei at the end of meiosis II. Alleles segregate during meiosis allowing new combinations to be formed by the fusion of gametes. Genetic variation is promoted by crossing over, random orientation of chromosomes and fusion of gametes from different parents.

Example 3.3.1.

The graphs show the schematic comparison of chromosome stages in meiosis I and II of rye (*Secale cereale*). The chromosome undergoes three stages: aggregation of chromatin into a chromosome filament; condensation in length, which involves a progressive increase in diameter; and separation of chromatids in segregation.

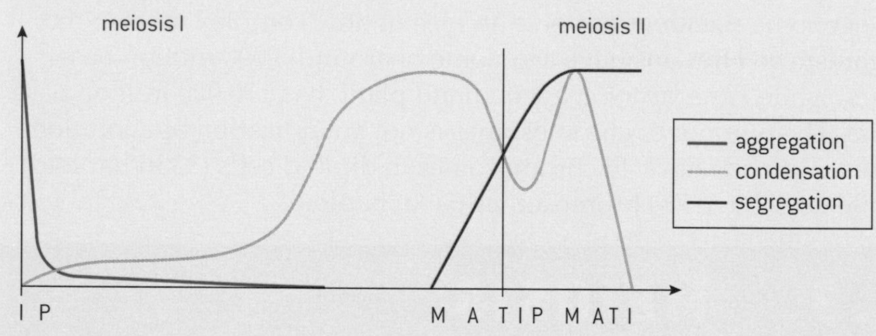

Key

I = interphase

P = prophase

M = metaphase

A = anaphase

T = telophase

a) State the stage at which the chromosomes are maximally condensed.

b) Suggest one reason there is no aggregation in meiosis II.

c) Explain chromosome condensation at each stage in meiosis I.

d) Compare and contrast chromosome segregation in meiosis I and meiosis II.

Solution

a) Metaphase.

b) Chromatin aggregates into chromosomes only in meiosis I, as in meiosis II chromosomes are already formed.

c) In interphase chromosomes are not condensed. Early prophase is when chromosomes start to condense and in mid prophase chromosomes can be distinguished. In late prophase chromosomes are partly separated and constrictions begin to form in mitosis. It is in metaphase when chromosomes are maximally condensed. While the chromosomes are still condensed, they segregate and move towards the poles during anaphase.

d) During anaphase I (in meiosis I), the homologous chromosome pairs are segregated at the poles—one chromosome from each pair migrates to an opposite pole. During anaphase II (meiosis II), each sister chromatid migrates to the opposite poles.

3.4 INHERITANCE

You should know:

- ✔ the inheritance of genes follows patterns.
- ✔ Mendel's experiments to explain inheritance.
- ✔ gametes are haploid so contain only one allele of each gene.
- ✔ the two alleles of each gene separate into different haploid daughter nuclei during meiosis.
- ✔ fusion of gametes results in diploid zygotes with two alleles of each gene that may be the same allele or different alleles.
- ✔ dominant alleles mask the effects of recessive alleles but codominant alleles have joint effects.
- ✔ many genetic diseases in humans can be due to autosomal recessive, dominant or codominant alleles.
- ✔ the inheritance of sex-linked genes.
- ✔ some genetic diseases and cancer are due to radiation and/or mutagenic chemicals.

You should be able to:

- ✔ describe the inheritance of ABO blood groups.
- ✔ explain the role sex-linked inheritance plays in red–green colour blindness and hemophilia.
- ✔ describe the inheritance of cystic fibrosis and Huntington's disease.
- ✔ describe the consequences of radiation at Hiroshima and Chernobyl.
- ✔ construct Punnett grids for monohybrid genetic crosses.
- ✔ compare predicted and actual outcomes of genetic crosses using real data.
- ✔ analyse pedigree charts to deduce the pattern of inheritance of genetic diseases.

Mendel's experiments with pea plants helped him to discover the laws of inheritance—the laws of segregation, independent assortment and dominance. Because each homologous chromosome migrates to a different pole during gamete formation, the alleles for each gene segregate from each other, so that each gamete carries only one allele for each gene (law of segregation). The homologous chromosomes migrate in a random way, therefore genes for different traits can segregate independently during the formation of gametes (law of independent assortment). Some alleles are dominant while others are recessive; an organism with at least one dominant allele will display the effect of the dominant allele (law of dominance).

In 1866, Gregor Johann Mendel, an Austrian monk and scientist, described the cross-breeding of pea plants to show the inheritance of seven characteristics. His results were used to establish the rules of inheritance, called the laws of Mendelian inheritance. In order to reach these laws, Mendel performed large numbers of crosses. Mendel's genetic crosses with pea plants generated numerical data. Many measurements with replicates were necessary to ensure reliability of the data obtained.

- **Segregation** is the separation of alleles into different nuclei.
- A **dominant** allele is one that produces its characteristic phenotype.
- A **recessive** allele only produces its characteristic phenotype when homozygous.
- **Codominant alleles** are those that are simultaneously expressed in the phenotype of the individual.
- **Sex linkage** is the phenotypic expression of an allele related to the sex chromosomes, usually the X chromosome.

Topic 1.6 introduces cell division.

Assessment tip

You may be asked to show genetic crosses in the exam. It is always better to use a Punnett grid to do so and follow annotation conventions. Alleles carried on X chromosomes should be shown as superscript letters on an upper case X, such as X^h. The expected notation for ABO blood group alleles is for the phenotype O, A, B and AB. For the genotype: ii for O group; I^AI^A or I^Ai for A; I^BI^B or I^Bi for B; and I^AI^B for AB.

Example 3.4.1.

What type of inheritance is shown in the pedigree?

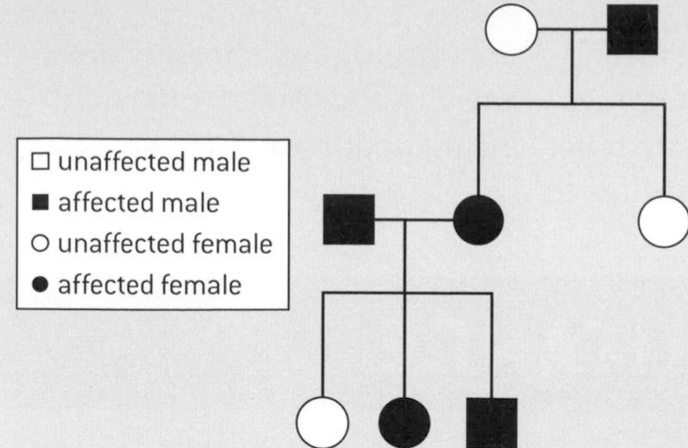

A. sex-linked recessive

B. autosomal dominant

C. multiple alleles

D. codominant alleles

Solution

The correct answer is **B** as it is showing an autosomal dominant inheritance. It is not sex-linked because if it was, the daughter in the F_1 pairing would not be affected, as her mother is unaffected, or both daughters would be affected if the mother were a carrier.
It is not showing multiple alleles or codominant alleles because the daughter in F_2 is not affected.

	X^B	Y
X^B	X^BX^B	X^BY
X^b	X^BX^b	X^bY

▲ **Figure 3.4.1.** Punnett grid

Colour-blindness is sex-linked recessive inheritance on the X chromosome. The Punnett grid shows the inheritance of colour-blindness in a family. (X^B shows the normal allele and X^b the colour-blind allele.) For females, there is a 50% of being a carrier and 50% chance of not being a carrier. For males, there is a 50% of being colour-blind and a 50% chance of being not colour-blind.

Assessment tip

In the exam you will be asked to construct Punnett grids for monohybrid genetic crosses. Remember to place the parents' alleles in the top and left-hand boxes and the offsprings' in the centre as a result of the possible combinations. Remember there is a 1/4 (25%) chance in each pregnancy for each one of the genotypes.

SAMPLE STUDENT ANSWER

The most common form of Stargardt disease is known to be autosomal recessive. Using a Punnett grid, deduce the probability of a child inheriting Stargardt disease if both the parents are carriers of the disease but do not have the disease themselves.

This answer could have achieved 0/3 marks:

		mother	
		x^h	x^h
father	x^h	x^hx^h	x^hx^h
	x^h	x^hx^h	x^hx^h

x^h – carriers

100% probability

▼ This student has shown the parents to be homozygous for the Stargardt allele so got the wrong percentage.

This answer could have achieved 3/3 marks:

Father (carrier) SⓈ
Mother (carrier) SⓈ ; S = normal allele, Ⓢ = Stargardt disease

		FATHER	
		S	Ⓢ
MOTHER	S	SS	SⓈ
	Ⓢ	SⓈ	ⓈⓈ

∴ probability of Stargardt disease = 25%

▲ The student drew the Punnett grid correctly. The gametes for the mother and father are shown. There are both dominant and recessive gametes because both parents are carriers. The four possible combinations of gametes are also shown. The answer is clear as 25% (or could have been 1/4) is written at the bottom.

3.5 GENETIC MODIFICATION AND BIOTECHNOLOGY

You should know:

- ✔ there are techniques for artificial manipulation of DNA, cells and organisms.
- ✔ gel electrophoresis separates proteins or fragments of DNA according to size.
- ✔ PCR can be used to amplify small amounts of DNA.
- ✔ DNA profiling involves comparison of DNA.
- ✔ genetic modification is carried out by gene transfer between species.
- ✔ clones are groups of genetically identical organisms, derived from a single original parent cell.
- ✔ many plant species and some animal species have natural methods of cloning.
- ✔ animals can be cloned at the embryo stage by breaking up the embryo into more than one group of cells.
- ✔ methods have been developed for cloning adult animals using differentiated cells.

You should be able to:

- ✔ describe DNA profiling in paternity and forensic investigations.
- ✔ explain gene transfer to bacteria via plasmids using restriction endonuclease and DNA ligase.
- ✔ evaluate the potential risks and benefits associated with genetic modification of crops.
- ✔ explain the production of cloned embryos produced by somatic-cell nuclear transfer.
- ✔ design an experiment to assess one factor affecting the rooting of stem-cuttings.
- ✔ analyse DNA profiles.
- ✔ analyse data on risks to monarch butterflies of Bt crops.

A polymerase chain reaction (PCR) is a technique by which small amounts of DNA can be amplified into large quantities. The DNA fragments obtained by this method can be separated by size in gel electrophoresis. The smaller molecules migrate further than the larger ones. A marker with known sizes of DNA fragments is used to compare sizes.

Replication, transcription and translation of DNA are studied in Topic 2.7.

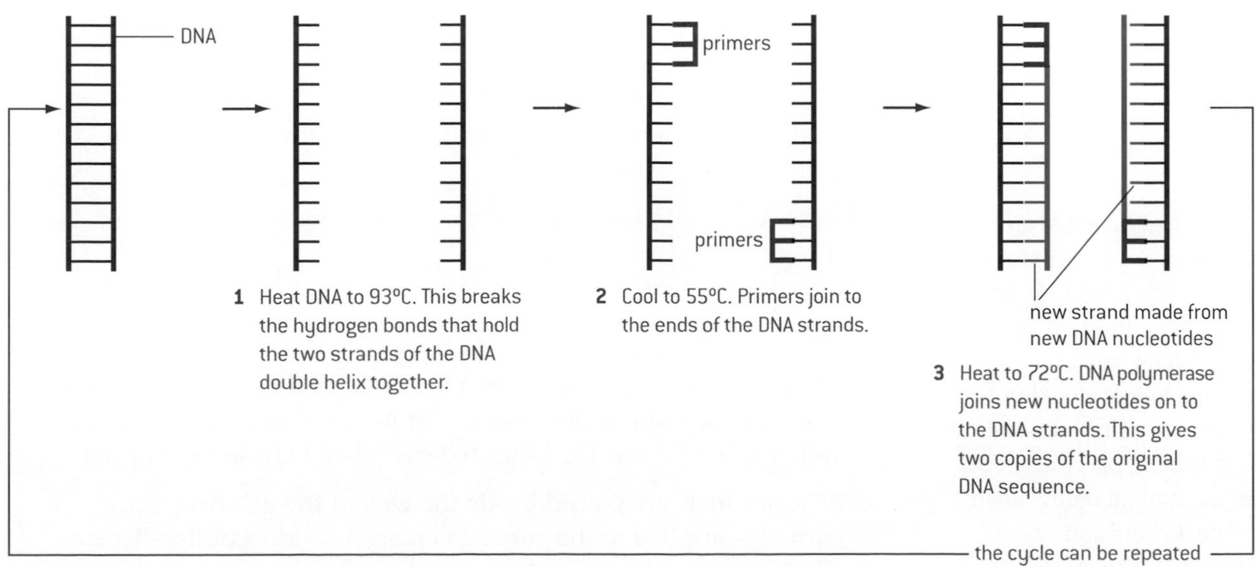

DNA
primers
primers
1 Heat DNA to 93°C. This breaks the hydrogen bonds that hold the two strands of the DNA double helix together.
2 Cool to 55°C. Primers join to the ends of the DNA strands.
new strand made from new DNA nucleotides
3 Heat to 72°C. DNA polymerase joins new nucleotides on to the DNA strands. This gives two copies of the original DNA sequence.
the cycle can be repeated

▲ **Figure 3.5.1.** Production of DNA molecules by PCR

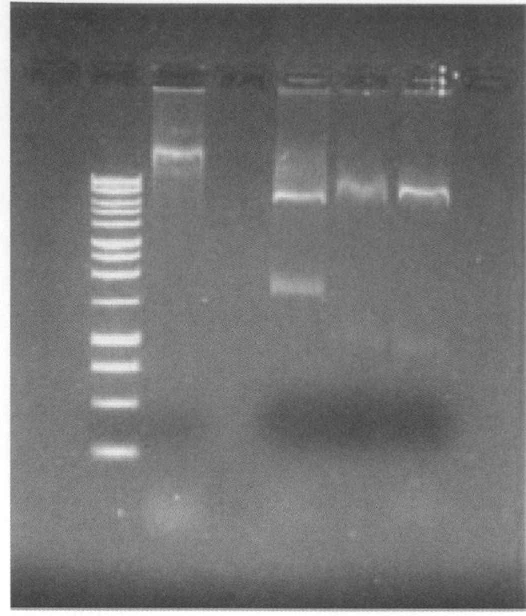

▲ **Figure 3.5.2.** Separation of DNA molecules by gel electrophoresis

- **Clones** are groups of genetically identical organisms, derived from a single original parent cell.
- **Endonucleases** are restriction enzymes that cut the DNA within the sequence.
- **DNA ligase** is an enzyme that joins DNA fragments by joining a sugar to a phosphate.

Example 3.5.1.

A DNA profiling was performed as a paternity test for the boy shown as DC. The first pair of maternal grandparents tested were both heterozygous for an allele (AB and AD) and the possible paternal grandparents were also heterozygous (AB and CD). Males are shown as squares and females as circles.

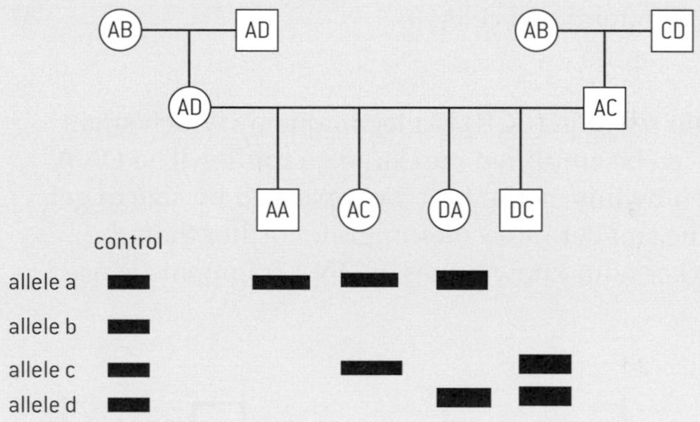

a) State the alleles present in the supposed father.

b) Suggest with a reason whether DC could have as maternal grandparents AB and AD.

c) Describe briefly how the DNA molecules were separated in this gel.

Solution

a) AC

b) DC could be the grandson of AB and AD. His mother is AD so she inherited one allele from her mother (A) and one from her father (D). Her son DC inherited the allele D from his mother.

c) PCR products are placed inside the well of the gel. An electric current is applied to the gel that is placed inside a buffer. Because DNA has a negative charge, the molecules migrate towards the positive pole. The speed at which they move is related to the size of the molecules. The larger particles move the least.

A major benefit of GM crops is increased yields. GM crops can be modified to include genes for food supplements such as vitamins, vaccines, stress resistance and longer shelf life. Where the introduced genes code for pest resistance, GM crops could also lead to a marked reduction in the use of pesticides and herbicides, which has many economic, environmental and health benefits.

Genetic engineering is the transfer of genetic information from one species to another. Usually what is transferred is a gene (a section of DNA). When the gene transfer is performed by bacteria, usually the gene is inserted in a plasmid. Plasmids usually have a gene of resistance to antibiotic or a gene to produce a fluorescent protein. The plasmid and the gene to be inserted are cut with the same restriction enzyme. This leaves "sticky ends" that allow them to be inserted into the plasmid. DNA ligase is then used to join these DNA sections. Once the gene is inserted into the plasmid, the transgenic bacteria are chosen by either growing them in the presence of antibiotics or by seeing their fluorescence.

Scientists must consider how genetically modified (GM) genes will move through food webs, in case there are toxic or allergic effects on non-target species. Gene transfer from GM plants to wild non-GM plants could also eventually lead to the extinction of wild plants and there are concerns the GM crops might become agricultural weeds and invade natural environments, compromising the biodiversity of these habitats. As some GM crops have pest resistance, this could impose intense selection pressure on pest populations to adapt to the resistance mechanism, leading to the development of super-pests that would be difficult to control.

Example 3.5.2.

A gene was transferred from one species to another. Why was this gene transfer possible?

A. Both species have the same genes

B. Both species have identical DNA

C. All species have the same number of chromosomes

D. All species have the same genetic code

Solution

The correct answer is **D**. Gene transfer can be done amongst species because they all have the same genetic code.

Practice problems for Topic 3

Problem 1

Explain how Cairns' technique is used for measuring the length of DNA molecules.

Problem 2

a) (i) Define codominant and recessive alleles.

(ii) Define locus and sex linkage.

b) Explain the inheritance of blood groups.

Problem 3

Explain how cloned embryos may be produced by somatic-cell nuclear transfer.

Problem 4

Explain the relationship between Mendel's law of segregation and independent assortment and meiosis.

4 ECOLOGY

4.1 SPECIES, COMMUNITIES AND ECOSYSTEMS

You should know:

- ✔ only sustainable communities will allow for continued survival of living organisms.
- ✔ species are groups of organisms that can potentially interbreed to produce fertile offspring.
- ✔ members of a species may be reproductively isolated in separate populations.
- ✔ most species have either an autotrophic or a heterotrophic method of nutrition.
- ✔ heterotrophs can be consumers, detritivores or saprotrophs.
- ✔ a community is formed by populations of different species living together and interacting with each other.
- ✔ a community forms an ecosystem by its interactions with the abiotic environment.
- ✔ autotrophs obtain inorganic nutrients from the abiotic environment.
- ✔ the supply of inorganic nutrients is maintained by nutrient cycling.
- ✔ ecosystems have the potential to be sustainable over long periods of time.

You should be able to:

- ✔ classify species as autotrophs, consumers, detritivores or saprotrophs.
- ✔ design sealed mesocosms to try to establish sustainability.
- ✔ analyse data obtained by quadrat sampling for association between two species using the chi-squared test (χ^2).
- ✔ interpret statistical significance in data.

Energy flow in ecosystems will be studied in Topic 4.2. The classification of biodiversity will be studied in Topic 5.3 and Topic C.1 will cover species and communities. Topic C.2 will broaden your knowledge of communities and ecosystems.

- **Autotrophs** are organisms that produce their own food.
- **Heterotrophs** are organisms that feed on other organisms.
- **Consumers** are heterotrophs that feed on living organisms by ingestion.
- **Detritivores** are heterotrophs that obtain organic nutrients from detritus by internal digestion.
- **Saprotrophs** are heterotrophs that obtain organic nutrients from dead organisms by external digestion.

Organisms need energy in order to live. Some organisms produce their own food while others need to feed on other organisms. Each species has one specific mode of nutrition. There are only a few species that have more than one mode of nutrition.

Example 4.1.1.

The diagram is a model representing different feeding interactions.

Which organisms could be saprotrophs?

A. V and W **C.** V only

B. W and Z **D.** Z only

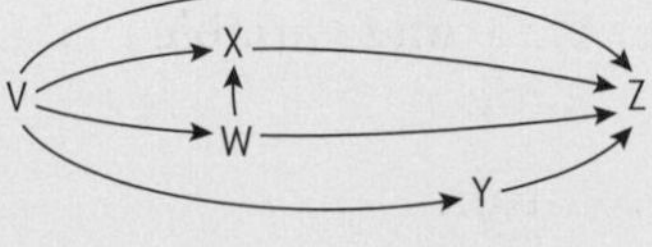

Solution

The answer is **B**. Saprotrophs are heterotrophs that obtain organic nutrients from dead organisms by external digestion. V is a producer, therefore an autotroph. Any other organisms could be a correct answer, as in this energy flow diagram the mode of feeding is not explicit.

Mesocosms are experimental systems that can be set up to study natural environments under controlled conditions. They can range in size from a small plastic bottle to a large tank. Mesocosms can be closed systems, allowing only energy exchange with the environment, or open systems, allowing exchange of materials as well. The advantage of a mesocosm is that many environmental conditions can be controlled and measured, the experiments can be replicated several times under the same conditions and it is easier to manipulate than in a free environment. The disadvantage is that it does not represent the real environment exactly, so you could reach false conclusions.

- **Species** are organisms that have common characteristics which are capable of producing fertile offspring.
- **Population** is a group of organisms of one species that interbreed and live in the same place at the same time.
- **Community** is formed by populations of different species living together and interacting with each other.
- **Ecosystem** is a community and its interactions with the abiotic environment.

The term sustainability refers to the capability of a system to endure and maintain itself. The three main challenges for sustainable ecosystems are finding sources of both nutrients and energy that will not run out and avoiding the accumulation of toxic wastes that will become hard to store or have an impact on the environment. The sustainable production of food uses renewable raw materials efficiently, avoids the use of environmentally unfriendly chemicals in their manufacture, uses clean and renewable energy and minimizes the production of waste. Because of the increase in our global population, it is becoming more important that we consider the sustainability of human activities.

Example 4.1.2.

A mesocosm experiment was set up to evaluate how the presence of the tiger fish (*Hoplias malabaricus*) affects the plankton structure of a lake in South America.

Twelve experimental units were set up. Each mesocosm was filled with groundwater, including sediment and free-floating plants and the plankton *Daphnia obtusa*. In some cases *Jenynsia multidentata,* a fish eaten by the tiger fish, was added with or without tiger fish of different sizes. The four different treatments were randomly assigned.

Control: plankton (*Daphnia obtusa*) (no fish)

JM: plankton (*Daphnia obtusa*) + *Jenynsia multidentata*

HMs: plankton (*Daphnia obtusa*) + *Jenynsia multidentata* + *Hoplias malabaricus* (small tiger fish)

HMb: plankton (*Daphnia obtusa*) + *Jenynsia multidentata* + *Hoplias malabaricus* (big tiger fish)

The abundance of plankton (*Daphnia obtusa*) remaining in the mesocosm was recorded as individuals per litre at different times.

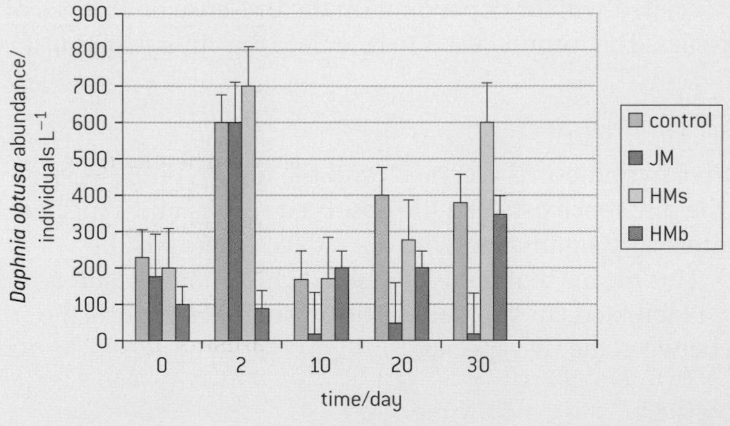

a) (i) State the amount of *Daphnia* in the JM mesocosm at the start of the experiment.

(ii) Calculate the percentage change in *Daphnia* in the JM mesocosm from the start to the end of the experiment.

b) State **one** variable that is not kept constant in these mesocosms.

Solution

a) (i) 180 ind L^{-1}

(ii) Amount of *Daphnia* at the start: 180 ind L^{-1}

Amount of *Daphnia* at the end: 20 ind L^{-1}

Assessment tip

Do not forget to provide the units.

Assessment tip

In markschemes, what is not expected in the answer will appear in brackets. In this case, because this is a "state" question, the reason is in brackets, as it is not expected in the answer.

First: work out the difference (increase) between the two numbers you are comparing.

Increase = New number – Original number

$= 20 - 180 = -160 \text{ ind } L^{-1}$

Then: divide the increase by the original number and multiply the answer by 100.

$$\% \text{ increase} = \frac{\text{Increase}}{\text{Original number}} \times 100 = \frac{-160}{180} \times 100 = -89\%$$

If your answer is a negative number, then this is a percentage decrease.

b) There are many correct answers to this question, as many variables are not kept constant inside each mesocosm. The answer should be only one variable, not a list of variables. Here are some possible answers: temperature (although all mesocosms are exposed to ambient temperature), pH (different metabolic activity in the mesocosm could change the amount of acids or alkalis produced, so this is variable across all mesocosms), nutrients (different organisms, so different in all mesocosms) and sunlight (as the plants could stop the rays from entering).

The chi-squared (χ^2) test can be used to determine whether there is a significant difference between the expected frequencies and the observed frequencies in one or more categories. For example, if you want to see if the distributions of two species are independent, you can use a chi-squared test. You should base your sampling on random numbers and in each quadrat record the presence or absence of the chosen species. The null hypothesis (also called H_0) is that the presence of one species is random in relation to the presence of the other species. The alternative hypothesis (also called H_a) is that the presence of one species is associated with or dependent on the presence or absence of the other species. The formula used to calculate the chi-squared test is:

$$\chi^2 = \sum \frac{(O - E)^2}{E}$$

The alternative hypothesis is accepted (and the null hypothesis is rejected) if the difference between the observed results and expected results is statistically significant (with a $p < 0.05$, where p is the probability). This means that if the calculated chi-squared result is higher than a tabulated chi-squared (critical value), it supports the association between the two species.

Example 4.1.3.

In a field trip, a student used a quadrat to count the presence of two species (1 and 2) in an ecosystem. In order to test whether there is association between the presence of one species and the presence of the second species, a chi-squared test was performed.

H_0 (null hypothesis) = The presence of species 1 is independent of the presence of species 2.

H_a (alternative hypothesis) = The presence of species 1 determines the presence of species 2.

a) The results for the observed numbers of organisms are shown in the table. Complete the table to:

i) identify the expected results if there is no association between the presence of each species

ii) calculate the chi-squared results.

b) Using the results and the chi-squared distribution table, determine whether there is an association or not between the presence of species 1 and the presence of species 2.

Solution

a)

Species	Number of organisms observed	Expected result if there is no association	$\frac{(\text{Observed}-\text{expected})^2}{\text{expected}}$
1 only	3	2.5	0.1
2 only	1	2.5	2.25
1 and 2	4	2.5	2.25
none	2	2.5	0.1
total	10	10	4.7

	Probability			
Degrees of freedom	**0.99**	**0.95**	**0.05**	**0.01**
1	0.000	0.004	3.841	6.635
2	0.020	0.103	5.991	9.210
3	0.115	0.352	7.815	11.345
4	0.297	0.711	9.488	13.277
5	0.554	1.145	11.070	15.086

b) The degrees of freedom in this study is n – 1, therefore 3 (as there are 4 possible scenarios of the presence or not of species 1 and 2). The critical value for a $p = 0.05$ is 7.815 (with a significance level of 5%). The calculated chi-squared is 4.7. Because the calculated value is smaller than the critical value, the alternative hypothesis is rejected and the null hypothesis accepted. This means there is no association between the presence of the two species.

4.2 ENERGY FLOW

You should know:

- ✔ ecosystems require a continuous supply of energy to fuel life processes and to replace energy lost as heat.
- ✔ most ecosystems rely on a supply of energy from sunlight.
- ✔ light energy is converted to chemical energy in carbon compounds by photosynthesis.
- ✔ chemical energy in carbon compounds flows through food chains by means of feeding.
- ✔ energy released from carbon compounds by respiration is used in living organisms and converted to heat.
- ✔ living organisms cannot convert heat to other forms of energy so it is lost from ecosystems.
- ✔ energy losses between trophic levels restrict the length of food chains and the biomass of higher trophic levels.

You should be able to:

- ✔ design representations of energy flow using pyramids of energy.
- ✔ analyse pyramids of energy.
- ✔ draw food chains.
- ✔ identify a food chain from a food web.
- ✔ explain the reason food chains have a limited length.

Species, communities and ecosystems are studied in Topic 4.1. In Topic 2.8 you can see cell respiration as a cause for the loss of energy in a food chain. In Topic 2.9 you can see photosynthesis as a mode of obtaining energy.

Theories can be used to explain natural phenomena; for example, the concept of energy flow explains the limited length of food chains. Because only around 10% of the energy in a trophic level can be used by the next trophic level, the food chains are limited in length.

Biomass in terrestrial ecosystems diminishes with energy along food chains due to the loss of carbon dioxide, water and other waste products, such as urea. The indigestible parts of organisms such as hair and nails are also lost from one link of the food chain to the next. Heat is lost through respiration. There is a continuous but variable supply of energy in the form of sunlight but the supply of nutrients in an ecosystem is finite and limited.

SAMPLE STUDENT ANSWER

The image shows a food web.

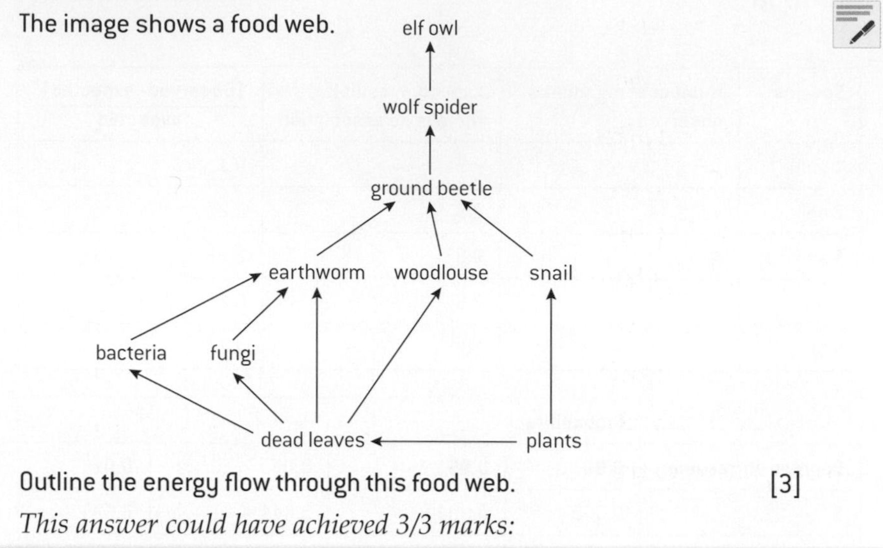

Outline the energy flow through this food web. [3]

This answer could have achieved 3/3 marks:

Plants (autotrophs) are the producers and convert light energy to chemical energy. This energy is used by the plant and lost as heat to the surroundings through the process of respiration. Primary consumers like snails consume plants. Only 10% of energy is passed on to the next trophic level as energy is lost as heat. Also, energy is lost as not all of the plant is eaten or digested.

▲ All answers are correct. The answer could have mentioned that as energy is lost at each trophic level, lengths of food chains or webs are restricted. The answer could have also added that detritivores or saprotrophs decay plant material that accumulates in the soil to obtain energy and that energy is used by organisms for metabolism.

Pyramids of energy are diagrams that represent the amount of energy converted to biomass at the different trophic levels.

Example 4.2.1.

The diagram is a trophic pyramid of energy (not shown to scale).

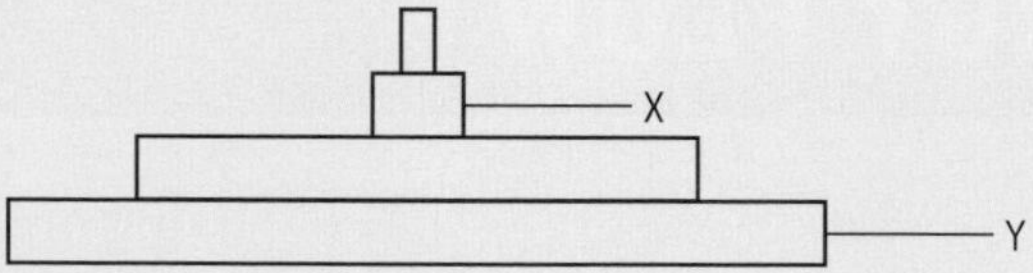

a) Label X and Y.

b) State the units used in a pyramid of energy.

c) Explain the loss of energy from one level to the next in a pyramid of energy.

Solution

a) X: second consumer

Y: producer

b) kJ m^{-2} yr^{-1} or kilojoules per metre squared per year. (The biomass can also be measured in joules or megajoules).

c) Autotrophs, mainly plants and algae in phytoplankton, release energy during photosynthesis. Plants are consumed by primary

consumers, which are then eaten by secondary consumers. Around 90% of the energy is lost between trophic levels. Most of the energy is lost as heat from respiration. The parts of organisms which cannot be digested are lost from one link of the food chain to the other as they are egested.

SAMPLE STUDENT ANSWER

Outline how the energy flow through food chains limits their length. [3]

This answer could have achieved 1/3 marks:

Food chains are often of about 3 to 4 organisms. This is because as energy flows, most of the energy is released or lost into the environment as heat. Only about 10% of the energy is passed from an energy level to the next level. Therefore, the length of the food chain is limited. Also, the energy available to the secondary consumers, which are often predators or carnivores, is very little. So, there could not be a lot of recorded energy that can be passed to the next consumer.

▲ The answer scored a mark for mentioning the small proportion (10%) of energy that can pass from one trophic level to the next.

▼ The answer did not score a mark for mentioning that energy is lost as heat because it does not specify that the heat loss occurs through respiration. The answer should have mentioned energy losses due to uneaten parts, or due to undigested parts that are lost through feces or egestion.

4.3 CARBON CYCLING

You should know:

- ✔ continued availability of carbon in ecosystems depends on carbon cycling.
- ✔ autotrophs convert carbon dioxide into carbohydrates and other carbon compounds.
- ✔ in aquatic ecosystems, carbon is present as dissolved carbon dioxide and hydrogencarbonate ions.
- ✔ carbon dioxide diffuses from the atmosphere or water into autotrophs.
- ✔ carbon dioxide is produced by respiration and diffuses out of organisms into water or the atmosphere.
- ✔ methane is produced from organic matter in anaerobic conditions by methanogenic archaeans and some diffuses into the atmosphere or accumulates in the ground.
- ✔ methane is oxidized to carbon dioxide and water in the atmosphere.
- ✔ peat forms when organic matter is not fully decomposed because of acidic and/or anaerobic conditions in waterlogged soils.
- ✔ partially decomposed organic matter from past geological eras was converted either into coal or into oil and gas that accumulate in porous rocks.
- ✔ carbon dioxide is produced by the combustion of biomass and fossilized organic matter.
- ✔ animals may have hard parts that are composed of calcium carbonate and can become fossilized in limestone.

You should be able to:

- ✔ estimate carbon fluxes due to processes in the carbon cycle.
- ✔ analyse data from air-monitoring stations to explain annual fluctuations.
- ✔ construct a diagram of the carbon cycle.
- ✔ explain the processes occurring in the carbon cycle.

The continued availability of carbon in ecosystems depends on carbon cycling. The cycle shows the processes that transfer carbon from one carbon sink to another. Carbon fluxes should be measured in gigatonnes, which are 1×10^{15} g (1,000,000,000,000,000 grams).

The structure of carbon compounds such as carbohydrates is studied in Topic 2.3. Carbon dioxide's role in photosynthesis will be studied in Topic 8.3. Carbon dioxide and climate change will be studied in Topic 4.4.

- A **carbon sink** (also known as carbon reservoir) is a reserve of carbon.
- A **carbon flux** is the transfer of carbon from one carbon sink to another.
- **Peat** is organic matter that is not fully decomposed due to acidic or waterlogged soils.
- **Fossil fuels** are partially decomposed organic matter from past geological eras that accumulate in porous rocks.
- **Limestone**, a rock composed of calcium carbonate, is mainly formed from sediments containing the remains of marine organisms such as corals and molluscs.

Example 4.3.1.

The diagram shows part of the carbon cycle.

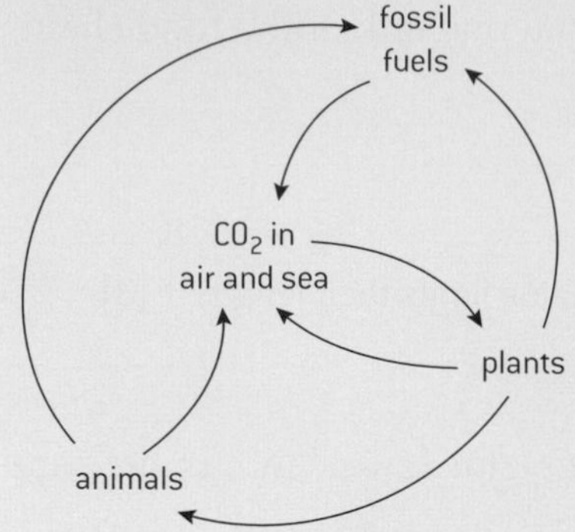

a) State **one** process by which CO_2 is transferred:

(i) from fossil fuels to the air

(ii) from plants to fossil fuels

(iii) from air to plants.

b) The carbon flux from plants and animals to CO_2 in the air and sea is about 120 gigatonnes per year. Explain the process by which this carbon is transferred to the air.

c) On the cycle, draw a box showing the carbon sink of decomposers and the carbon flux due to decomposition leading to the carbon sink of decomposers.

d) Describe the formation of peat.

Solution

a) (i) combustion

(ii) fossilization

(iii) photosynthesis

b) Carbon dioxide is produced by plants and animals by the process of respiration. Respiration involves many chemical reactions that convert food into energy. The reactions are catabolic, as they involve the breakdown of glucose to produce energy in the form of adenosine triphosphate (ATP). This process occurs in the mitochondria of cells. This carbon dioxide diffuses out of cells and passes into the atmosphere, or into the water in the case of aquatic plants. In aquatic habitats carbon dioxide dissolves forming hydrogencarbonate ions.

c) *A box (added anywhere in the cycle) labelled decomposers with arrows from animals and from plants to decomposers (usually in soil).*

d) Waterlogged soils do not have oxygen, as all air gaps are full of water. In these soils anaerobic respiration occurs, resulting in acid conditions. Saprotrophic bacteria and fungi are unable to function in acidic conditions, so full decomposition is inhibited, resulting in peat formation.

SAMPLE STUDENT ANSWER

Increasing carbon dioxide concentration in the atmosphere leads to acidification of the ocean. This in turn reduces the amount of dissolved calcium carbonate. A study was undertaken to investigate the effect of increasing the concentration of atmospheric carbon dioxide on the calcification rate of marine organisms. Calcification is the uptake of calcium into the bodies and shells of marine organisms. The study was undertaken inside Biosphere-2, a large-scale closed mesocosm. The graph shows the results of the data collection.

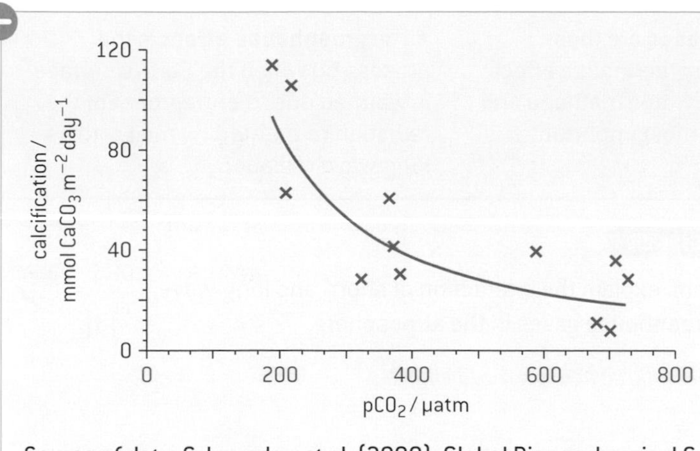

Source of data: C. Langdon et al. (2000), Global Biogeochemical Cycles, 14(2), pp. 639–654

a) State the relationship between atmospheric carbon dioxide and calcification rates. [1]

This answer could have achieved 1/1 marks:

As carbon dioxide concentration increases, the calcification rates decrease exponentially.

▲ Although what is shown in the graph on the X axis is the partial pressure of carbon dioxide, the concentration of carbon dioxide was accepted as an answer.

b) Outline **one** way in which reef-building corals are affected by increasing atmospheric carbon dioxide. [2]

This answer could have achieved 1/2 marks:

Reef-building corals require carbonate ions to help build (deposits in) their shells. If atmospheric CO_2 is high, more carbon dioxide is taken into the water and the CO_2 reacts with water to form H_2CO_3 which dissociates into H^+ and HCO_3^-, further decreasing the carbonate ions available for the reef-building corals to deposit.

▲ The answer scored the mark for mentioning the fact that the carbonate ion availability is decreased with increased CO_2 levels.

▼ The answer does not mention that an increase in CO_2 increases global temperature (CO_2 is a greenhouse gas). Higher ocean temperatures (the same applies for acidification) leads to rejection of zooxanthellae, therefore increasing coral bleaching.

4.4 CLIMATE CHANGE

You should know:

- ✔ concentrations of gases in the atmosphere affect climates experienced at the Earth's surface.
- ✔ carbon dioxide and water vapour are the most significant greenhouse gases.
- ✔ methane and nitrogen oxides have less impact.
- ✔ the impact of a gas depends on its ability to absorb long-wave radiation as well as on its concentration in the atmosphere.
- ✔ warmed Earth emits longer-wavelength radiation (heat).
- ✔ longer-wave radiation is absorbed by greenhouse gases that retain the heat in the atmosphere.
- ✔ global temperatures and climate patterns are influenced by concentrations of greenhouse gases.
- ✔ there is a correlation between rising atmospheric concentrations of carbon dioxide since the start of the industrial revolution 200 years ago and average global temperatures.
- ✔ recent increases in atmospheric carbon dioxide are largely due to increases in the combustion of fossilized organic matter.

You should be able to:

- ✔ explain the effect of greenhouse gases on climate change.
- ✔ analyse data on coral reefs from increasing concentrations of dissolved carbon dioxide.
- ✔ analyse data showing correlations between global temperatures and carbon dioxide concentrations on Earth.
- ✔ evaluate claims that human activities are not causing climate change.

The properties of water and how these control temperature are studied in Topic 2.2. The production of carbon dioxide in respiration is shown in Topic 2.8 and the use of carbon dioxide in photosynthesis is discussed in Topic 2.9.

Claims, such as the claims that human activities are causing climate change, need to be assessed. Scientific evidence for warming of the climate system is unequivocal. The global temperature of the Earth's surface has increased 1.8°C since 1880. Many scientific models have been used to explain these changes. It is likely that this trend is partially due to human activity, as it is unlikely that these changes occurred only from natural internally generated variability of the climate system. Nevertheless, there are many uncertainties that need to be tested, especially those related to variability due to natural causes.

- **Greenhouse gases** are those that produce the greenhouse effect (carbon dioxide, water, methane and nitrogen are the most important).
- The **greenhouse effect** is the process by which the Earth's surface is warmed due to entrapment of the radiation re-emitted by the Earth as long-wave radiation.

SAMPLE STUDENT ANSWER

Using the diagram, explain the interaction of short- and long-wave radiation with greenhouse gases in the atmosphere. [3]

This answer could have achieved 3/3 marks:

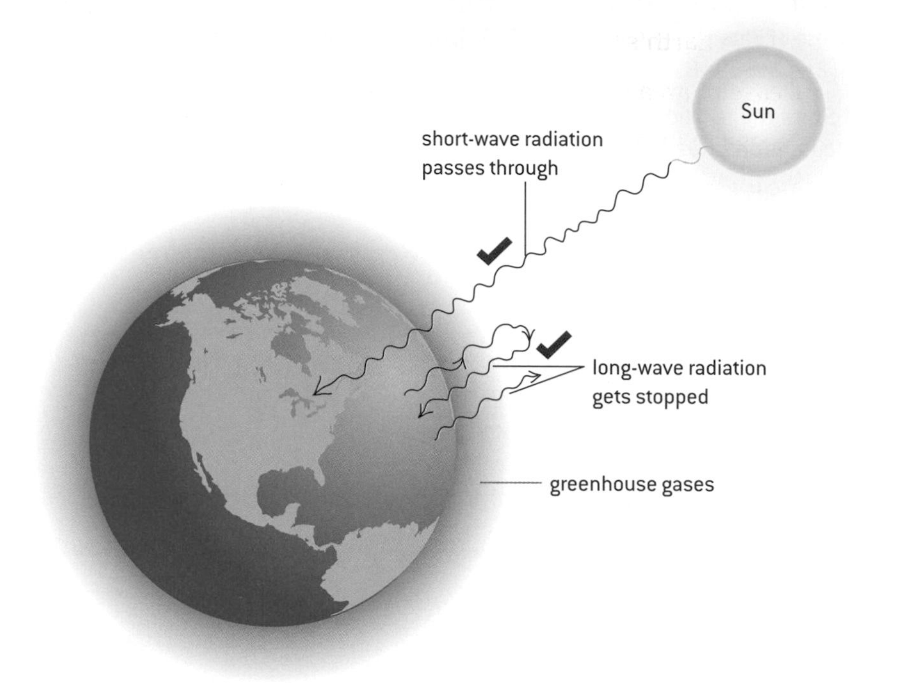

Greenhouse gases include gases such as CO_2, methane, water vapour, etc. Short-wave radiations from the rays of the Sun penetrate these gases in the atmosphere, reaching the surface of the Earth and thus delivering heat energy. However, when they reflect from the surface, they become long-wave radiation which is absorbed by the gases and reflected back to the Earth or trapped in the atmosphere. This causes the Earth to become warmer.

▲ The answer scored full marks for mentioning that the reflected radiation is long-wave radiation and is reflected as heat. Long-wave radiation is unable to pass through the greenhouse gases. Short-wave radiation (as UV) passes through the greenhouse gases to the Earth's surface.

Methane is an important greenhouse gas. It is produced from organic matter in anaerobic conditions. Methanogenic archaeans produce methane and water from carbon dioxide and hydrogen. They can also produce methane from acetate. Some methane diffuses into the atmosphere or accumulates in the ground. Methane is oxidized to carbon dioxide and water in the atmosphere, but this takes many years.

There are two strategies to use to identify the correct answer to a multiple choice question: 1. pick the correct answer directly, 2. eliminate all the wrong answers.

Example 4.4.1.

The enhanced greenhouse effect has caused global warming. What is likely to occur in the Arctic if the Earth's surface temperature rises?

A. Increase in area of ice habitat

B. Decrease in numbers of pest species and pathogens

C. Decreased rates of decomposition of detritus

D. Increased amount of predators from temperate regions

Solution

The correct answer is **D**. All the other answers are the opposite of what happens, as in **A** there is a decrease in ice areas and in **B** and **C** they all increase due to global warming.

Example 4.4.2.

What gases do methanogenic archaeans use to produce methane that is liberated into the atmosphere?

A. Carbon dioxide, hydrogen and oxygen

B. Carbon dioxide, ethanol and hydrogen

C. Carbon monoxide or acetate and ozone

D. Carbon dioxide and hydrogen or acetate

Solution

The correct answer is **D** as methanogenic archaeans can produce methane from the reaction of carbon dioxide and hydrogen or from acetate.

Practice problems for Topic 4

Problem 1

Outline the potential energy value of earthworms as primary consumers, grown on household food waste such as fruit and salad leftovers, as a potential source of food.

Problem 2

a) Describe the production of methane by bacteria.

b) Explain the reason carbon dioxide is a more significant greenhouse gas than methane.

Problem 3

Discuss the processes in the carbon cycle that affect concentrations of carbon dioxide and methane in the atmosphere and the consequences for climate change.

5 EVOLUTION AND BIODIVERSITY

5.1 EVIDENCE FOR EVOLUTION

You should know:

- ✔ evolution occurs when heritable characteristics of a species change.
- ✔ fossil records provide evidence for evolution.
- ✔ selective breeding of domesticated animals shows that artificial selection can cause evolution.
- ✔ evolution of homologous structures by adaptive radiation explains similarities in structure when there are differences in function.
- ✔ populations of a species can gradually diverge into separate species by evolution.
- ✔ continuous variation across the geographical range of related populations supports the concept of gradual divergence.

You should be able to:

- ✔ explain the processes involved in evolution.
- ✔ analyse data on development of melanistic insects in polluted areas.
- ✔ compare the pentadactyl limbs of mammals, birds, amphibians and reptiles with different methods of locomotion.

The inheritance of genetic traits is studied in Topic 3.4 and Topic 5.2 discusses how natural selection leads to evolution.

- **Evolution** is the change in characteristics that can be inherited by different generations of a population.
- **Homologous structures** are those inherited from a common ancestor.
- **Analogous structures** have similar structures but a different ancestry.
- **Convergent evolution** is the development of similar features in unrelated species, for example, wings in birds and insects or the streamlined bodies of fish and cetaceans.

There is overwhelming evidence for the evolution of life on Earth. The fossil record provides evidence for evolution. This includes not only direct evidence such as preserved bones, shells, teeth, and pine cones but also indirect evidence such as footsteps, feces and pollen.

Comparative anatomy has shown that certain structural features are basically similar in different species. This homology may be regarded as evidence of evolution from a common ancestor. The forelimbs of tetrapod vertebrates have similar structures, as they all have a five-finger pattern (pentadactyl), but are adapted for different purposes. For example bats use their wings to fly while moles use their front limbs to dig. Homologous structures showing adaptations to different conditions or lifestyles are examples of divergent evolution.

Diversity of life has evolved and continues to evolve by selection. Natural selection increases the frequency of characteristics that make individuals better adapted and decreases the frequency of other characteristics leading to changes within the species. These changes in species can lead to evolution. Artificial selection can also cause evolution, as can be seen in selective breeding of domesticated animals.

- **Divergent evolution** describes the evolution of new species from a single parent species in response to evolutionary triggers, for example Darwin's finches.
- **Adaptive radiation** is a rapidly occurring type of divergent evolution in which the newly diversified species are adaptively specialized to a particular environmental niche.
- **Speciation** is the creation of new species from one existing species, often as a result of divergent evolution.

Example 5.1.1.

What evidence for evolution can be seen in domesticated animals?

A. Ancestors of domesticated animals cannot be found in the fossil record.

B. Variation in domesticated animals is due to sexual reproduction.

C. Animals adapt to the environment but these characteristics are not inherited.

D. Differences between breeds show that artificial selection can cause species to change.

Solution

The correct answer is **D**. Artificial selection is much faster than natural selection. The fossils of ancestors of domesticated animals have been found, so **A** is incorrect. Although answer **B** is correct, this is not evidence for evolution for domesticated animals only. Answer **C** is not related to evolution, as characteristics acquired during the lifetime of an individual are not heritable.

SAMPLE STUDENT ANSWER

Outline how reproductive isolation can occur in an animal population. [3]

This answer could have achieved 1/3 marks:

Reproductive isolation can for example occur through a change in mating behaviour. The different behaviour can be passed on to the offspring up to a point, where individuals practicing the old behaviour won't mate with those who do the new one any more. Speciation has occurred as the two groups won't interbreed anymore.

▲ This answer scored a mark for outlining the process of mating behaviour—a type of behavioural isolation

▼ Reproductive isolation can be sympatric or allopatric. The other methods the answer could have mentioned are the temporal isolation, geographic isolation, or isolation shown by plants with polyploidy

Assessment tip

If the question is worth 3 marks, you will need to outline at least three methods.

5.2 NATURAL SELECTION

You should know:

- ✔ diversity of life has evolved and continues to evolve by natural selection.
- ✔ natural selection can only occur if there is variation among members of the same species.
- ✔ mutation, meiosis and sexual reproduction cause variation between individuals in a species.
- ✔ adaptations are characteristics that make an individual suited to its environment and way of life.
- ✔ species tend to produce more offspring than the environment can support.
- ✔ individuals that are better adapted tend to survive and produce more offspring while the less well adapted tend to die or produce fewer offspring.
- ✔ individuals that reproduce pass on characteristics to their offspring.
- ✔ natural selection increases the frequency of characteristics that make individuals better adapted and decreases the frequency of other characteristics, leading to changes within the species.

You should be able to:

- ✔ describe adaptations to environments.
- ✔ explain the processes involved in natural selection.
- ✔ explain the process of speciation.
- ✔ analyse data respecting the changes in beaks of finches on Daphne Major.
- ✔ explain evolution of antibiotic resistance in bacteria.

The inheritance of traits is studied in Topic 3.4. The characteristics of species are studied in Topic 4.1 and the evidence for evolution in Topic 5.1.

Theories are used to explain natural phenomena, for example, the theory of evolution by natural selection can explain the development of antibiotic resistance in bacteria. Methicillin-resistant *Staphylococcus aureus* (MRSA) is a bacterium that causes mild infections on the skin, like sores or boils. It can also cause more serious skin infection or if it enters the bloodstream it can reach the lungs or the urinary tract. Though most MRSA infections aren't serious, some can be life-threatening. Because it is resistant to commonly used antibiotics, it is tougher to treat than most strains of *Staphylococcus aureus*. This resistance has arisen by the excessive use of antibiotics which selects for the most resistant strains by competitive exclusion.

There are three levels at which the evolutionary process acts. First is the origin of genetic novelties caused by variation. Second is the ordering of those variations in moulding the genetic structures of populations into new frequencies and third is the fixation of the diversity already attained on the preceding two levels. The first is attained by mutations and recombinations, the second by natural selection and the third level is obtained by isolating mechanisms.

- **Variations** are differences in a species due to mutation, meiosis and sexual reproduction.
- **Adaptations** are characteristics that make an individual suited to its environment and way of life.
- **Natural selection** is a theory involving mechanisms that contribute to the selection of individuals that reproduce. It can lead to the evolution of species.

Variation in species arises from recombination of genetic material. This can happen in different ways. Crossing over and independent assortment of homologous chromosomes during meiosis I produce different combinations of gametes. The combination of male and female gametes when fertilization occurs also produces variation. Mutations are random and usually neutral but can be beneficial or harmful. These variations lead to adaptations which can aid a population's chance of survival and reproduction.

Example 5.2.1.

What increases variation in a population?

I. Mutation II. Meiosis III. Sexual reproduction

A. I only

B. I and II only

C. II and III only

D. I, II and III

Solution

The correct answer is **D**, as they all promote variation within a population. Mutations can cause a change in the DNA of organisms, changing their genotype. Mutations occurring in gametes are inherited. In sexually reproducing organisms, variation occurs during meiosis by crossing over of genes between homologous chromosomes and the random orientation in the migration of chromatids. Another source of variation results from the random fusion of female and male gametes.

SAMPLE STUDENT ANSWER

Some lice live in human hair and feed on blood. Shampoos that kill lice have been available for many years but some lice are now resistant to those shampoos. Two possible hypotheses are:

Hypothesis A	Hypothesis B
Resistant strains of lice were present in the population. Non-resistant lice died with increased use of anti-lice shampoo and resistant lice survived to reproduce.	Exposure to anti-lice shampoo caused mutations for resistance to the shampoo and this resistance is passed on to offspring.

Discuss which hypothesis is a better explanation of the theory of evolution by natural selection. [3]

This answer could have achieved 2/3 marks:

Hypothesis A is better as it follows the theory of natural selection. Whereby certain organisms possessed the resistance to the shampoo, probably through mutation. Due to the use of anti-lice shampoo, there was a fight for survival in which the resistant lice who were better adapted to live in such environment survived, while those who were not, died. The resistant lice thus were able to reproduce and passed on this trait to the offspring (which hypothesis B mentions).

Assessment tip

As this is a discussion, valid statements for both hypotheses can be mentioned. For example, variation is necessary for natural selection to occur, and in this case both hypotheses include variation in the population of lice, resistant and non-resistant.

▲ One mark was awarded for hypothesis A as the better one as the resistance (mutation) to the shampoo would be present in the population initially and not caused by the shampoo. The other mark was for saying that the frequency of the best adapted increases and resistance is passed on through reproduction.

▼ The answer fails to mention that characteristics acquired during the lifetime of the individual might not be inherited as not all mutations are heritable.

Darwin's finches, inhabiting the Galapagos archipelago, are a good example of speciation and adaptive radiation. Their common ancestor arrived on the Galapagos about two million years ago. Fifteen of the currently recognized species evolved from this common ancestor. Changes in the size and form of the beak have enabled different species to utilize different food resources, especially seeds of different sizes. The environment on the small island called Daphne Major is harsh so only a few organisms can survive. Here, the medium ground finch, *Geospiza fortis,* subsists mainly on small seeds but members of the *G. fortis* population with larger beaks are able to tackle the bigger seeds of a plant called *Tribulus cistoides*. In 1982, a few members of a larger species of finch, *Geospiza magnirostris,* flew onto the island. Birds of this species have bigger beaks, allowing them to crack the large seeds of *T. cistoides* easily. The population of *G. magnirostris* grew, until there were enough to compete with the *G. fortis* for *T. cistoides* seeds. During a drought, there was little availability of the large seeds, so the *G. fortis* finches with smaller beaks had an advantage over the *G. magnirostris* finches with bigger beaks, as they could feed on smaller seeds. This trait was then passed down to the next generation, showing a strong shift towards smaller beak size among *G. fortis*.

Assessment tip

The term trait is accepted instead of allele or gene, as it is the phenotypic expression of the characteristic.

(a) *G. fortis (large beak)*

(b) *G. fortis (small beak)*

(c) *G. magnirostris*

▲ **Figure 5.2.1.** Darwin's finches

5.3 CLASSIFICATION OF BIODIVERSITY

You should know:

- ✔ species are named and classified using an internationally agreed system—the binomial system.
- ✔ the binomial system is used to name newly discovered species.
- ✔ taxonomists classify species using a hierarchy of taxa.
- ✔ organisms are classified into three domains: Eukarya, Archaea and Eubacteria.
- ✔ the main taxa used to classify eukaryotes are kingdom, phylum, class, order, family, genus and species.
- ✔ in a natural classification, the genus and accompanying higher taxa consist of all the species that have evolved from one common ancestral species.
- ✔ taxonomists sometimes reclassify groups of species when new evidence shows that a previous taxon contains species that have evolved from different ancestral species.
- ✔ natural classifications help in identification of species and allow the prediction of characteristics shared by species within a group.

You should be able to:

- ✔ determine the classification of one plant and one animal species from domain to species level.
- ✔ state recognition features of Bryophyta, Filicinophyta, Coniferophyta and Angiospermophyta.
- ✔ state recognition features of Porifera, Cnidaria, Platyhelminthes, Annelida, Mollusca, Arthropoda and Chordata.
- ✔ state recognition features of birds, mammals, amphibians, reptiles and fish.
- ✔ construct a dichotomous key for use in identifying specimens.

The characteristics of species, communities and ecosystems are studied in Topic 4.1. The evidence for evolution is covered in Topic 5.1. The characteristics of plant species are related to the presence of vascular tissues. Transport in plants will be studied in Topic 9.1 and Topic 9.2.

- Eukaryotes are **classified** from higher hierarchy to lower as kingdom, phylum, class, order, family, genus and species.
- The **binomial system** of classification uses *"Genus species"* to name organisms.

Natural classifications help in identification of species and allow the prediction of characteristics shared by species within a group. Sometimes taxonomists reclassify groups of species when new evidence shows that a previous taxon contains species that have evolved from different ancestral species. For example, sequencing of the 16S rRNA bases (part of the 30S ribosome of prokaryotes) has resulted in the reclassification of many bacterial species. The binomial system is used to avoid the confusion which often arises from the use of common names. The genus and species names are normally underlined or written in italics. The genus name begins with an upper case (capital) letter and the species name with a lower case (small) letter; for example, *Homo sapiens* is the scientific name for humans.

Taxonomists classify organisms by assigning them to groups based on the presence or absence of shared features. The features used to assign plants to four different phylums include:

- Bryophyta: Non-vascular land plants including liverworts, mosses and hornworts. They do not produce fruits, pollen or seeds. They have rhizoids instead of roots. Fertilization of gametes depends on the presence of water as it occurs outside the organism.
- Filicinophyta: Vascular plants including ferns. Have roots, stems and leaves with xylem and phloem. Do not produce pollen and have no ovaries. Reproduction is by spores on the underside of leaves.

- Coniferophyta: Vascular plants including evergreen pine trees. Have roots, stems and leaves with xylem and phloem. Pollen and ovules are produced in cones, where seeds are produced. They do not produce flowers.
- Angiospermophyta: Vascular flowering plants such as roses. They have roots, stems and leaves with xylem and phloem. Pollen and ovules are produced in flowers, where fruits containing seeds are produced.

Assessment tip

Other characteristics can be used to construct the key, but always try to keep it simple.

Example 5.3.1.

Construct a simple dichotomous key to classify plants.

Solution

1. a. Produce seeds .... go to 2
   b. Do not produce seeds .... go to 3
2. a. Seeds enclosed in ovary .... **Angiospermophyta**
   b. Seeds not enclosed in ovary ....**Gymnospermophyta**
3. a. Has vascular tissue ....**Filicinophyta**
   b. Does not have vascular tissue ....**Bryophyta**

- **Invertebrates** are animals that have no vertebral column, such as the Arthropoda.
- **Vertebrates** are animals that have a vertebral column. All vertebrates have a spinal cord (notochord). These are classified as Chordata, along with a few primitive species that have a notochord but no vertebral column.

Features used by taxonomists to classify invertebrates as members of the following six phyla include:

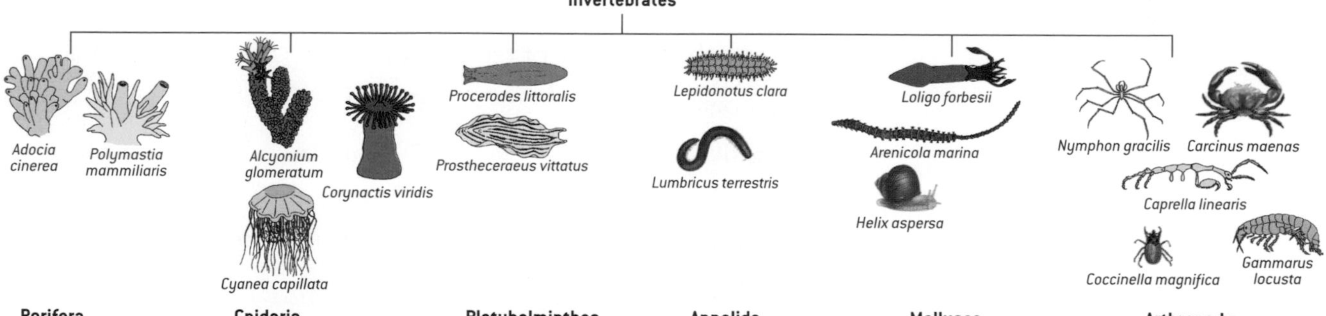

Porifera	Cnidaria	Platyhelminthes	Annelida	Mollusca	Arthropoda
• no mouth or anus	• radial symmetry	• bilateral symmetry	• bilateral symmetry	• bilateral symmetry	• bilateral symmetry
• pores on surface	• sting cells	• mouth	• mouth and anus	• mouth and anus	• mouth and anus
• spicules	• tentacles	• no skeleton	• blood vessels	• soft body	• jointed legs
	• calcium exoskeleton		• no skeleton	• hard shell	• exoskeleton

▲ **Figure 5.3.1.** The six phyla of invertebrate taxonomy

SAMPLE STUDENT ANSWER

Lice are wingless insects that belong to the phylum Arthropoda. State **two** characteristics that identify lice as members of the Arthropoda. [2]

This answer could have achieved 1/2 marks:

1. have head, legs and exoskeleton.
2. have bilateral symmetry.

▼ Although it is correct, the mark for exoskeleton is not given because if there is a list, only the first answer is considered and "head" is too vague to score a mark. The answer could have mentioned jointed appendages, segmented body, or mouth and anus

▲ The answer scored one mark for bilateral symmetry

Dichotomous keys can be used to identify organisms. A series of questions need to be answered in order to deduce to which group an organism belongs.

Features used by taxonomists to classify chordate animals are shown in Figure 5.3.2.

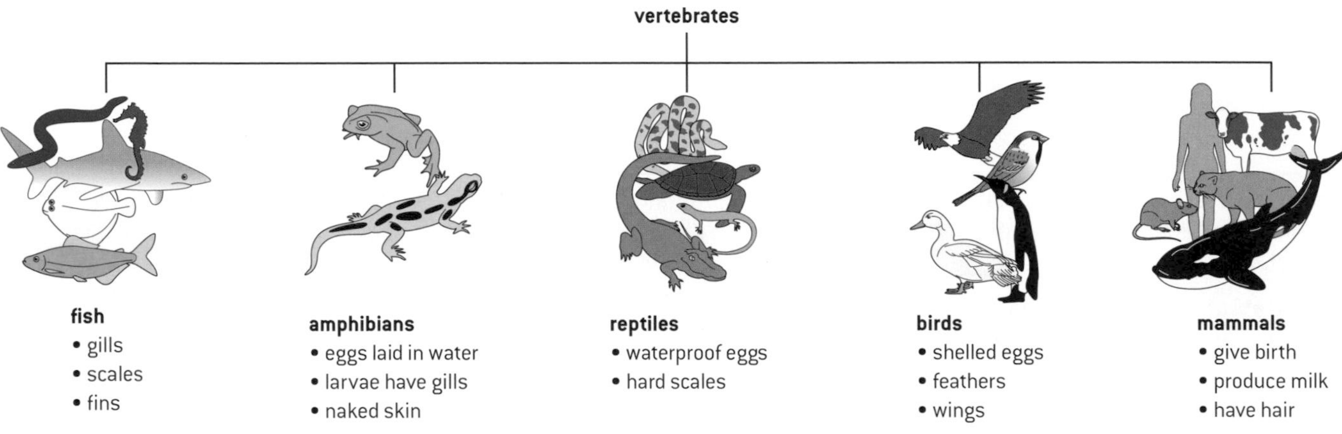

▲ **Figure 5.3.2.** Features for classifying vertebrates

Example 5.3.2.

Which is a characteristic of reptile skin?

A. soft moist

B. dry scaly

C. hairy

D. feathery

Solution

The answer is **B**, as **A** is for amphibians, **C** is for mammals and **D** is for birds.

Classifying the more than 2 million known species is important in facilitating meaningful communication among scientists and society in general. Traditional taxonomic identification based on observable traits is rapidly being replaced by molecular approaches. Classification based on genetic or protein sequences has two advantages over observation of morphology. Genetically distinct species which cannot be visually distinguished are easily recognized by differences in their DNA or protein sequences and there is no human error when identifying species by their DNA. However, because it is difficult to obtain genetic sequences during a field trip, classifying organisms using external features is the most advantageous approach when not in a laboratory.

5.4 CLADISTICS

You should know:

- ✔ a clade is a group of organisms that have evolved from a common ancestor.
- ✔ base or amino acid sequences are evidence for which species are part of a clade.
- ✔ sequence differences accumulate gradually—there is a positive correlation between the number of differences between two species and the time since they diverged from a common ancestor.
- ✔ traits can be analogous or homologous.
- ✔ cladograms are tree diagrams that show the most probable sequence of divergence in clades.
- ✔ evidence from cladistics has shown that classifications of some groups based on structure did not correspond with the evolutionary origins of a group or species.

You should be able to:

- ✔ construct cladograms including humans and other primates.
- ✔ analyse cladistic evidence that led to the reclassification of the figwort family.
- ✔ analyse cladograms to deduce evolutionary relationships.

Evidence for which species are part of a clade can be obtained from the base sequence of a gene or the corresponding amino acid sequence of a protein. In 1977 Carl Woese, studying the sequences of 16S ribosomal RNA, realized that Archaea had a separate line of evolutionary descent from bacteria. This determined the new classification of domains.

DNA and protein sequences are usually shown in a format called FASTA. The similarities in DNA sequences can be shown using the software BLASTn and protein sequences using BLASTp. Sequence alignments can be performed in BLAST or ClustalW.

A cladogram is a tree diagram used in cladistics to show relationships between organisms. A cladogram uses lines that branch off in different directions ending at a clade. A clade is a group of organisms with a last common ancestor. At each node a splitting event occurs. The node therefore represents the end of the ancestral taxonomic group (taxon), and the stems represent the newly formed species that split from the ancestor.

The inheritance of traits is studied in Topic 3.4. The evidence for evolution and how evolution took place are studied in Topic 5.1 and Topic 5.2, respectively. Classification of species is studied in Topic 5.3.

• A **trait** is an identifiable, heritable phenotypic characteristic of an organism, such as blue eyes, or red or white flowers. It is used in the description of the observable characteristics expressed by an allele.

Plant families have been reclassified as a result of evidence from cladistics. In 2001, Richard Olmstead and Pat Reeve studied the DNA sequences of three chloroplast genes through PCR sequencing of 65 taxa of the aquatic angiosperms in the Scrophulariaceae, commonly known as the figwort family. They found that the groups previously included in this family were not all descendants of a common ancestor; rather, they were five different groups.

Example 5.4.1.

The cladogram shows the evolutionary relationship between some animals.

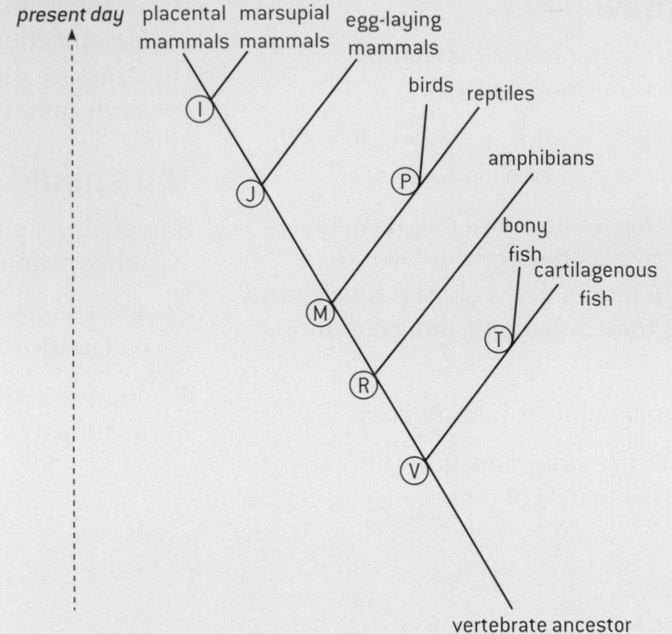

a) State the organisms that are closely related to reptiles.

b) Identify the group of organisms that is most distantly related to fish.

c) Identify the node (bifurcation) where fish diverged.

Solution

a) birds **b)** mammals **c)** V

Example 5.4.2.

The FASTA sequences for cytochrome c proteins in humans (*Homo sapiens*) and mice (*Mus musculus*) were aligned using BLASTp.

Score	Expect	Method	Identities	Positives	Gaps
199 bits(507)	2e-73	Compositional matrix adjust.	96/105(91%)	99/105(94%)	0/105 (0%)

```
Query    1  MGDVEKGKKIFIMKCSQCHTVEKGGKHKTGPNLHFGRKTGQAPGYSYTAANKNKGIIW  60
            MGDVEKGKKIF+ KC+QCHTVEKGGKHKTGPNLHFGRKTGQA G+SYT ANKNKGI W

Subject  1  MGDVEKGKKIFVQKCAQCHTVEKGGKHKTGPNLHFGRKTGQAPGFSYTDANKNKGITW  60

Query   61  GEDTLMEYLENPKKYIPGTKMIFVGIKKKEERADLIAYLKKATNE  105
            GEDTLMEYLENPKKYIPGTKMIF GIKKK ERADLIAYLKKATNE

Subject 61  GEDTLMEYLENPKKYIPGTKMIFAGIKKKGERADLIAYLKKATNE  105
```

a) State the number of amino acids in the studied protein that are different in these organisms.

b) Suggest a reason cytochrome c is useful to construct cladograms.

c) Yeast (*Saccharomyces cerevisiae*) cytochrome c protein has 64 differences from human cytochrome c. Identify whether mice or yeast are closer to humans in the cladogram.

Solution

a) 9

b) Cytochrome c is an evolutionarily conserved protein. Substitutions within the primary structure of cytochrome c are relatively constant over time, which makes characterizing cytochrome c a useful molecular clock.

c) Mice are closer, as the more differences there are, the earlier they separated; therefore yeast separated earlier.

Practice problems for Topic 5

Problem 1

Beak shape in Darwin's finches is emblematic of natural selection and adaptive radiation. Representatives of four distinct beak morphologies of finches that feed on different parts of cacti are: small blunt in *Geospiza fuliginosa*, medium blunt in *Geospiza fortis*, large blunt in *Geospiza magnirostris* and medium pointed in *Geospiza scandens*. The length-to-width ratios of the beaks were measured and are shown in the scatter diagram.

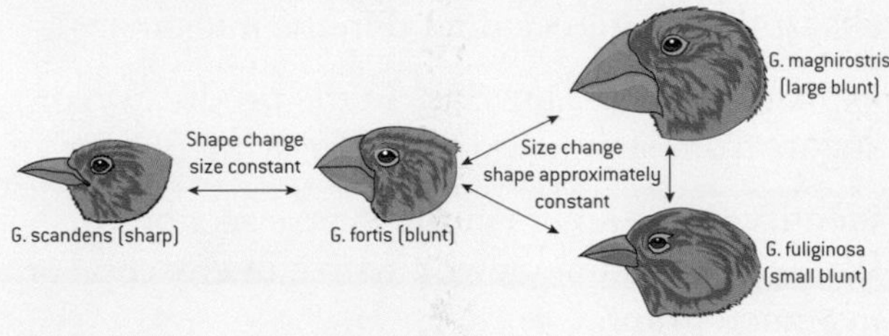

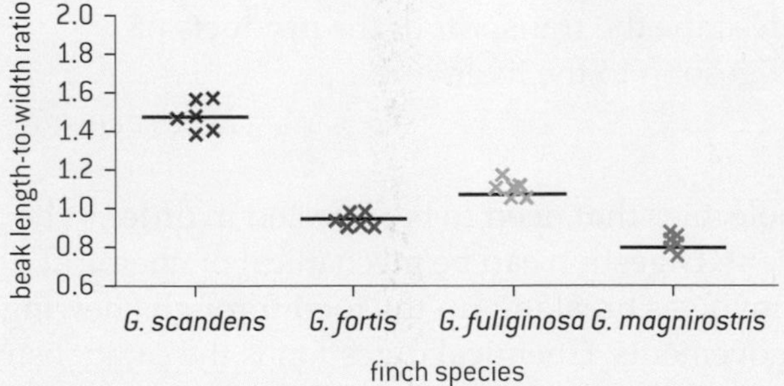

a) State the median beak length-to-width ratio in *G. fortis*.

b) Compare the beak shape of *G. scandens* with that of *G. magnirostris*.

c) Explain how the shapes of the beaks allow different feeding habits.

d) Using the example of finches, suggest how populations can gradually diverge into separate species by evolution.

Problem 2

Explain how natural selection can lead to evolution.

Problem 3

Complete the table to compare and contrast the following invertebrates. Use a tick (✓) to show the feature is present.

	Phylum					
Feature	**Porifera**	**Cnidaria**	**Platyhelminthes**	**Annelida**	**Mollusca**	**Arthropoda**
Articulated legs						
Bilateral symmetry						
Outer shell						

Problem 4

The image shows part of a cladogram.

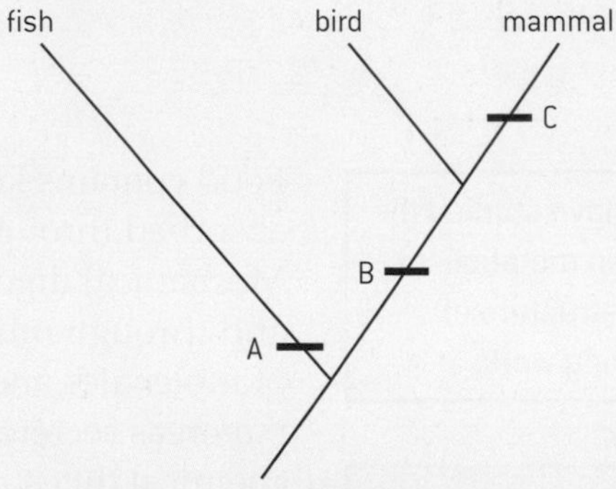

a) Using the cladogram, suggest one diagnostic feature that characterizes the given groups of vertebrates at A, B and C.

b) State the name of the domain to which these organisms belong.

6 HUMAN PHYSIOLOGY

6.1 DIGESTION AND ABSORPTION

You should know:

- ✔ the structure of the wall of the small intestine allows it to move, digest and absorb food.
- ✔ contraction of circular and longitudinal muscle of the small intestine mixes the food with enzymes and moves it along the gut.
- ✔ the pancreas secretes enzymes into the lumen of the small intestine.
- ✔ enzymes digest most macromolecules in food into monomers in the small intestine.
- ✔ villi increase the surface area of epithelium over which absorption is carried out.
- ✔ villi absorb monomers formed by digestion as well as mineral ions and vitamins through different membrane transport methods.

You should be able to:

- ✔ draw an annotated diagram of the digestive system.
- ✔ explain the processes occurring in the small intestine that result in the digestion of starch.
- ✔ describe and make annotated diagrams of the structure of the villus.
- ✔ explain how molecules are absorbed in the villus.
- ✔ describe the use of dialysis tubing to model absorption of digested food in the intestine.
- ✔ explain the use of amylase in the production of sugars from starch and in the brewing of beer.
- ✔ identify tissue layers in transverse sections of the small intestine viewed with a microscope or in a micrograph.
- ✔ describe the transport of the products of digestion to the liver.

In Topic 2.1 you have studied the molecules involved in metabolism and in Topic 2.5 the structure of enzymes and how they work.

- **Digestion** is the process by which foods are transformed into soluble molecules.
- **Absorption** is the process by which monomers enter the cells of the intestine wall.
- **Amylase** is the enzyme that digests starch, a polysaccharide, into disaccharides.
- **Lipase** is the enzyme that digests lipids such as triglycerides into fatty acids and glycerol.
- **Proteases** are enzymes that digest proteins into peptides and then into amino acids.

Food contains large molecules that need to be digested in order to be absorbed through the gut. Digestion can be mechanical or chemical. Mechanical digestion involves breaking up the food through chewing and through muscle movements. Chemical digestion is the catabolism of molecules and includes the use of acids or alkalis and enzymes. The pancreas secretes enzymes into the lumen of the small intestine for chemical digestion.

Example 6.1.1.

a) Label the diagram of the digestive system.

b) Describe the digestion of starch in humans.

c) Explain the reason lipase is not involved in the digestion of starch.

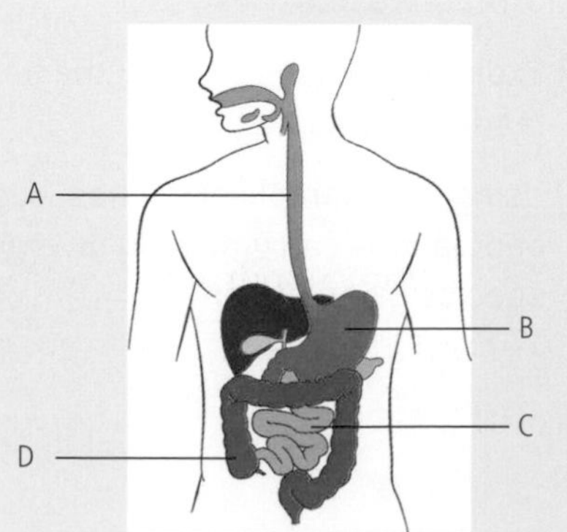

Solution

a) A: esophagus; B: stomach; C: small intestine; D: large intestine

b) Starch is first digested in the mouth. The teeth chew the starchy food and the tongue helps to mix it with saliva. The starch starts to be digested with salivary amylase, which breaks down some of the starch molecules into maltose. Starch is a polysaccharide while maltose is a disaccharide. The food is then taken to the stomach and from there to the small intestine. The small intestine has circular and longitudinal muscles that help move the food along the gut. The pancreatic amylase is secreted into the lumen of the small intestine. It continues the digestion of starch molecules into maltose. Maltose is then digested by maltase into glucose molecules (monomers). Glucose can now be absorbed by the intestinal villi to enter the blood.

c) Lipase is an enzyme that digests lipids. Specifically it breaks down triglycerides into fatty acids and glycerol. Enzymes are specific—their active site will usually accept only one substrate. Lipase is no good for digesting starch because starch does not fit its active site.

Assessment tip

You can be asked questions on how starch, glycogen, lipids and nucleic acids are digested into monomers. Remember that cellulose remains undigested.

- **Villi** are small finger-like projections of the intestinal wall facing the lumen.
- **Microvilli** are small projections found on the surface of the epithelial cells of the villi.
- A **lacteal** is the lymphatic vessel found in each villus which aids lipid absorption.
- The **mucosa** is the layer of epithelial tissue where food absorption occurs.
- **Circular and longitudinal muscles** contract in different directions, allowing the movement of food through the gut.

The small intestine is lined with villi. Villi absorb monomers formed by digestion as well as mineral ions and vitamins through different methods. Villi increase the surface area of epithelium over which absorption is carried out. Each villus has a central lacteal into which lipids are absorbed and a network of capillaries into which the rest of the useful substances from digestion are absorbed.

Example 6.1.2.

a) State **one** type of molecule absorbed into the lacteal.

b) State the route of amino acids into the villus.

c) State **one** function of:

(i) circular muscle (ii) mucosa.

Solution

a) fatty acids

b) from the lumen of the small intestine to capillaries in the villus

c) (i) contract to move food along the gut

(ii) absorption of foods

Example 6.1.3.

The micrograph shows epithelial cells of a villus.

a) On the micrograph label:

(i) a mitochondrion

(ii) microvilli

(iii) a nucleus.

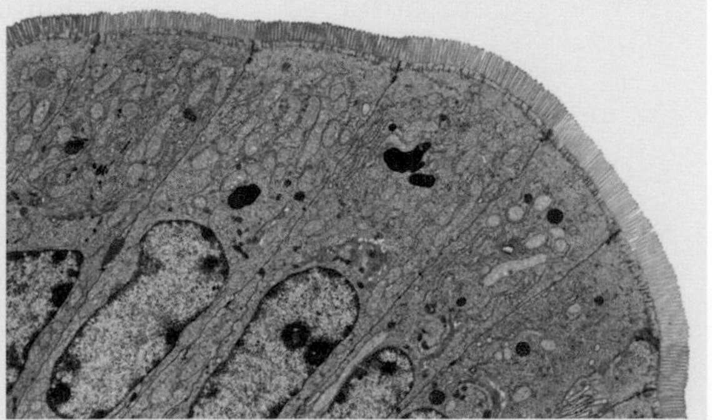

The structure and function of the intestine wall is difficult to study, as mechanisms to do so are invasive. Models can be used as representations of the real world. Dialysis tubing can be used to model absorption in the intestine because it is selectively permeable.

b) Suggest **one** adaptation of the epithelial cells of the villus to absorption other than many mitochondria for energy.

Solution

a) Labels as follows: microvilli are shown yellow at the top of the cell, mitochondria are shown orange and occur throughout the cell and the nucleus is shown blue.

b) Villi have many microvilli that increase the surface area for absorption. They also contain basal channels that help in the absorption of substances. Enzymes expressed on the surface of intestinal epithelium ensure close proximity of the food molecules to the villi for absorption. The blood capillaries close to the epithelium increase absorption.

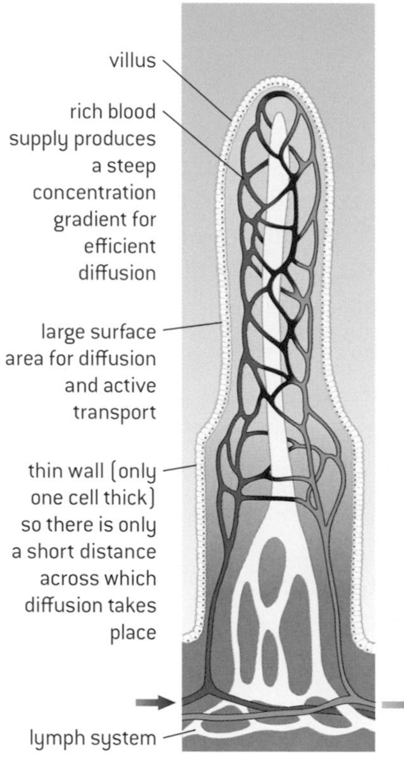

▲ **Figure 6.1.1.** Adaptations of the villus to absorption

Once digested, sugars or polysaccharides have become monosaccharides, lipids such as triglycerides have become fatty acids and glycerol, and proteins have become amino acids.These monomers can now be absorbed by the epithelial cells of the villi. Glycerol is absorbed by simple diffusion, as it can pass through the cell membrane. Fatty acids are absorbed by facilitated diffusion. Glucose and amino acids are absorbed by co-transport with sodium. This process itself does not require energy, but to maintain the sodium gradient, the sodium–potassium pump must move the sodium from one side of the cell to the other through active transport. This does require energy in the form of ATP. Most water-soluble vitamins are absorbed by simple diffusion.

• **Sodium co-transport** is the process by which sodium is transported outside the cell by active transport to allow the entrance of other substances such as glucose into the cell by co-transport with the sodium ions.

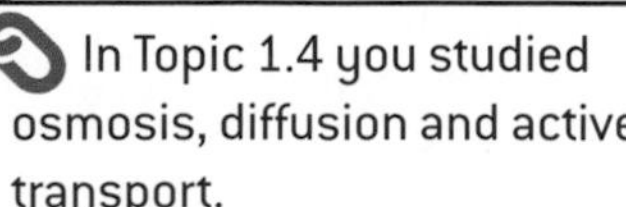

In Topic 1.4 you studied osmosis, diffusion and active transport.

Example 6.1.4.

How is glycerol absorbed into the epithelium of the intestine wall?

A. Simple diffusion

B. Osmosis

C. Facilitated diffusion

D. Endocytosis

Solution

Glycerol is absorbed by simple diffusion, so the correct answer is **A**. Glycerol is soluble in lipids, therefore it can pass through the bilipid layer of the cell membrane without the need for a channel.

6.2 THE BLOOD SYSTEM

You should know:

- ✔ the blood system continuously transports substances to cells and simultaneously collects waste products.
- ✔ arteries carry blood at high pressure from the ventricles to the tissues of the body.
- ✔ arteries have muscle cells and elastic fibres in their walls which maintain blood pressure between pump cycles.
- ✔ blood flows through tissues in capillaries.
- ✔ capillaries have permeable walls that allow exchange of materials between cells in the tissues and the blood in the capillary.
- ✔ veins collect blood at low pressure from the tissues of the body and return it to the atria of the heart.
- ✔ valves in veins and the heart ensure circulation of blood by preventing backflow.
- ✔ there is a separate circulation for the lungs.
- ✔ the sinoatrial node initiates the heartbeat, acting as a pacemaker.
- ✔ the sinoatrial node sends out an electrical signal that stimulates contraction as it is propagated through the walls of the atria and then the walls of the ventricles.
- ✔ the heart rate can be increased or decreased by impulses brought to the heart through two nerves from the medulla of the brain.
- ✔ adrenaline increases the heart rate to prepare for vigorous physical activity.

You should be able to:

- ✔ identify blood vessels as arteries, capillaries or veins from the structure of their walls.
- ✔ explain how the structure of arteries, veins and capillaries are suited to their function.
- ✔ sketch diagrams of the chambers and valves of the heart and the blood vessels connected to it.
- ✔ explain the double circulation of blood in the heart.
- ✔ identify and describe the function of valves.
- ✔ explain the function of the sinoatrial node and describe the nervous and hormonal control of the heart rate.
- ✔ describe pressure changes in the left atrium, left ventricle and aorta during the cardiac cycle.
- ✔ describe William Harvey's discovery of the circulation of the blood with the heart acting as the pump.
- ✔ explain causes and consequences of occlusion of the coronary arteries.
- ✔ discuss social implications of coronary heart disease.

Blood flows twice through the heart, once to be oxygenated in the lungs and a second time for circulation to the rest of the body.

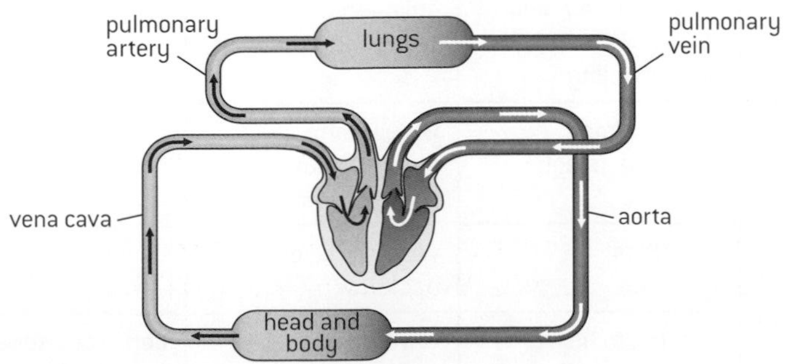

▲ **Figure 6.2.1.** Double circulation in the heart

> In Topic 2.2 you studied the structure of water and how some molecules can dissolve in it to be transported. In Topic 2.3 you have learned about carbohydrates and lipids, which are transported in blood. In Topic 6.4 you will study gas exchange occurring between blood and lung tissues. In Topic 6.6 you will study the hormones that can be carried in blood and how they help in homeostasis and reproduction.

The heart and blood vessels make up the circulatory system. There are three main types of blood vessels: arteries that carry blood away from the heart, veins that carry blood towards the heart and capillaries that connect both these vessels.

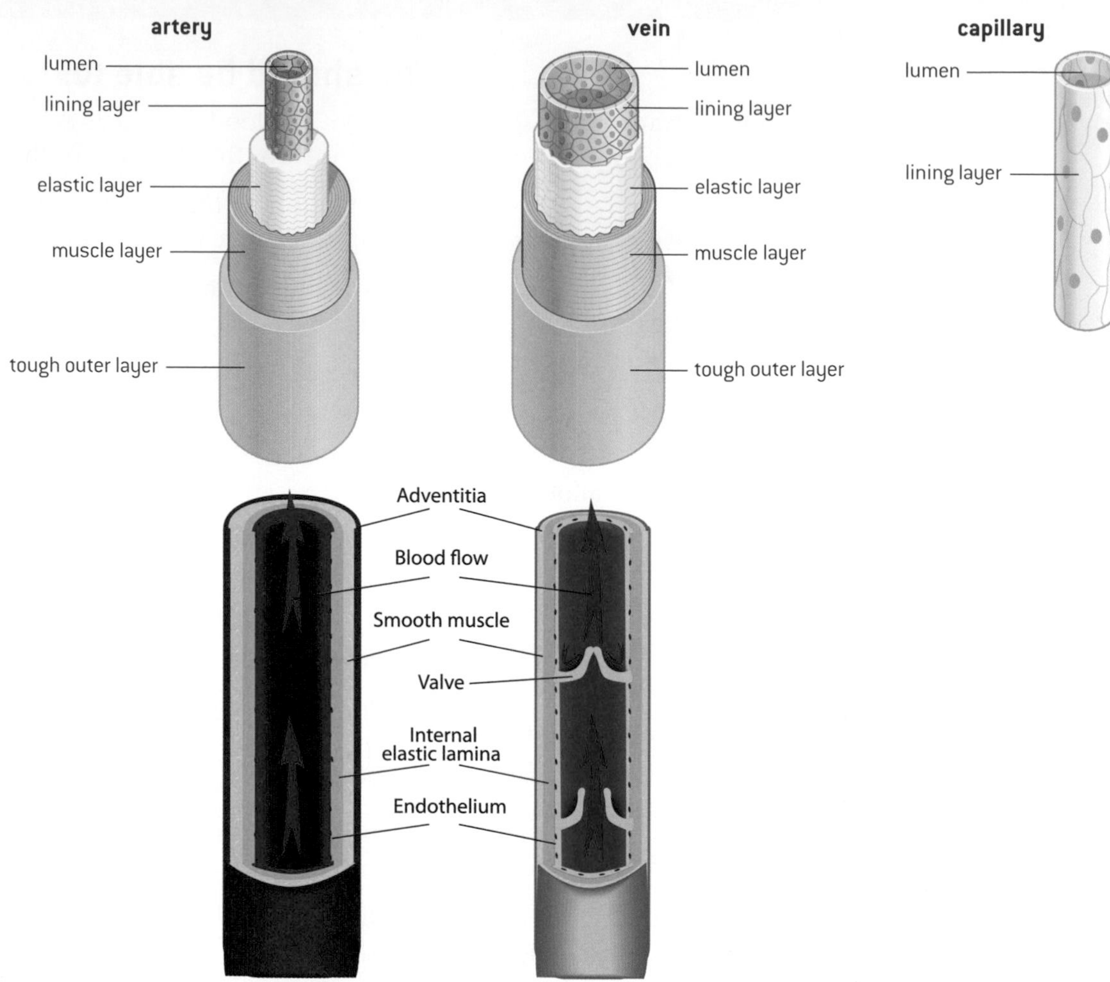

▲ **Figure 6.2.2.** Blood vessels

Theories are regarded as uncertain. In 1628 William Harvey described how blood flow in the body was unidirectional, with valves stopping the blood from backflowing, and that blood returned to the heart to be recycled by pumping. With this, he overturned theories developed by the ancient Greek philosopher Galen on the movement of blood in the body.

Example 6.2.1.

Compare and contrast the structure of arteries, veins and capillaries.

Solution

	Arteries	Veins	Capillaries
Walls	thick walls to withstand high pressure	thin wall to allow muscles to exert pressure on them	wall one layer of cells thick to allow diffusion of substances
Fibres in wall	collagen and elastic fibres in outer layer to give wall strength and flexibility	thin outer layer of collagen and elastic fibres for protection	no fibres
Lumen	narrow lumen to maintain high pressure	wide lumen so greater volume of blood can pass	narrow lumen to fit in small spaces
Valves	no valves	valves to avoid backflow	no valves
Direction	away from heart	towards heart	connect arteries to veins
Pores	no pores	no pores	pores to allow lymphocytes to leave

Assessment tip

You do not need to answer the question in the form of a table, but it helps to ensure you are really comparing and contrasting.

The pressure in the different chambers of the heart changes according to whether the ventricles are contracting (systole), or relaxing after a contraction (diastole). A whole heartbeat cycle can be seen on a heart pressure graph. The first part of a heartbeat can be seen as the first parabolic curve where the atria are contracting (atrial systole) and the second, larger curve, where the ventricles are contracting (ventricular systole). In healthy adults you can hear two normal heart sounds in each heartbeat. These are often described as a "lub", the closing of the atrioventricular (AV) valves, and a "dub", the closing of the semilunar valves.

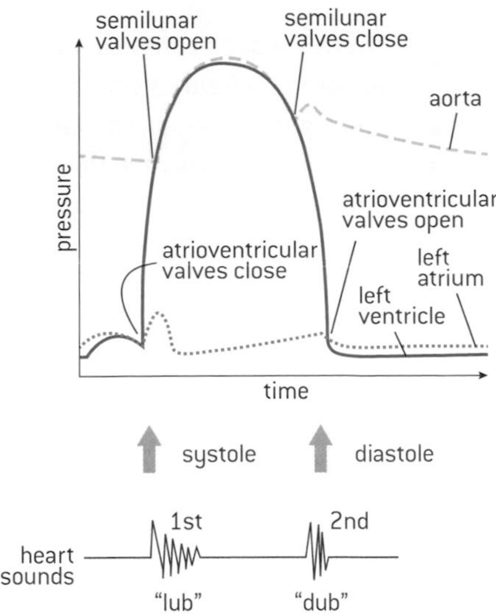

▲ **Figure 6.2.3.** Cardiac cycle

Example 6.2.2.

a) Figure 6.2.4 shows a human heart.

Label A to D in the diagram of the heart.

b) (i) Explain the reason the pressure in the ventricle increases during atrial contraction.

(ii) The pressure in the aorta is around 80 mmHg just before the semilunar valves open. Predict the pressure in the ventricle at the time these valves open.

(iii) Describe the change in pressure in the left ventricle after the semilunar valves close.

Solution

a) A: aorta; B: pulmonary artery; C: left ventricle wall; D: vena cava.

b) (i) During atrial contraction, the atrioventricular (AV) valve is open; therefore the blood is flowing from the left atrium into the left ventricle, increasing the pressure.

(ii) The pressure must be just above 80 mmHg to allow the blood flowing from the ventricles into the aorta. The ventricular pressure must be high enough to open the valves, letting blood out of the heart.

(iii) The pressure in the left ventricle decreases, as blood has left the ventricle and entered the aorta to be carried to the rest of the body. The atria are filling with blood, but because the AV (mitral) valve is closed, no blood is flowing into the ventricle.

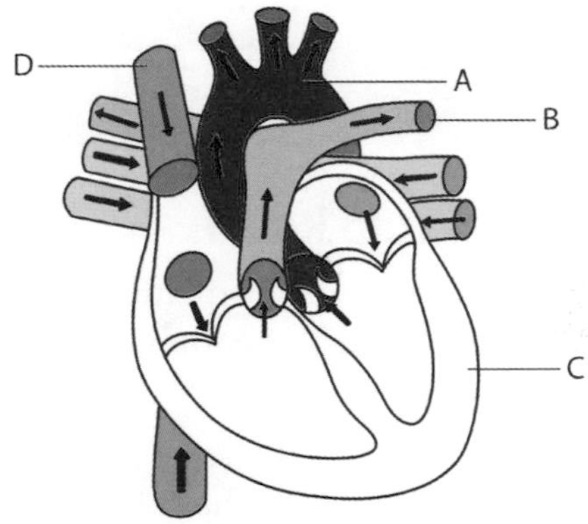

▲ **Figure 6.2.4.** The human heart

Assessment tip

Part (b) in example 6.2.2 is referring to the left ventricle, but as the same happens in both ventricles simultaneously, no marks will be taken off if you just refer to the ventricles instead of the left ventricle.

- The **medulla oblongata** is the part of the brain involved in the control of the heart rate.
- **Adrenaline** is a hormone that accelerates the heart rate.

There are social implications of coronary heart disease. Diets rich in glucose and *trans* fats can lead to obesity, which at the same time increases the chances of atherosclerosis. High blood pressure is also a cause of coronary heart disease. Smoking, stress, certain drugs and a diet rich in salt can contribute to increasing blood pressure. Coronary heart disease can lead to disabilities; therefore it is important to keep a healthy lifestyle to reduce your chances of developing it.

The heartbeat is initiated by a group of specialized muscle cells in the right atrium called the sinoatrial node. The sinoatrial node acts as a pacemaker, sending out an electrical signal that stimulates contractions as it is propagated through the walls of the atria and then the walls of the ventricles. The heart rate can be affected by nervous and hormonal control. Two nerves link the heart to the medulla oblongata of the brain. One of these nerves carries messages to increase the heart rate, the other to decrease the heart rate. Adrenaline is a hormone that increases the heart rate to prepare for vigorous physical activity.

Atherosclerosis is caused by the formation of a plaque in the arteries called atheroma. The atheroma is mainly formed by low density lipoprotein (LDL) cholesterol and white blood cells. It clogs the arteries, impeding the flow of blood. If the artery occluded (blocked) is the coronary artery, the heart can suffer damage due to lack of oxygen and nutrients reaching the heart muscle tissue. If this happens the heart cannot contract properly. This damage can lead to a heart attack or an infarct (an area of dead tissue).

6.3 DEFENCE AGAINST INFECTIOUS DISEASE

You should know:

- ✔ skin and mucous membranes form a primary defence against pathogens that cause infectious disease.
- ✔ clotting factors are released from platelets allowing cuts in the skin to be sealed by blood clotting.
- ✔ the cascade results in the rapid conversion of fibrinogen to fibrin by thrombin.
- ✔ ingestion of pathogens by phagocytic white blood cells gives non-specific immunity to diseases.
- ✔ production of antibodies by lymphocytes in response to particular pathogens gives specific immunity.
- ✔ antibiotics block processes that occur in prokaryotic cells but not in eukaryotic cells.
- ✔ viruses lack a metabolism and cannot therefore be treated with antibiotics.
- ✔ some strains of bacteria have evolved with genes that confer resistance to antibiotics and some strains of bacteria have multiple resistance.

You should be able to:

- ✔ explain mechanisms of defence against pathogens.
- ✔ describe the process of blood clotting and explain the causes and consequences of blood clot formation in coronary arteries.
- ✔ outline the production of antibodies by lymphocytes.
- ✔ outline how lymphocytes act as memory cells.
- ✔ explain the action of antibodies against pathogens.
- ✔ explain the evolution of antibiotic-resistant bacteria.
- ✔ describe Florey and Chain's experiments to test penicillin on bacterial infections in mice.
- ✔ describe the effects of HIV on the immune system and methods of transmission.

In Topic 5.2 you have seen how natural selection can result in bacterial resistance to antibiotics.

Pathogens are organisms that cause disease. The human body has structures and processes that resist the continuous threat of invasion by pathogens. Skin and mucous membranes form a primary defence against pathogens that cause infectious disease, not allowing organisms to enter the body. To avoid organisms entering through a wound, clotting factors are released from platelets allowing cuts in the skin to be sealed by blood clotting. The cascade results in the rapid conversion of fibrinogen to fibrin using the enzyme thrombin. Fibrin is a protein that makes a mesh joining the platelets, stopping the blood from flowing through the wounds.

Example 6.3.1.

What sequence of reactions occurs during blood clotting?

A. Fibrinogen is broken down into trypsin by fibrin.

B. Fibrin is broken down into fibrinogen by trypsin.

C. Fibrinogen is broken down into fibrin by thrombin.

D. Fibrin is broken down into fibrinogen by thrombin.

Solution

The correct answer is **C**. Fibrinogen is a globular, soluble protein found in blood plasma. In the event of a cut, an enzyme that is found in an inactive form is transformed into active thrombin. Thrombin breaks down part of the soluble fibrinogen protein, now transforming it into a smaller but insoluble fibrous protein called fibrin. Molecules of fibrin form cross-linking bonds resulting in a large polymer which attaches to platelets, and thus clots blood.

An understanding of immunity has led to the development of vaccinations and treatment of cancer. In 2018, George P. Smith and Gregory Winter both became Nobel Laureates in Chemistry. They were inspired by the power of evolution and used the same principles—genetic change and selection—to develop proteins to solve medical problems. They used bacteriophages to evolve new proteins which can be used in the pharmaceutical industry for the production of antibodies that neutralize toxins, counteract autoimmune diseases or cure metastatic cancer.

Once organisms have entered the tissues, the immune system activates a series of mechanisms to defend the body. Ingestion of pathogens by phagocytic white blood cells gives non-specific immunity to diseases. Production of antibodies by lymphocytes in response to particular pathogens gives specific immunity. Antibodies are proteins that can attach to pathogens, destroying them or flagging them for other lymphocytes to engulf them by phagocytosis. Lymphocytes mature to become antibody-producing cells called plasma cells (or B cells). Memory cells are a clone of plasma cells that remember how to produce particular antibodies, and therefore are very useful if there is a second encounter with the same pathogen.

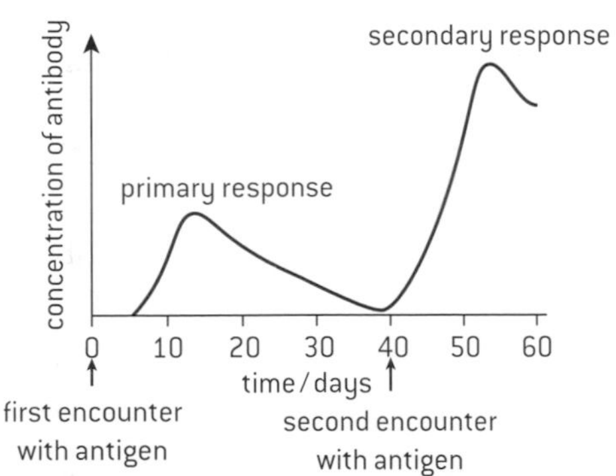

▲ **Figure 6.3.2.** Production of antibodies

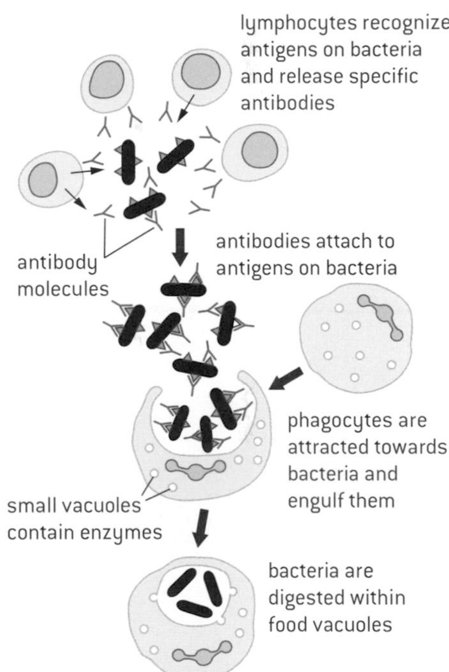

▲ **Figure 6.3.1.** Specific immune response

- **Lymphocytes** are white blood cells involved in the production of antibodies.
- **Plasma cells** are mature lymphocytes that can produce antibodies in the primary response.
- An **antigen** is part of the pathogen (usually a protein) recognized by the immune system.
- An **antibody** is a protein that attaches to the antigen to destroy or flag the pathogen.
- **Memory cells** produce antibodies if a pathogen carrying a specific antigen is re-encountered. This is called the secondary response.
- **HIV** is a virus that causes the disease AIDS.

In Topic 11.1 you will cover antibody production in more detail, along with how vaccinations work.

Assessment tip

In an eight-mark question you should mention many proteins, not just explain how one type of protein is made.

▲ The student scored one mark for saying that lymphocytes create antibodies.

▼ The student failed to mention clotting factors and fibrin as a protein that permits blood clotting. Although antibodies are mentioned, the student does not mention their specificity, the role of plasma cells in producing large amounts of specific antibodies or how memory cells retain the ability to produce specific antibodies.

SAMPLE STUDENT ANSWER

Some blood proteins are involved in the defence against infectious diseases. Explain the roles of **named** types of proteins in different defence mechanisms. [8]

This answer could have achieved 1/8 marks:

Lymphocytes are phagocytes that are involved in defence against infectious diseases. When the infection occurs, the bacteria carry antigens on their surface, which the lymphocytes could recognize. The lymphocytes will then find a phagocyte that is responsible for the antigen. There are different types of antigens. The lymphocyte would have to multiply cells that defend from different bacteria. When the lymphocyte is found, it will now begin rapid cell division to increase the number of lymphocytes creating antibodies. The phagocyte would then engulf the bacteria and digest it. The lymphocytes are now able to recognize similar antigens on bacteria, so if a similar or the same antigen enters the body the lymphocyte would be able to quickly make phagocytes to destroy the bacteria.

Antibiotics are chemicals produced by living organisms such as fungi or which can be produced synthetically. They recognize antigens that are on the pathogen, usually a part of a protein or a glycoprotein of the cell wall or membrane. Antibiotics block processes that occur in prokaryotic cells but not in eukaryotic cells. Some block the formation of the cell wall or plasma membrane, others inhibit replication or transcription of the microorganism's genetic material. Some strains of bacteria have evolved with genes that confer resistance to antibiotics, and some strains have multiple resistance. Viruses lack a metabolism, so they cannot be treated with antibiotics.

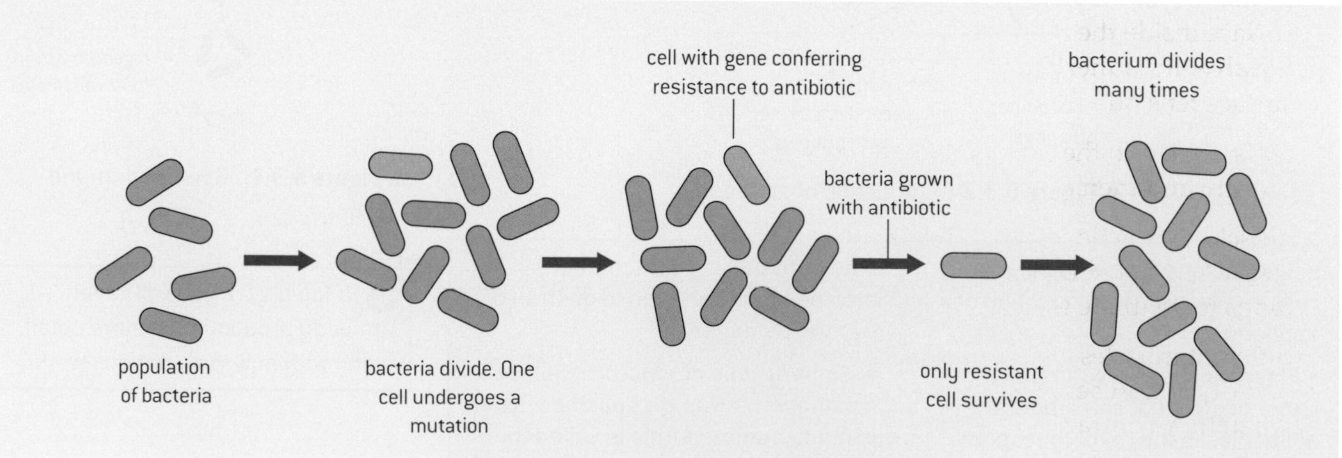

▲ **Figure 6.3.3.** Antibiotic resistance in bacteria

Human immunodeficiency virus (HIV) acts on the immune system, reducing the number of active lymphocytes (or T cells) and causing a loss of the ability to produce antibodies, leading to the development of acquired immune deficiency syndrome (AIDS).

Example 6.3.2.

How does HIV affect the immune system?

A. Reduces the number of phagocytes

B. Reduces the number of active lymphocytes

C. Increases the amount of antibiotics

D. Increases the amount of clotting factors

Solution
The correct answer is **B**. HIV is a virus that destroys active lymphocytes or T helper cells. These cells are involved in the immune reaction against pathogens; therefore the person infected with HIV will have a poor response to any other disease.

Florey and Chain's tests on the safety of penicillin would not be compliant with current protocol on testing. After testing penicillin on eight mice, Florey and Chain decided to test it on humans. Fortunately it proved to be effective in most cases and penicillin started to be produced in large quantities. Nowadays more testing must take place before a drug can be used on humans. Drugs must first be tested on animals, then on healthy volunteers, only then on consenting infected patients. Randomized controlled trials which are double-blind, placebo-controlled trials must be performed to eliminate bias.

6.4 GAS EXCHANGE

You should know:

- ✔ ventilation maintains concentration gradients of oxygen and carbon dioxide between the air in alveoli and the blood flowing in adjacent capillaries, so gas exchange occurs passively.
- ✔ type I pneumocytes are extremely thin alveolar cells that are adapted to carry out gas exchange.
- ✔ type II pneumocytes secrete a solution containing surfactant that creates a moist surface inside the alveoli to prevent the sides of the alveolus adhering to each other by reducing surface tension.
- ✔ air is carried to the lungs in the trachea and bronchi and then to the alveoli in bronchioles.
- ✔ muscle contractions cause the pressure changes inside the thorax that force air in and out of the lungs to ventilate them.
- ✔ different muscles are required for inspiration and expiration because muscles do work only when they contract.

You should be able to:

- ✔ draw the respiratory system.
- ✔ explain lung ventilation including the muscles involved.
- ✔ draw a diagram to show the structure of an alveolus and an adjacent capillary.
- ✔ describe the structure of alveoli and the function of pneumocytes.
- ✔ explain how gas exchange occurs in the alveoli and how they are suited to their function.
- ✔ outline causes and consequences of lung cancer and emphysema and discuss the social consequences of these diseases.
- ✔ design experiments to monitor ventilation in humans at rest and after mild and vigorous exercise.

In Topic 1.4 you studied membrane transport. Oxygen and carbon dioxide are transported across membranes in the alveoli to and from the blood system you studied in Topic 6.2. Cancer is excessive cell division, which you studied in Topic 1.6.

Ventilation is the movement of air into and out of the lungs. It is also called breathing. The lungs are actively ventilated to ensure that gas exchange can occur passively. Ventilation maintains concentration gradients of oxygen and carbon dioxide between air in alveoli and blood flowing in adjacent capillaries. It is performed by four different sets of muscles. The external and internal intercostal muscles, and the diaphragm and abdominal muscles are examples of antagonistic muscle action. In inhalation (breathing in), the external intercostal muscles and the diaphragm contract, expanding the thoracic cavity, allowing air to flow into the lungs. In exhalation (breathing out), these two sets of muscles relax. In contrast, during inhalation, the internal intercostal muscles and the abdominal muscles are relaxed. They contract only during exhalation. Ventilation can be monitored either by simple observation and simple apparatus or by data logging with a spirometer or chest belt and pressure meter.

- **Ventilation rate** is the amount of air breathed in a period of time.
- **Tidal volume** is the amount of air breathed in one breath.

Example 6.4.1.

The diagram shows the respiratory system.

a) (i) Label A to D.

(ii) On the diagram identify a bronchus and bronchioles.

b) Outline the function of B.

c) State the process occurring at E.

Solution

a) (i) A: ribs; B: diaphragm; C: lungs; D: windpipe or trachea.

(ii) Bronchus between trachea and bronchioles; bronchiole between bronchus and alveoli.

b) The function of B is to contract to allow inspiration. When the diaphragm contracts, the rib cage gets larger, reducing the pressure in the thoracic cavity. This allows air to rush into the lungs.

c) At E, in the alveoli, gaseous exchange occurs.

SAMPLE STUDENT ANSWER

The lung volume was recorded as a student breathed into a spirometer. The air was exhaled over a soda lime chamber that absorbs carbon dioxide to avoid suffocation.

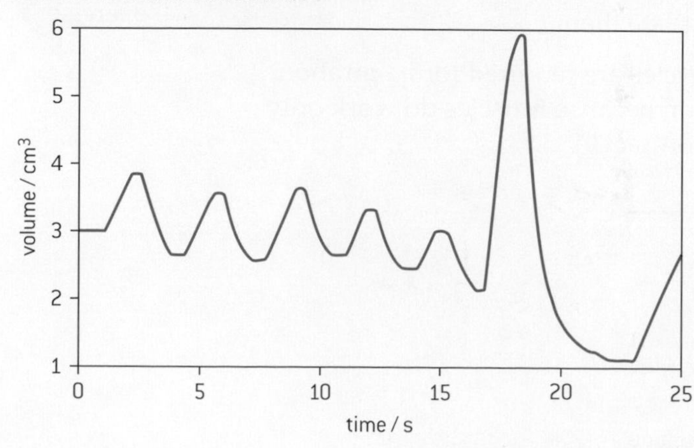

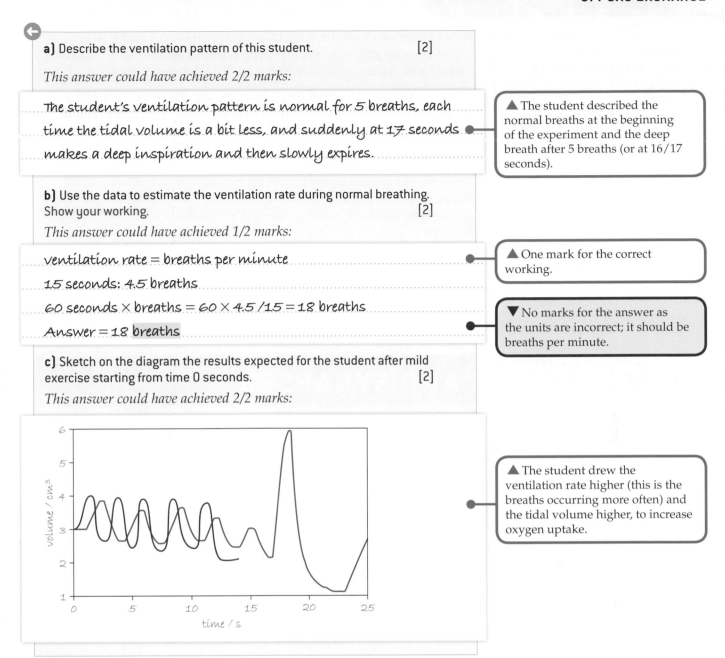

a) Describe the ventilation pattern of this student. [2]

This answer could have achieved 2/2 marks:

The student's ventilation pattern is normal for 5 breaths, each time the tidal volume is a bit less, and suddenly at 17 seconds makes a deep inspiration and then slowly expires.

▲ The student described the normal breaths at the beginning of the experiment and the deep breath after 5 breaths (or at 16/17 seconds).

b) Use the data to estimate the ventilation rate during normal breathing. Show your working. [2]

This answer could have achieved 1/2 marks:

ventilation rate = breaths per minute

15 seconds: 4.5 breaths

60 seconds × breaths = 60 × 4.5 / 15 = 18 breaths

Answer = 18 breaths

▲ One mark for the correct working.

▼ No marks for the answer as the units are incorrect; it should be breaths per minute.

c) Sketch on the diagram the results expected for the student after mild exercise starting from time 0 seconds. [2]

This answer could have achieved 2/2 marks:

volume / cm^3 (1, 2, 3, 4, 5, 6)

time / s (0, 5, 10, 15, 20, 25)

▲ The student drew the ventilation rate higher (this is the breaths occurring more often) and the tidal volume higher, to increase oxygen uptake.

Gaseous exchange occurs in the alveoli of the lungs. The alveoli are adapted to this function. They are moist, are one cell thick and have capillaries close by.

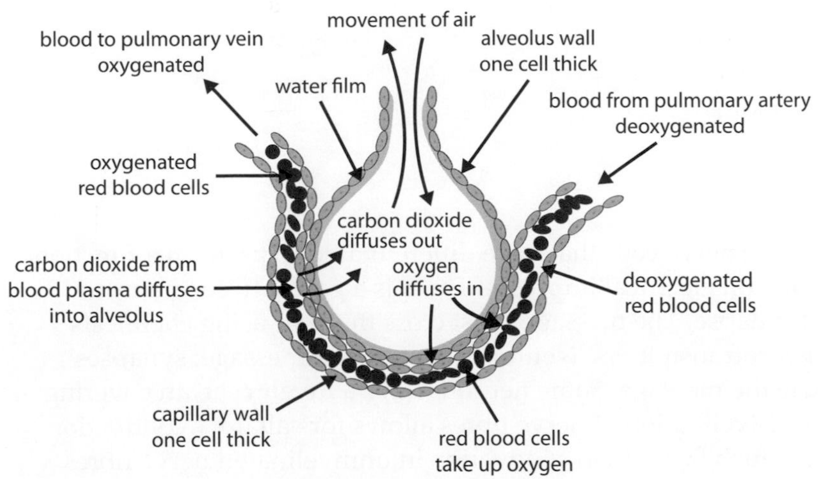

▲ **Figure 6.4.1.** Gaseous exchange in the alveoli

Cancer is uncontrolled cell division. For further information on cell division see Topics 3.2, 3.3 and 10.1.

- **Type I pneumocytes** are flattened epithelial cells allowing gaseous exchange.
- **Type II pneumocytes** are cuboidal epithelial cells producing surfactant.
- **Lung cancer** is the presence of tumours growing in the lung tissue.
- **Emphysema** is the destruction of alveolar walls by elastase leading to impaired gaseous exchange.

Type I pneumocytes, which are numerous, very large and flat, are where gaseous exchange occurs. Type II pneumocytes, which are cube shaped and occur amongst the type I cells, produce surfactant.

Cancer is a disease in which some of the cells of the body multiply out of control and spread into surrounding tissue. Lung cancer can be caused by smoking, carcinogens such as asbestos and air pollution. The consequence is the formation of tumours causing pain, breathing difficulties and fatigue. It can progress to metastasis of the tumour and death. Smoking causes lung cancer because cigarette smoke contains carcinogens that damage the cells that line the lungs. Emphysema is a progressive disease of the lungs caused by destruction of lung tissue involved in gaseous exchange. The walls of the alveoli become thicker and the air sacs larger. This is caused by the presence of elastase, an enzyme produced by phagocytes that digests elastin, a protein that allows alveoli to stretch. Smoking increases the number of phagocytes in the alveoli, therefore increasing the number of elastase. Gaseous exchange is impaired, causing shortness of breath and possibly leading to death.

6.5 NEURONS AND SYNAPSES

You should know:

- ✔ neurons transmit electrical impulses.
- ✔ myelination of nerve fibres allows for saltatory conduction.
- ✔ neurons pump sodium and potassium ions across their membranes to generate a resting potential.
- ✔ an action potential consists of depolarization and repolarization of the neuron.
- ✔ nerve impulses are action potentials propagated along the axons of neurons.
- ✔ propagation of nerve impulses is the result of local currents that cause each successive part of the axon to reach the threshold potential.
- ✔ synapses are junctions between neurons and between neurons and receptor or effector cells.
- ✔ when presynaptic neurons are depolarized they release a neurotransmitter into the synapse.
- ✔ a nerve impulse is initiated only if the threshold potential is reached.

You should be able to:

- ✔ draw the structure of a myelinated neuron.
- ✔ explain electrical impulses across neurons.
- ✔ explain an action potential.
- ✔ explain the effect of myelination on conduction.
- ✔ analyse oscilloscope traces showing resting potentials and action potentials.
- ✔ describe chemical synapses.
- ✔ explain the propagation of nerve impulses across synapses.
- ✔ explain the secretion and reabsorption of acetylcholine by neurons at synapses.
- ✔ explain the blocking of synaptic transmission at cholinergic synapses in insects by binding of neonicotinoid pesticides to acetylcholine receptors.

In Topic 1.4 you studied the different methods of membrane transport.

Neurons are nerve cells that have differentiated to perform a function. They transmit electrical impulses. There is a gap between neurons called a synapse. The message must cross this gap using chemicals called neurotransmitters. Neurons transmit the message; synapses modulate the message. Some neurons have a myelin sheath covering the axon. Myelination of nerve fibres allows for saltatory conduction, which is much faster than conduction in unmyelinated nerve fibres.

Example 6.5.1.

The diagram shows a motor neuron.

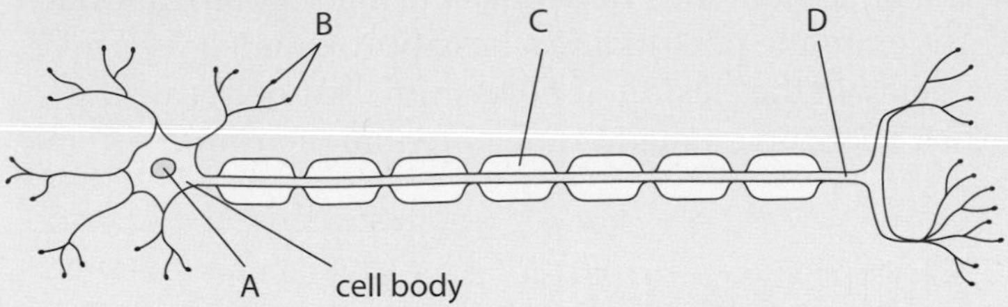

Which label shows the myelin sheath?

Solution
The correct answer is **C**.

- **Membrane potential** is the difference in voltage between the outside and inside of the cell membrane.
- **Resting potential** is the membrane potential as long as there is no perturbance; it is around −70 mV.
- **Action potential** is the depolarization and repolarization of the neuron allowing impulse transmission.
- **Depolarization** is the opening of sodium ion channels, increasing the membrane potential.
- **Repolarization** is the closing of sodium ion channels and opening of potassium ion channels restoring the low membrane potential.
- **Hyperpolarization** is when the membrane potential is at its lowest. This is caused by the delay in closing of the potassium ion channels.

Nerve impulses are action potentials propagated along the axons of neurons. Propagation of nerve impulses is the result of local currents that cause each successive part of the axon to reach the threshold potential. Membrane potentials in neurons can be measured by placing electrodes on each side of the membrane. The potentials can be displayed using an oscilloscope.

The resting potential is the potential across the membrane caused by the exchange of three sodium ions moving out of the cell and two potassium ions into the cell by the sodium–potassium pump, with the use of one ATP molecule. An action potential consists of depolarization and repolarization of the neuron.

Example 6.5.2.

Ions move across the plasma membrane of a neuron during an action potential. The oscilloscope trace shows voltage changes generated in a neuron during three action potentials.

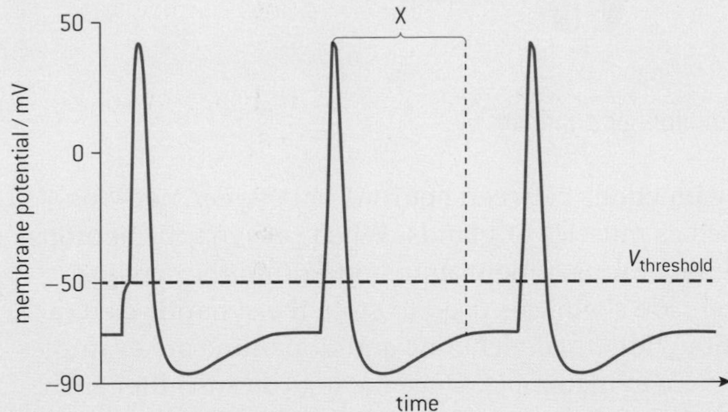

a) Describe depolarization.

b) Explain the movement of ions which causes the voltage changes observed during the interval labelled X on the graph.

Solution

a) Depolarization is caused by the opening of the sodium channels, allowing sodium ions to diffuse into the neurone down a concentration gradient.

Assessment tip

The question refers to a section labelled X, so do not answer referring to what occurs before or after the section labelled X, as no marks will be given or marks can be lost due to contradictions.

b) At the peak the sodium (Na^+) channels close and the potassium (K^+) channels open. The membrane potential is approximately +45 mV. The potassium ions flow out, causing repolarization to occur. The membrane potential reaches approximately −70 mV. There is a delay in the closing of potassium channels, causing hyperpolarization, and a membrane potential of around −78 mV.

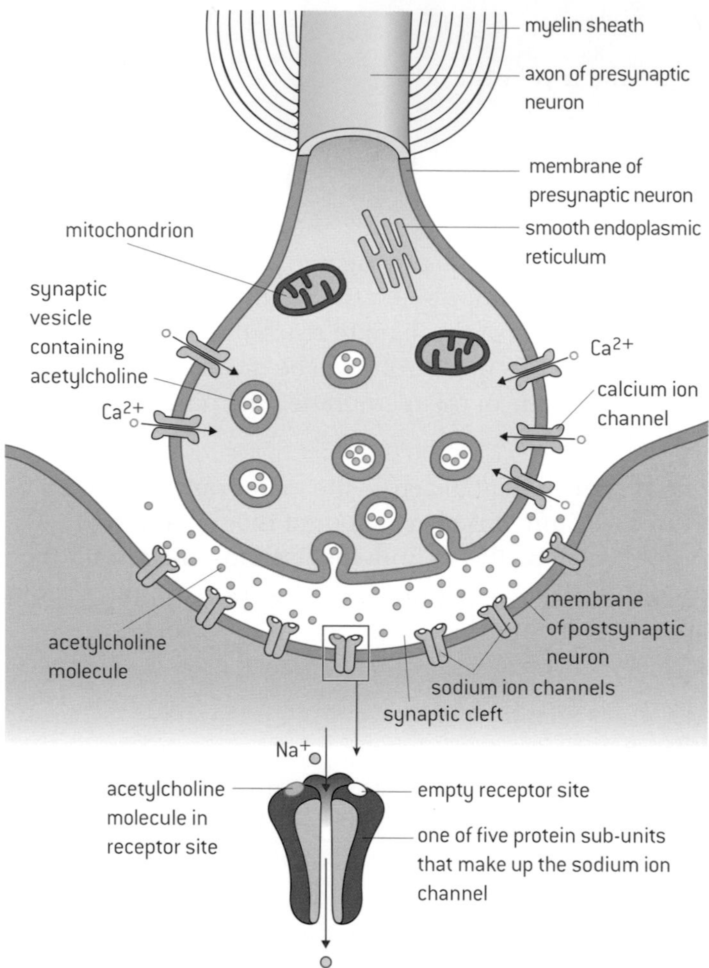

▲ **Figure 6.5.1.** Cholinergic synapse

Synapses are the junctions between neurons or between neurons and effectors such as muscles or glands. When presynaptic neurons are depolarized they release a neurotransmitter into the synapse. Neurotransmitters are chemicals used to cross the synaptic cleft (a gap around 20 nm wide). Acetylcholine and dopamine are examples of neurotransmitters in mammals. Once the neurotransmitter such as acetylcholine reaches the postsynaptic neuron, it attaches to a receptor, inducing the action potential by the opening of sodium ion channels. The acetylcholine is then broken down by the enzyme acetylcholinesterase and the resting potential is recovered.

Neonicotinoids are synthetic chemicals used as insecticides. They have a similar composition to nicotine. They bind irreversibly to the acetylcholine receptor in cholinergic synapses in the central nervous system of insects. Because the receptors are blocked, acetylcholine is no longer capable of binding to the postsynaptic membrane, inhibiting synaptic transmission. This causes paralysis and death of insects.

6.6 HORMONES, HOMEOSTASIS AND REPRODUCTION

You should know:

✔ hormones are used when signals need to be widely distributed.

✔ insulin and glucagon are secreted by α and β cells of the pancreas respectively to control blood glucose concentration.

✔ thyroxin is secreted by the thyroid gland to regulate the metabolic rate and help control body temperature.

✔ leptin is secreted by cells in adipose tissue and acts on the hypothalamus of the brain to inhibit appetite.

✔ melatonin is secreted by the pineal gland to control circadian rhythms.

✔ a gene on the Y chromosome causes embryonic gonads to develop as testes and secrete testosterone.

✔ testosterone causes pre-natal development of male genitalia and both sperm production and development of male secondary sexual characteristics during puberty.

✔ estrogen and progesterone cause pre-natal development of female reproductive organs and female secondary sexual characteristics during puberty.

✔ the menstrual cycle is controlled by negative and positive feedback mechanisms involving ovarian and pituitary hormones.

You should be able to:

✔ explain the hormonal control of blood glucose levels.

✔ outline causes and treatment of type I and type II diabetes.

✔ explain the regulation of metabolic rate by thyroxin.

✔ explain the effect of leptin on appetite.

✔ outline testing of leptin on patients with clinical obesity and reasons for the failure to control the disease.

✔ describe how hormones are used in a variety of therapies such as replacement therapies.

✔ explain the action of melatonin.

✔ outline the causes of jet lag and use of melatonin to alleviate it.

✔ explain the action of testosterone.

✔ draw and annotate diagrams of the male and female reproductive systems to show the names of structures and their functions (front and side views).

✔ explain the roles of FSH, LH, estrogen and progesterone in the menstrual cycle.

✔ outline the use in IVF of drugs to suspend the normal secretion of hormones, followed by the use of artificial doses of hormones to induce superovulation and establish a pregnancy.

✔ describe William Harvey's investigation of sexual reproduction in deer.

Hormones are chemicals secreted by glands. They travel through blood, affecting target organs. Hormones are used when signals need to be widely distributed. Their effect is usually long lasting. For example, insulin and glucagon are secreted by the α and β cells of the pancreas, respectively, to control blood glucose concentration.

- **Insulin** induces the uptake of glucose from blood to be stored in the liver as glycogen when glucose levels are high.
- **Glucagon** induces the breakdown of glycogen into glucose to be released when glucose levels are low in blood.
- **Thyroxin** is secreted by the thyroid gland to regulate the metabolic rate and help control body temperature.
- **Leptin** is secreted by cells in adipose tissue and acts on the hypothalamus of the brain to inhibit appetite.
- **Melatonin** is secreted by the pineal gland to control circadian rhythms. It promotes sleep when it is dark. It is used to alleviate the effects of jet lag.

Developments in scientific research follow improvements in apparatus—William Harvey was hampered in his observational research into reproduction by lack of equipment. William Harvey failed to solve the mystery of sexual reproduction because effective microscopes were not available when he was working. The microscope was invented 17 years after his death. Fusion of gametes and subsequent embryo development remained undiscovered till many years later.

SAMPLE STUDENT ANSWER

a) Glands are organs that secrete and release particular chemical substances. Melatonin is an important hormone secreted in the pineal gland in the brain. Describe its role in mammals. [2]

This answer could have achieved 2/2 marks:

Melatonin controls the circadian rhythms in mammals as it sets the pattern of sleep by inducing sleep when it is night time (and dark) and waking the mammal up during the daytime when it is bright.

▲ The student scored one mark for mentioning that melatonin regulates the circadian rhythm (or biological clock). The other mark is for saying that it affects the sleep–wake cycles. The other marks that could be obtained are for the fact that production is controlled by the amount of light detected in the retina and that there is a high level of secretion during the night and a low level during the day; this means secretion is directly proportional to night-time duration.

b) Describe how the hormone leptin helps to prevent obesity. [3]

This answer could have achieved 2/3 marks:

The hormone leptin signals to the brain that we are full and do not need to eat any more. An experiment on mice showed that those with leptin inhibitors became obese. If leptin is given to a human, he/she is more likely to feel full having eaten less. Therefore he/she is less likely to eat too much, retain too many calories and become obese. Studies conducted on humans with increased leptin have not been successful so far, although slight weight losses have been noticed.

▲ The student scored one mark for saying that leptin inhibited appetite and another mark for saying it reduces food intake.

▼ The student could have scored a mark for mentioning that leptin is produced in the adipose tissue and acts on the hypothalamus of the brain. The higher the amount of adipose tissue, the greater the amount of leptin produced. In humans, obese people have leptin resistance; therefore for them leptin is no help in losing weight.

Diabetes is a medical condition in which blood glucose levels are higher than normal, even after fasting. Glucose can be detected in urine. Type I diabetes is an autoimmune disease that causes the destruction of the β cells of the islets of Langerhans of the pancreas. This causes the body to lack insulin, therefore raising the level of glucose in blood. It is treated with insulin injections. Type II diabetes is caused by resistance to insulin. It can be caused by obesity.

SAMPLE STUDENT ANSWER

Outline the cause of type I diabetes in humans. [1]

This answer could have achieved 1/1 marks:

It is caused by an inability to produce insulin. B cells do not function properly.

▲ The answer scored one mark for mentioning the inability to produce insulin.

▼ Although the student did score full marks, to avoid confusion, it would have been better to specify that the B cells were pancreatic, as there are blood immune B cells too.

This answer could have achieved 0/1 marks:

It's the result of genetic transfer from the parents to the offspring. It is a genetic development with the onset during childhood.

▼ This answer did not score a mark because it is implying that type I diabetes is inherited, when it should have said it is an autoimmune condition involving destruction of beta/β cells, thus reducing the production of insulin.

A gene on the Y chromosome found only in males causes embryonic gonads to develop as testes and secrete testosterone. Testosterone causes pre-natal development of male genitalia and both sperm production and development of male secondary sexual characteristics during puberty.

Example 6.6.1.

The diagram shows a section through the male reproductive system. Which structure represents the prostate gland?

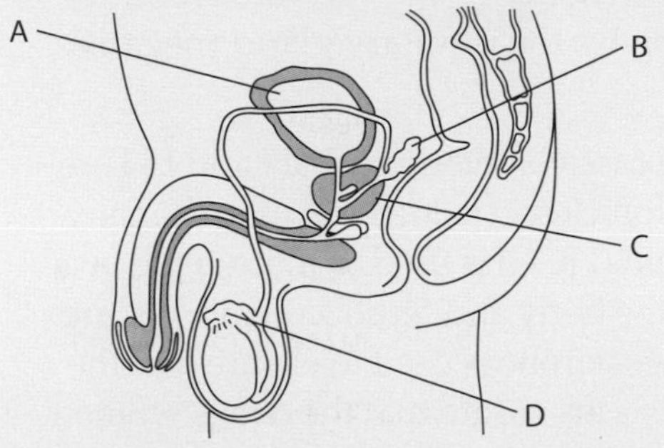

Solution
The correct answer is **C**. **A** shows the bladder which is part of the urinary system and where urine is stored. **B** shows the seminal vesicles that produce seminal fluid and **D** the epididymis where sperm are stored.

- **Testes** are the male reproductive organs that produce sperm and testosterone.
- The **scrotum** holds the testes.
- The **epididymis** is a duct where sperm are stored until ejaculation.
- **Sperm ducts** transfer the sperm from the epididymis to the urethra during ejaculation.
- **Seminal vesicles** and the **prostate gland** secrete fluid that makes semen.
- The **penis** is the organ that penetrates the vagina during sexual intercourse.

Estrogen and progesterone cause pre-natal development of female reproductive organs and female secondary sexual characteristics during puberty.

Example 6.6.2.

The diagram shows the female reproductive system.

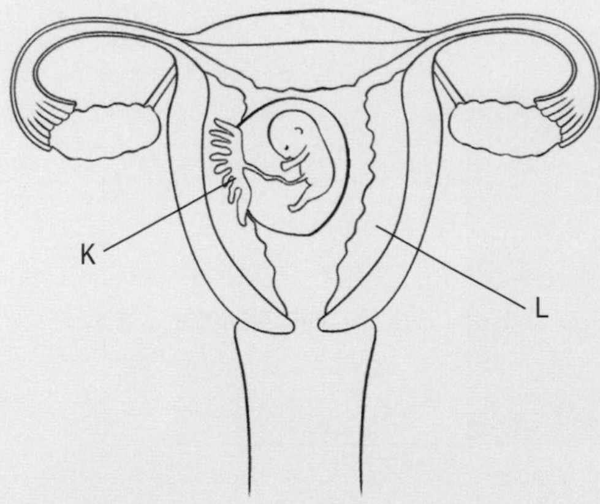

a) Which structures do K and L identify?

	K	L
A	endometrium	uterine wall
B	placenta	endometrium
C	amnion	placenta
D	fetus	uterine wall

Solution
The correct answer is **B**.

- **Ovaries** are the organs where eggs, estrogen and progesterone are produced.
- **Oviducts** are the canals that receive the eggs at ovulation and are the site of fertilization.
- The **uterus** is the organ where the embryo is implanted after fertilization.
- The **cervix** is the lower part of the uterus. Its muscles dilate to provide a birth canal.
- The **vagina** is the duct joining the vulva with the uterus allowing sexual intercourse.
- The **vulva** protects the internal parts of the female reproductive system.

- **FSH** stimulates the development of follicles and the secretion of estrogen by the follicle wall.
- **Estrogen** stimulates the repair and thickening of the endometrium.
- **LH** stimulates the development of follicles, leading to ovulation.
- **Progesterone** promotes the thickening and maintenance of the endometrium.

The menstrual cycle is controlled by negative and positive feedback mechanisms involving ovarian and pituitary hormones. Follicle stimulating hormone (FSH) stimulates the development of follicles and secretion of estrogen by the follicle wall. Estrogen stimulates the repair and thickening of the endometrium after menstruation. High levels of estrogen inhibit the secretion of FSH (negative feedback) and stimulate luteinizing hormone (LH) secretion. LH stimulates the completion of meiosis and allows ovulation. LH also promotes the development of the wall of the follicle after ovulation into the corpus luteum which secretes estrogen (positive feedback) and progesterone. Progesterone promotes the thickening and maintenance of the endometrium. It also inhibits FSH and LH secretion by the pituitary gland (negative feedback).

In vitro fertilization (IVF) is used in cases of infertility. FSH and LH are used to stimulate development of follicles, causing ovulation of many ova. Human chorionic gonadotropin (HCG) is used to mature the ova. Fertilization is performed outside the body and fertilized embryos are then implanted in the woman. Progesterone is used to ensure that the uterus lining is maintained. Scientists are aware that the drugs women take in fertility treatment pose potential risks to health; nevertheless this method is widely used throughout the world in treating infertile couples.

Practice problems for Topic 6

Problem 1

In an experiment on food digestion and absorption, a student placed amylase and starch inside a dialysis tube. The dialysis tube was then placed in a test tube containing a solution of water and maltase. The tube was left for three hours. The amount of glucose was measured outside the dialysis tube.

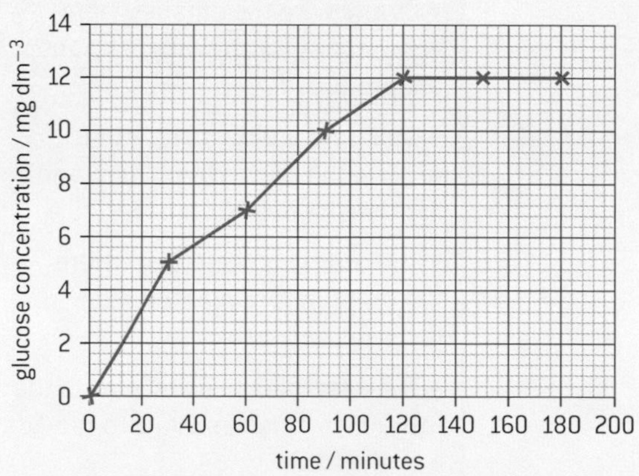

a) State the reason a dialysis tube is used in this experiment.

b) State the glucose concentration outside the dialysis tube 1 hour after the start of the experiment.

c) Explain the source of glucose:

(i) outside the dialysis tube

(ii) inside the dialysis tube.

Problem 2

a) Describe William Harvey's discovery of the circulation of the blood with the heart acting as the pump.

b) Explain what is meant by the double circulation of blood through the heart.

Problem 3

The bar graph shows the life expectancy of non-smokers and smokers.

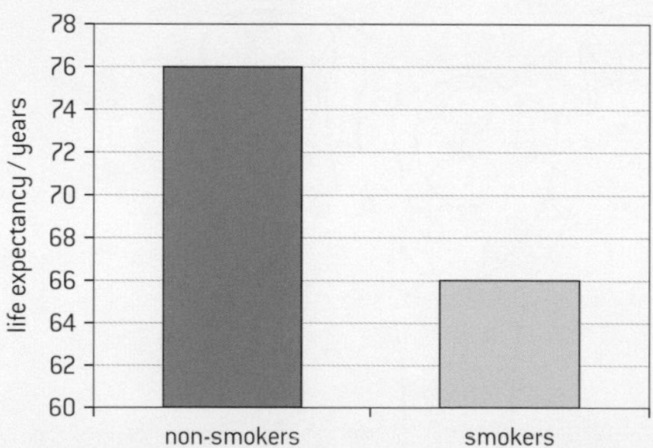

a) (i) Calculate the percentage difference in the life expectancy of smokers and non-smokers.

(ii) Suggest whether this data supports the hypothesis that smoking adversely affects life expectancy.

b) Explain gaseous exchange in the lungs.

c) Discuss the social consequences of lung cancer and emphysema.

Problem 4

Botulin or Botox is a neurotoxic protein produced by the bacterium *Clostridium botulinum*. Intramuscular administration of botulin inhibits the release of the neurotransmitter acetylcholine at the neuromuscular junction from presynaptic motor neurons causing muscle paralysis. It is used in cosmetics to reduce wrinkling of skin. In a study to see its side effects, ten brown rats (*Rattus norvegicus*) were injected with different doses of botulin. Their muscles were examined under the microscope for atrophy and classified according to severity as normal, mild and severe.

	Muscular atrophy (n = 10)		
Dose of botulin / ng kg^{-1} day^{-1}	**Normal (no atrophy)**	**Mild**	**Severe**
0	10	0	0
1	3	7	0
3	0	10	0
9	0	2	8

a) (i) State the number of rats that had atrophied muscles with a dose of 1 ng per kilogram of mass per day.

(ii) Identify the lowest dose that caused all rats to have some atrophy in their muscles.

b) Suggest whether the results support the hypothesis that botulin causes muscle atrophy.

The body mass of rats was recorded every five days during 20 days and compared with that of control rats.

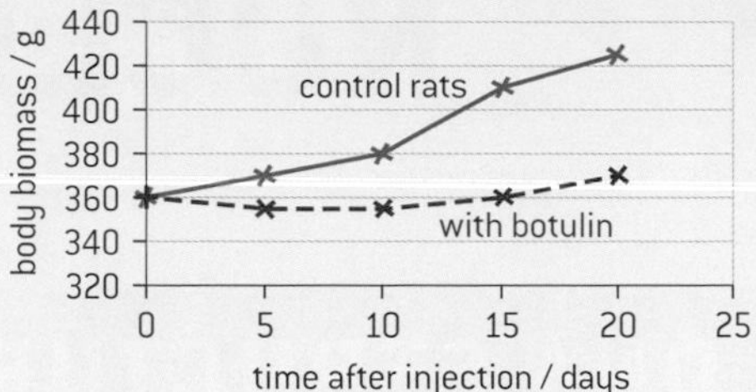

c) Compare and contrast the effect of botulin with time on the body mass of botulin-treated rats and untreated control rats.

d) Evaluate all the data on the effect of botulin on rats.

Problem 5

Explain how an electrical impulse is propagated along a myelinated neuron.

7 NUCLEOTIDES (AHL)

7.1 DNA STRUCTURE AND REPLICATION

You should know:

- ✔ the structure of DNA is ideally suited to its function.
- ✔ nucleosomes help to supercoil the DNA.
- ✔ DNA structure suggested a mechanism for DNA replication.
- ✔ DNA polymerases can start adding nucleotides only at the 3′ end of a primer.
- ✔ DNA replication is continuous on the leading strand and discontinuous on the lagging strand.
- ✔ DNA replication is carried out by a complex system of enzymes.
- ✔ some regions of DNA do not code for proteins but have other important functions.

You should be able to:

- ✔ explain DNA replication (only in prokaryotes), describing the function of helicase, DNA gyrase, single strand binding proteins, DNA primase and DNA polymerases I and III in replication.
- ✔ describe the function of non-coding regions of DNA such as regulators of gene expression, introns, telomeres and genes for tRNAs.
- ✔ explain the use of nucleotides containing dideoxyribose to stop DNA replication in the preparation of samples for base sequencing.
- ✔ describe Rosalind Franklin's and Maurice Wilkins' investigation of DNA structure by X-ray diffraction.
- ✔ analyse results of the Hershey and Chase experiment providing evidence that DNA is the genetic material.
- ✔ analyse the association between protein and DNA within a nucleosome using molecular visualization software.
- ✔ explain how tandem repeats are used in DNA profiling and analyse profile electrophoretic gels.

In Topic 2.6 you have already studied the basic structure of DNA and RNA. Here you will see more detail and study the replication of DNA.

DNA consists of a double strand of nucleotides in the form of a helix. In eukaryotes it is associated with histone proteins forming nucleosomes. This allows the DNA to be condensed. A nucleosome consists of a core of four pairs of histone proteins with around 150 base pairs of DNA coiled around the proteins. A short section of naked DNA connects one nucleosome to the next. An additional histone protein molecule binds the DNA to the core particle.

Rosalind Franklin's X-ray diffraction studies provided crucial evidence that DNA is a double helix. Rosalind Franklin and Maurice Wilkins used novel techniques with a high resolution X-ray camera to obtain very clear images of diffraction patterns from DNA. They were able to deduce that the DNA molecule was helical in shape. This helped Watson and Crick figure out the structure of DNA.

Alfred Hershey and Martha Chase devised an experiment that provided evidence that DNA is the genetic material and not proteins as was believed at the time. In their experiment, they cultured viruses that contained proteins with radioactive (^{35}S) sulfur and they separately cultured viruses that contained DNA with radioactive (^{32}P) phosphorus. They infected bacteria separately with the two types of viruses. They separated the non-genetic component by centrifugation and measured the radioactivity in the pellet (where the infected bacteria where) and the supernatant (where the virus coats were). They found that most radioactivity was found in the infected bacteria when using radiolabelled DNA.

Bacteriophages are viruses that create a channel across the bacterial cell membrane and wall to transfer the viral genome DNA to the host cell cytoplasm, initiating infection in the bacterium.

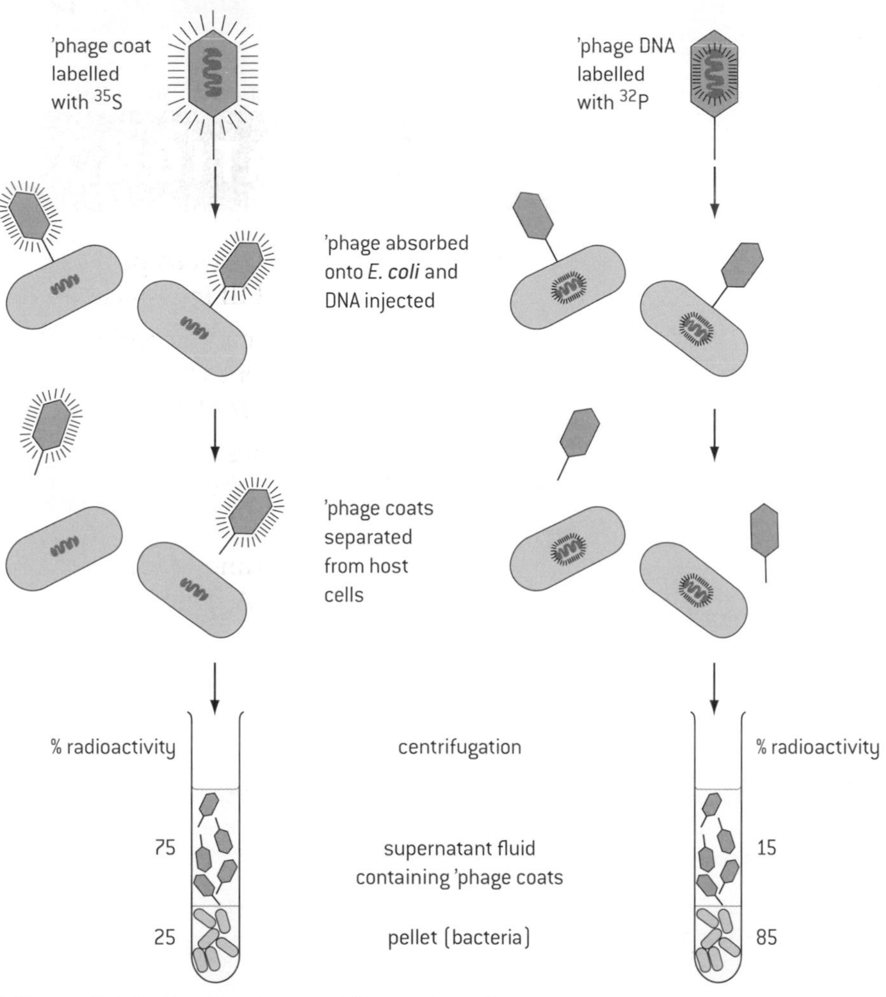

▲ **Figure 7.1.1.** The Hershey and Chase experiment

Example 7.1.1.

Scientists wanted to study the importance of calcium in channelling of bacteriophage DNA to the host bacterium *Bacillus subtilis*. The DNA of a bacteriophage was marked with radioactive phosphorus and used to infect bacteria, with and without calcium being present. The amount of radioactivity was recorded in the supernatant (left) and the percentage of infected bacteria calculated (right).

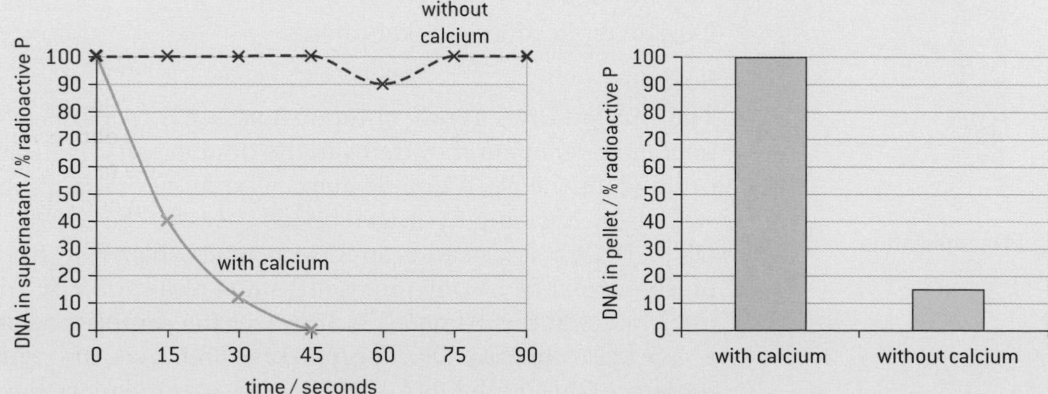

a) (i) State the percentage radioactive phosphorus in the supernatant at 30 seconds for infection with and without calcium.

(ii) Explain the results obtained in **(i)**.

b) Suggest whether the results obtained in both graphs support the hypothesis that calcium is necessary for infection.

c) Discuss the results in term of the Hershey–Chase theory of genetic inheritance.

Solution

a) (i) 12 % with calcium and 100% without calcium.

(ii) With calcium the radioactive phosphorus of the viral DNA enters the bacterium and therefore disappears from the supernatant. In the infection occurring without calcium, the bacteriophages cannot make the channel for the DNA to enter the bacterium, therefore the viral DNA remains in the supernatant.

b) The data supports the hypothesis, as there is no or little infection when calcium is missing, as seen in both graphs.

c) The Hershey–Chase experiments provided evidence that DNA is the genetic material and not proteins. In their experiment, they cultured viruses that contained proteins with radioactive sulfur and other viruses that contained DNA with radioactive phosphorus. They observed that the radioactivity in the pellet was due to phosphorus and that in the supernatant was due to sulfur. In the experiment on *Bacillus subtilis*, the same results were obtained for phosphorus; it was in the pellet only when infection was possible (with calcium). When infection was not possible (without calcium), the DNA did not enter the bacterial cells and was found in the supernatant; very little was found in the pellet.

Replication of DNA is needed for the duplication of the genetic material in advance of cell division by mitosis or meiosis. In eukaryotes it produces the sister chromatids in chromosomes. DNA replication is carried out by a complex system of enzymes.

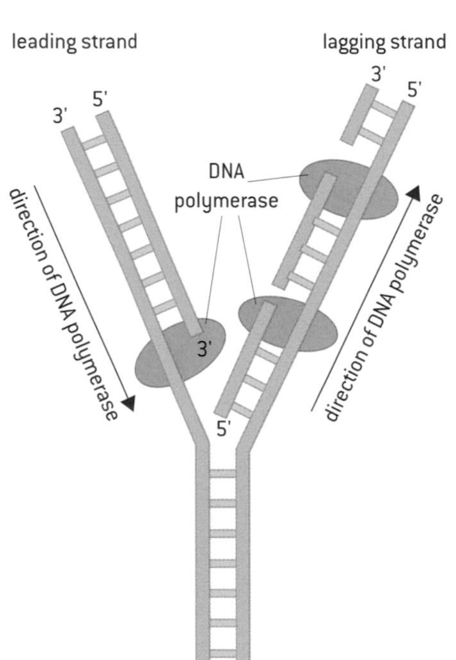

▲ **Figure 7.1.2.** DNA replication

- **DNA gyrase** (topoisomerase II) relieves the strain on the coil, ready for helicase to start working.
- **Helicase** breaks hydrogen bonds between strands, allowing the helix to uncoil.
- **Single strand binding proteins** avoid re-binding of complementary base pairs.
- **DNA primase** starts adding RNA primer at the 5′ end.
- **DNA polymerase III** adds complementary nucleotides at the primer in the 5′–3′ direction.
- **DNA polymerase I** removes the primer.
- **DNA ligase** seals the nicks.
- **DNA polymerase I** and **DNA polymerase III** proofread the new DNA.

Example 7.1.2.

Explain replication in eukaryotes. [8]

Solution

The enzyme DNA gyrase or topoisomerase II prepares DNA for uncoiling, relieving the strains in the double helix. Helicase uncoils or unwinds the DNA double helix, separating or unzipping the strands by breaking hydrogen bonds between the two strands of DNA. DNA primase adds an RNA primer which is a short length of RNA. Next DNA polymerase III starts replication by adding nucleotides at the primer. It synthesizes the complementary strand in a 5′ to 3′ direction. DNA polymerase I removes the primer by replacing RNA with DNA. DNA ligase seals the nicks linking sections of replicated DNA, or Okazaki fragments, in the lagging strand. Once replication is completed, DNA polymerase I and DNA polymerase III proofread for mistakes.

Non-coding regions are regions of DNA that do not code for proteins but have other important functions such as regulators of gene expression, introns, telomeres and genes for tRNAs. Regulators of gene expression can be areas of the DNA such as the promoter, where the RNA polymerase joins to start transcription. Many promoters require activator proteins to start transcription and can be down-regulated by repressor proteins. Introns are sections found in eukaryotic DNA that are transcribed to mRNA but are edited out before translation. Telomeres are the extremes of chromosomes and protect the DNA from damage. There are 170 to 570 genes coding for tRNAs in eukaryotes.

- **Regulators of gene expression** are non-coding regions of DNA that increase or decrease gene expression.
- **Introns** are sections of mRNA that are edited out (spliced) before translation.
- **Telomeres** are the non-coding DNA extremes of chromosomes.
- **Genes for tRNAs** are DNA sections that are transcribed to RNA molecules that act as tRNAs.

Example 7.1.3.

Which are examples of non-coding regions of DNA?

I telomeres II promoter III exons

A. I only **C.** I and III only

B. I and II only **D.** I, II and III

Solution

The correct answer is **B**, as exons are the coding regions.

Example 7.1.4.

Tandem repeats occur in DNA when a pattern of one or more nucleotides is repeated and remain adjacent to each other. Tandem repeats can show us patterns that help identify the inherited traits of an individual.

In a paternity test, a PCR was performed on each organism tested and the amplified material was run through electrophoresis. The results are shown in the gel.

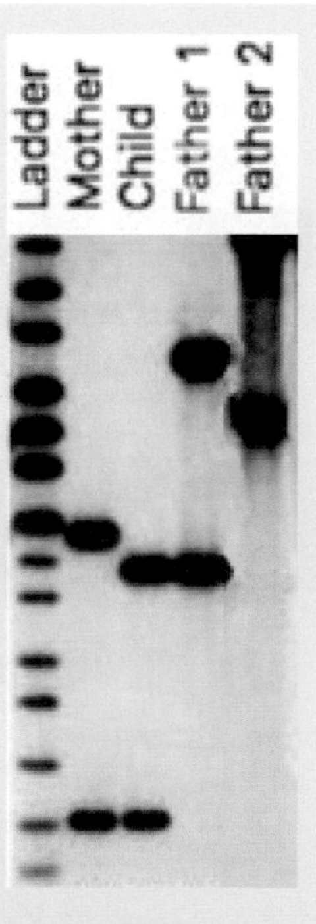

a) (i) State the most probable father of the child.

(ii) Explain the choice in **(i)**.

b) Describe the process of electrophoresis.

Solution

a) (i) Father 1.

(ii) Father 1 has one band at the same height as the child that is not present in the mother nor in father 2. This means that both the child and father 1 share a common tandem repeat.

b) In electrophoresis, the products obtained from the PCR reaction (short nucleotide sequences of DNA) are placed in a well of an agarose gel. The gel is placed in a buffer and an electric current is applied. The DNA fragments will move towards the positive pole as DNA is negative due to the phosphate groups. The smaller fragments move more than the larger fragments, therefore the fragments are separated according to their size. A ladder of fragments of known size is placed in a well in order to compare sizes.

Tandem repeats in coding regions can cause diseases such as Huntington's disease. If they occur in non-coding regions they are usually harmless, although diseases such as the fragile X syndrome are caused by tandem repeats in non-coding regions. When a PCR is performed, if there are tandem repeats, fragments of different sizes will be obtained. These are directly proportional to the number of repetitions; this means that if a tandem repeat appears six times, the fragments produced in a PCR will be double the length of one that is repeated three times.

Sequencing is used to determine the DNA nucleotide order. Frederick Sanger determined the sequence termination method, using nucleotides containing dideoxyribose to stop DNA replication in preparation of samples for base sequencing.

Stages 1–2: DNA strand chopped up, mixed with primer, bases, DNA polymerase + terminator bases

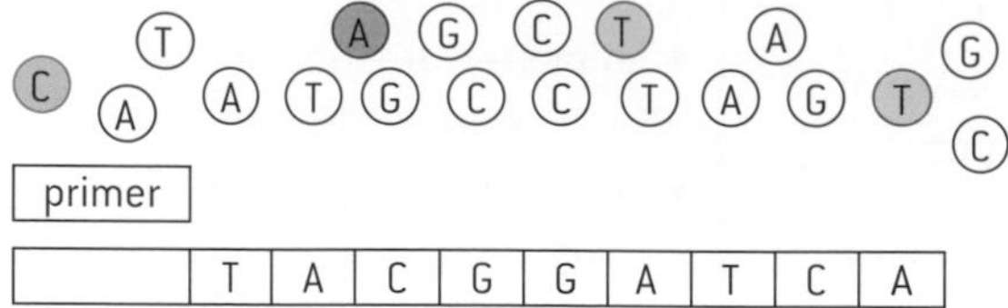

Stage 3: Each time a terminator base is added a strand terminates until all possible chains produced

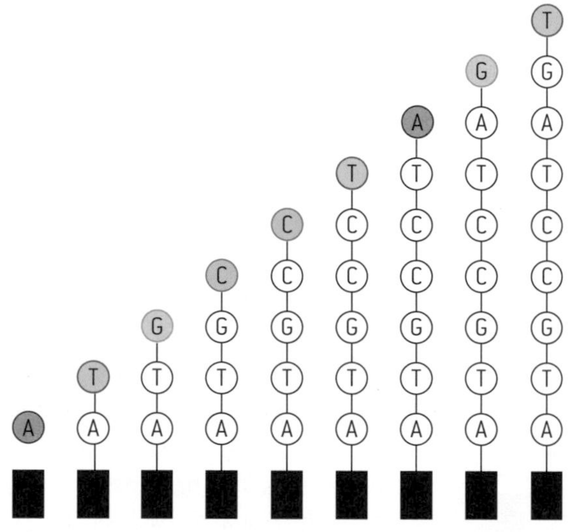

Stage 4: Readout from capillary tubes: DNA fragments separated by electrophasis in capillary tubes

▲ **Figure 7.1.3.** Sequencing

7.2 TRANSCRIPTION AND GENE EXPRESSION

You should know:

✔ gene expression is regulated by proteins that bind to specific base sequences in DNA.

✔ the environment of a cell and of an organism have an impact on gene expression.

✔ in transcription information stored as a code in DNA is copied onto mRNA.

✔ transcription occurs in a 5′ to 3′ direction.

✔ nucleosomes help to regulate transcription in eukaryotes.

✔ eukaryotic cells modify mRNA after transcription.

✔ splicing of mRNA increases the number of different proteins an organism can produce.

You should be able to:

✔ explain gene expression and its regulation.

✔ explain the process of transcription and the nucleosome regulation of it.

✔ explain post-transcriptional changes to RNA and how they affect gene expression.

✔ explain alternative splicing.

✔ describe the promoter as an example of non-coding DNA with a function.

✔ analyse changes in the DNA methylation patterns.

Gene expression is the process by which a gene product is made using information in the genes. These gene products are usually proteins, but can also be tRNA or small RNAs that can act as gene expression regulators. Although the genome of an organism is the same in all its cells, only some genes are expressed in different tissues. For example, insulin is expressed only in the α cells of the pancreas. This is because gene expression is regulated. Gene expression can be modulated by transcriptional initiation, RNA processing and post-translational modification of proteins.

In Topic 2.7 you studied DNA replication, transcription and translation and Topic 7.1 looked at replication in more detail.

Example 7.2.1.

Explain the control of gene expression in eukaryotes.

Solution

The mRNA conveys genetic information from DNA to the ribosomes, where it guides polypeptide production. Gene expression requires the production of specific mRNA through transcription. Most genes are turned off, therefore not being transcribed at any one time. This means that gene expression is regulated and some genes are expressed only at certain times in certain cells of different tissues. Cell differentiation involves changes in gene expression. Transcription factors, which are usually proteins, can increase (or decrease) transcription. Such proteins may prevent or enhance the binding of RNA polymerase. Nucleosomes limit the access of transcription factors to DNA, thereby regulating gene transcription. Gene expression can be regulated by post-transcriptional modification such as mRNA splicing. DNA methylation or protein acetylation appear to control gene expression as epigenetic factors. Some DNA methylation patterns are inherited, for example the one that determines X chromosome silencing. Introns may contain positive or negative gene regulators. Hormones or the chemical environment of a cell can also affect gene expression.

Transcription is the synthesis of mRNA, the first step of gene expression. The RNA polymerase binds the DNA at the promoter site. The two DNA strands are separated due to the breaking of the hydrogen bonds between complementary nucleotides. RNA polymerase adds the 5′ end of the free RNA nucleotide to the 3′ end of the growing mRNA molecule. The temporary hydrogen bonds between DNA and RNA are broken, liberating the mRNA. The mRNA either remains in the nucleus or leaves via a nuclear pore.

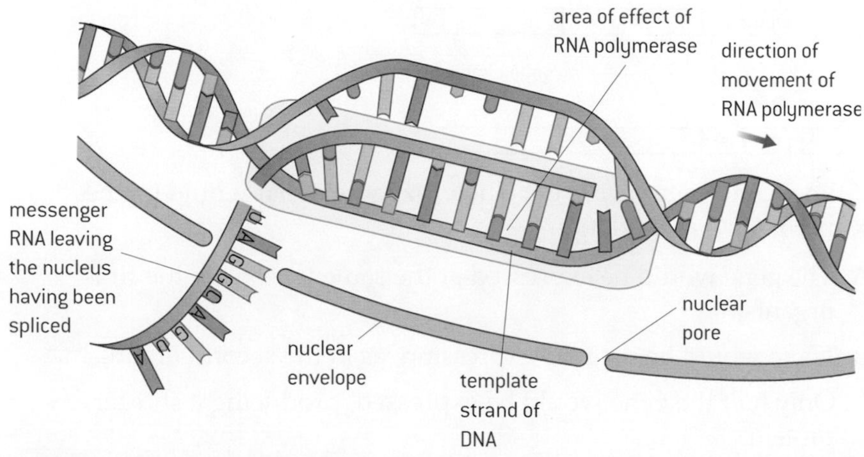

▲ **Figure 7.2.1.** DNA transcription

It is now well known that there are epigenetic factors that can modify gene expression. There is mounting evidence that the environment can trigger heritable changes in epigenetic factors. The nature versus nurture debate concerning the relative importance of an individual's innate qualities versus those acquired through experience is still under discussion.

Example 7.2.2.

Which statement applies to transcription in eukaryote cells but not in prokaryote cells?

A. Initiation of transcription requires a promoter sequence of DNA.

B. Uracil replaces thymine in RNA.

C. It occurs in the nucleus and is followed by translation in the cell cytoplasm.

D. RNA polymerase transcribes DNA to produce a strand of RNA.

Solution

The correct answer is **C**, as prokaryotes do not have a nucleus. In eukaryotes, translation usually occurs outside the nucleus.

The promoter is essential in the regulation of gene expression. It is a non-transcribed sequence of DNA usually found upstream of the gene. If the RNA polymerase cannot join the promoter on the DNA, the replication into mRNA will not occur. Gene expression can be inhibited by the joining of a repressor to the promoter. In other occasions, an activator can enhance the binding of the RNA polymerase to the promoter, increasing gene expression.

NH2, HN, O, N, H — DNA methyltransferase → NH2, CH3, HN, O, N, H

▲ **Figure 7.2.2.** DNA methylation

Methylation is the addition of a methyl group ($-CH_3$) to a nucleotide, usually a cytosine found in a CpG island. CpG islands are areas of DNA in or close to the promoters where there is a cytosine nucleotide followed by a guanine nucleotide in a linear sequence in the 5′ to 3′ direction (5′-C-phosphate-G-3′). Normally the cytosines of these islands are not methylated, but if they are, this affects transcription inversely.

Example 7.2.3.

The diagram shows the methylation pattern of the DNA promoters of homologous chromosomes in two homozygote organisms.

Organism 1

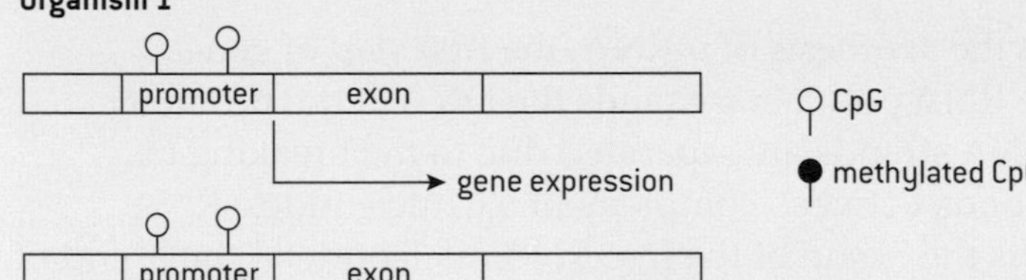

Organism 2

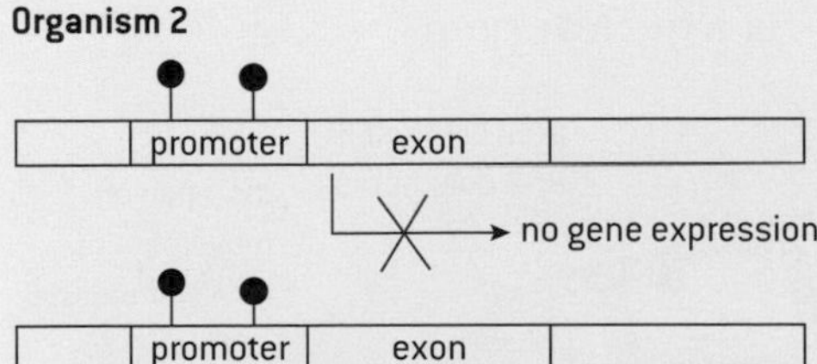

If the second organism were a heterozygote, what would be the result of CpG methylation?

A. The gene would be expressed at the same level as in the first organism.

B. There would be no gene expression, as in the second organism.

C. Only half the gene would be expressed, producing a shorter protein.

D. Only the methylated strand would be expressed.

Solution

The correct answer is **A**. If one chromosome has the promoter unmethylated, then that strand can be transcribed to RNA. In this case methylation is inhibiting transcription, therefore the strand that is not methylated can be transcribed. Usually only one chromosome is used as a template for replication, therefore the amount of gene expression will not change.

In eukaryotes the mRNA suffers post-transcription modifications such as splicing, polyadenylation of the 3′ end and capping of the 5′ end. Splicing is the removal of introns by special proteins and enzymes. If alternative sequences are used as introns, different peptides can be obtained from the same gene (alternative splicing).

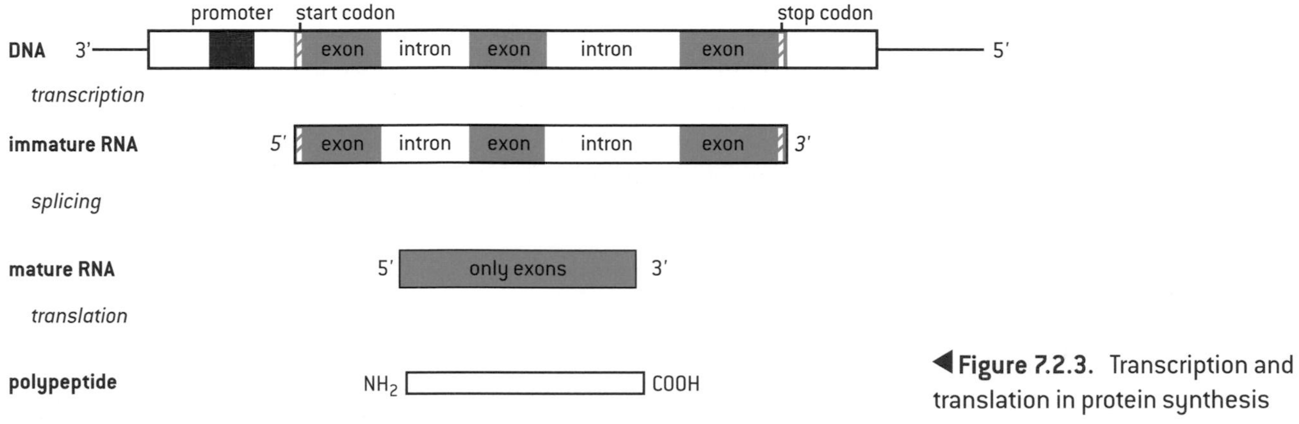

◀ **Figure 7.2.3.** Transcription and translation in protein synthesis

7.3 TRANSLATION

You should know:

- ✔ information transferred from DNA to mRNA is translated into an amino acid sequence.
- ✔ initiation of translation involves assembly of the components that carry out the process.
- ✔ synthesis of the polypeptide involves a repeated cycle of events.
- ✔ disassembly of the components follows termination of translation.
- ✔ free ribosomes synthesize proteins for use primarily within the cell.
- ✔ bound ribosomes synthesize proteins primarily for secretion or for use in lysosomes.
- ✔ translation can occur immediately after transcription in prokaryotes due to the absence of a nuclear membrane.
- ✔ the sequence and number of amino acids in the polypeptide is the primary structure.
- ✔ the secondary structure is the formation of alpha helices and beta pleated sheets stabilized by hydrogen bonding.
- ✔ the tertiary structure is the further folding of the polypeptide stabilized by interactions between R groups.
- ✔ the quaternary structure exists in proteins with more than one polypeptide chain.

You should be able to:

- ✔ explain translation.
- ✔ explain tRNA-activation by enzymes through phosphorylation and amino acid binding.
- ✔ describe protein structure.
- ✔ explain the reason polar and non-polar amino acids are relevant to the bonds formed between R groups.
- ✔ describe how quaternary structure may involve the binding of two or more peptides or of a prosthetic group to form a conjugated protein.
- ✔ identify polysomes in electron micrographs of prokaryotes and eukaryotes.
- ✔ describe how molecular visualization software can be used to analyse the structure of eukaryotic ribosomes and a tRNA molecule.

In Topic 2.7 you studied DNA replication, transcription and translation. In Option B you can study bioinformatics.

Assessment tip

Answers should be written inside the answer box as continuations below the box or to the side may not be seen. If you need to continue the answer on another sheet, then you should state this clearly where the answer in the box ends. Continuation answers should be fully labelled, for example, 2(a)(ii), not just (ii).

Information transferred from DNA to mRNA is translated into an amino acid sequence. Synthesis of the polypeptide involves a repeated cycle of events. Initiation of translation involves assembly of the components that carry out the process. Firstly the tRNA is activated by phosphorylation in the cytoplasm. A tRNA-activating enzyme attaches a specific amino acid to a determined tRNA, using ATP for energy. To begin the process of translation, an mRNA molecule binds to the small ribosomal subunit at an mRNA binding site. An initiator tRNA molecule carrying methionine binds at the start codon (AUG). The large subunit of the ribosomes then joins the small subunit. It has three sites, the A (aminoacyl) site, P (peptidyl) site and E (exit) site. The initiator tRNA is in the P site of the ribosome. The next codon signals another tRNA to bind at the A site. A peptide bond is formed between the amino acids in the P and A sites. The ribosome translocates three bases along the mRNA (in the 5′–3′ direction), moving the tRNA in the P site to the E site. This tRNA is now free in the cytoplasm to be activated again. The tRNA with the appropriate anticodon will bind to the next codon and occupy the vacant A site. This will continue till a stop codon (UAG, UGA or UAA) is reached. The disassembly of the components follows termination of translation.

SAMPLE STUDENT ANSWER

Outline the roles of the different binding sites for tRNA on ribosomes during translation. [4]

This answer could have achieved 4/4 marks:

Ribosomes have 3 binding sites; A, P and E. At any time, only 2 tRNA can bind to the site. Translation happens from 5′ to 3′. During translation, an mRNA containing the amino acid methionine, corresponding to the start codon AUG arrives at the P site of the ribosome. Another tRNA with the amino acid corresponding to its anticodon attaches at the A site. As the mRNA moves along the ribosome, the tRNA moves from the P site to the E site, leaving the ribosome. The tRNA from the A site moves to the P site and a peptide bond forms between amino acids. The binding sites are present on the large subunits of the ribosome.

▲ The student scores the marks for saying that the ribosome has A, P and E binding sites. The second and third marks are for saying that met-tRNA binds to the start codon, in the P site. The fourth mark is for saying that the peptide bond is formed between the amino acid of the A site and the polypeptide at the P site. Another mark could have been given for including that the E binding site is where the tRNA exits the ribosome.

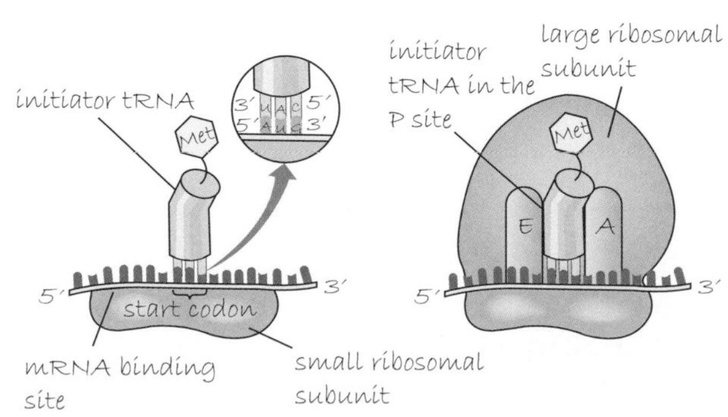

Ribosomes are the site of translation. Free ribosomes synthesize proteins that are used primarily within the cell while bound ribosomes synthesize proteins primarily for secretion or for use in lysosomes. Translation can occur immediately after transcription in prokaryotes due to the absence of a nuclear membrane.

Example 7.3.1.

The diagram shows the structure of a polysome.

a) Identify X and Y.

b) Indicate with an arrow the direction of the mRNA movement in this diagram.

c) (i) State where polysomes can be found in eukaryotic cells.

(ii) Suggest a reason why the presence of polysomes enhances translation.

Solution

a) X: polypeptide and Y: ribosome

b) Arrow shown from left to right.

c) (i) Free in the cytoplasm or attached to rough endoplasmic reticulum.

(ii) Many ribosomes can simultaneously read one mRNA chain to synthesize the same protein. This allows for the formation of a greater amount of the same protein in a shorter period of time.

The use of computers has enabled scientists to make advances in bioinformatics applications such as locating genes within genomes and identifying conserved sequences.

For example, scientists can diagnose diseases by comparing DNA sequences. Von Willebrand disease is the most common hereditary blood-clotting disorder in humans. It arises from a deficiency in a protein required for platelet adhesion during clotting, the von Willebrand factor. The gene for this factor is found on the short arm of chromosome 12 and has 52 exons. Patients whose DNA sequence for this clotting factor is different to the norm could be liable to clotting problems and suffer severe bleeding. A simple genetic sequencing test and a BLAST comparison enable diagnosis of the disease before the symptoms occur.

- The **primary structure** of proteins or polypeptides is the sequence and number of amino acids.
- The **secondary structure** is the formation of alpha helices and beta pleated sheets stabilized by hydrogen bonding.
- The **tertiary structure** is the further folding of the polypeptide into globular or fibrous forms, stabilized by interactions between R groups. The bonds maintaining tertiary structure are disulfide bridges, hydrophobic interactions, electrostatic attractions and hydrogen bonds.
- A **quaternary structure** exists in proteins with more than one polypeptide chain. It can have other molecules such as a prosthetic group.

Proteins can have different structures. Hemoglobin is a globular protein that has a quaternary structure. It is formed by four polypeptides that have alpha helix secondary structures. The four prosthetic groups in hemoglobin are the same, one heme group joined to each polypeptide. The heme group is a molecule containing iron, which oxidizes allowing the transport of oxygen in red blood cells.

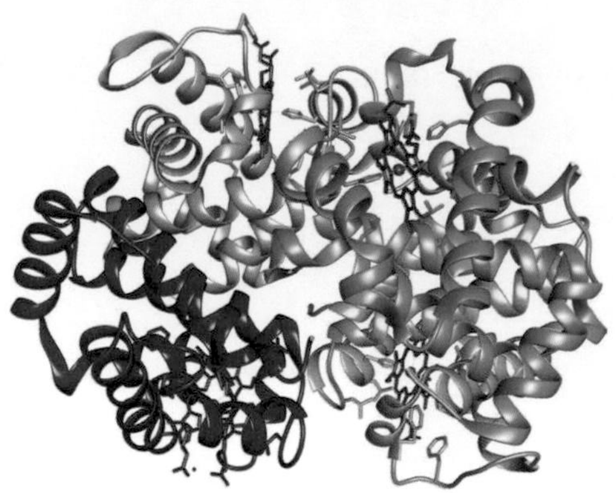

▲ **Figure 7.3.1.** Quaternary structure of hemoglobin showing the four polypeptides and four heme groups (red)

Practice problems for Topic 7

Problem 1

Explain the use of nucleotides containing dideoxyribose to stop DNA replication in the preparation of samples for base sequencing.

Problem 2

Acetylation and methylation are epigenetic changes that can occur in peptides. The epigenetic differences were recorded in twins at the ages of 4 years and 40 years.

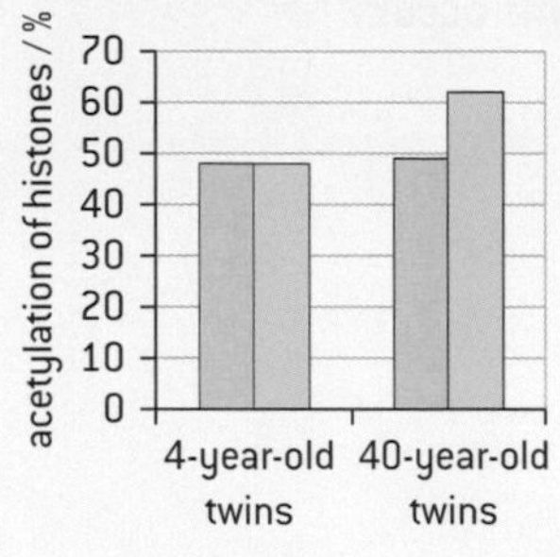

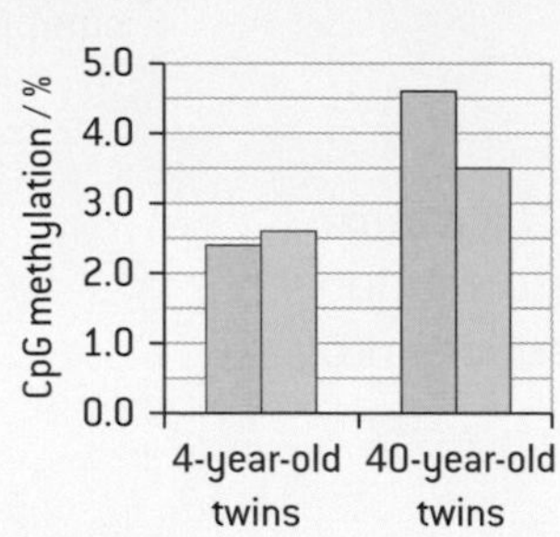

a) (i) Describe the histone acetylation patterns of the twins.

(ii) Explain how the acetylation of histones affects gene expression.

b) (i) Suggest a reason for the differences in CpG methylation patterns between the twins.

(ii) Explain how the methylation of DNA can affect gene expression.

Problem 3

Explain how a DNA sequence can eventually determine the primary structure of a protein.

8 METABOLISM, CELL RESPIRATION AND PHOTOSYNTHESIS (AHL)

8.1 METABOLISM

You should know:

- ✔ metabolic reactions are regulated in response to the cell's needs.
- ✔ metabolic pathways consist of chains and cycles of enzyme-catalysed reactions.
- ✔ enzymes lower the activation energy of the chemical reactions that they catalyse.
- ✔ enzyme inhibitors can be competitive or non-competitive.
- ✔ metabolic pathways can be controlled by end-product inhibition.

You should be able to:

- ✔ explain the need for metabolic pathways and how enzymes work in metabolic processes.
- ✔ explain the end-product inhibition of the pathway that converts threonine to isoleucine.
- ✔ describe how databases are used to search for potential new anti-malarial drugs.
- ✔ calculate and plot rates of reaction from raw experimental results.
- ✔ distinguish different types of inhibition from graphs at specified substrate concentration.

Enzymes are globular proteins that lower the activation energy of the chemical reactions they catalyse. Metabolic pathways are enzyme-mediated biochemical reactions that lead to the synthesis or breakdown of natural substrates into products within a cell or tissue. The glycolysis reactions in respiration and the Calvin cycle in photosynthesis are examples of metabolic pathways.

In Topic 2.5 you studied how enzymes work. In Topic 2.7 and Topic 7 you studied DNA replication, transcription and translation that lead to gene expression.

Example 8.1.1.

a) Explain how the active site of an enzyme helps to lower activation energy.

b) Compare and contrast the primary and secondary structure of an enzyme.

Solution

a) The active site is formed by the amino acids responsible for the catalytic activity of enzymes. It is an area of the protein where the substrate enters and is specific for each substrate.
The enzyme lowers the energy needed for the reaction to occur by bringing the substrates closer to join in an anabolic reaction or by straining the molecules in a catabolic reaction.

b) Both the primary and secondary structures help determine the final structure of the protein. The primary structure is the number and sequence of amino acids in the protein. These are joined by peptide bonds. The secondary structure is the folding pattern of the primary structure. It is held by hydrogen bonds, which are weaker than the peptidic bonds. The secondary structure can be alpha-helices or beta sheets. Where there is no structure it is called loops.

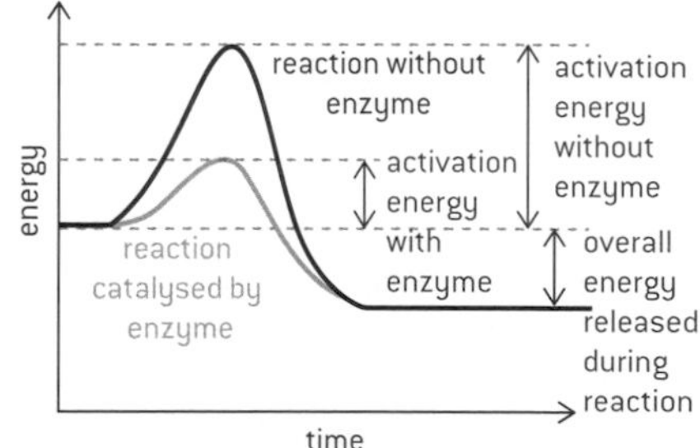

▲ **Figure 8.1.1.** Enzymes lower activation energy

Assessment tip

For questions that ask you to **compare and contrast**, remember to include at least one difference and one similarity.

Assessment tip

Remember that questions are often based on more than one topic. This question is linked to Topic 7, so you can add information from that topic too.

Developments in bioinformatics, such as the interrogation of databases, have facilitated research into metabolic pathways. Databases such as Kegg have curated most metabolic pathways, the enzymes involved and their inter-relationships.

- A **rate of reaction** is the speed at which a chemical reaction takes place. It can be measured according to the concentration (or amount) of product formed in a unit of time or as the concentration (or amount) of substrate that disappears in a unit of time.
- An **inhibitor** is a molecule that binds to an enzyme, decreasing its catalytic activity.
- A **competitive inhibitor** joins at the active site. Its binding is affected by the substrate concentration.
- A **non-competitive inhibitor** joins away from the active site. Its binding is not affected by the substrate concentration.

Malaria is a disease caused by the parasite *Plasmodium*, that leads to around 2 million deaths per year. Increased resistance of malaria parasites to drug treatment requires the development of new therapies. *Plasmodium* nucleic acid synthesis metabolic pathways are different from those of their human hosts. Thus, targeting these metabolic pathways provides a promising route for novel drug development. Bioinformatic analysis has provided an improved route to inhibitor design targeted to specific enzymes, including those of nucleic acid synthesis metabolism. Scientists can use experimental information as well as training artificial neural networks to predict the outcome of novel drugs on different targets.

Enzyme inhibitors can be competitive or non-competitive. Competitive inhibitors compete with the substrate for the active site. This means that increasing the substrate concentration reduces the inhibition and eventually the reaction is similar to that without the inhibitor. Conversely, non-competitive inhibitors bind to the enzyme in a different place from the active site, so increasing the substrate concentration will not affect the inhibition. Many enzyme inhibitors have been used in medicine. For example, ethanol and fomepizole have been used to act as competitive inhibitors for antifreeze poisoning.

SAMPLE STUDENT ANSWER

The graph shows the relationship between the reaction rate and substrate concentration in the presence and absence of a competitive inhibitor.

no inhibitor
with inhibitor
reaction rate
substrate concentration

a) Explain the effect of the competitive inhibitor on the reaction rate. [2]

This answer could have achieved 0/2 marks:

The competitive inhibitor increases the rate of the reaction. This means that a certain point in time the substrate concentration with the presence of a competitive inhibitor will be higher than the substrate concentration in the absence of the competitive inhibitor at the same point in time.

▼ The student confuses rate of reaction and substrate concentration. The answer should have mentioned that the competitive inhibitor slows the reaction rate as it competes for the active site. The competitor has a similar shape or composition to the substrate, binding to the enzyme in a reversible way. As the substrate concentration increases, more substrate binds to the active site than the competitor and the reaction rate increases. The reaction rate then reaches the maximum plateau, the same as with no inhibitor.

b) On the graph, sketch the relationship between the reaction rate and substrate concentration in the presence of a non-competitive inhibitor. [2]

This answer could have achieved 2/2 marks:

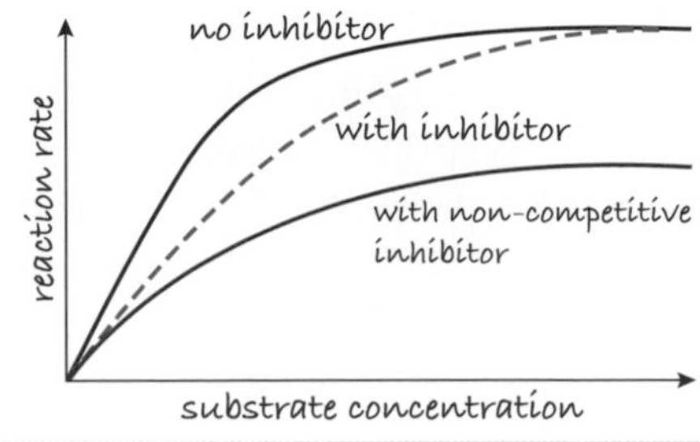

▲ The student drew the correct line, starting at the same point as the one without an inhibitor but always with a lower rate of reaction and making a plateau at a lower point.

Metabolic reactions are regulated in response to the cell's needs. Metabolic pathways consist of chains and cycles of enzyme-catalysed reactions. Because enzymes lower the activation energy of the chemical reactions that they catalyse, the reactions can occur at a higher rate. Metabolic pathways can be controlled by end-product inhibition. This means the initial reaction is inhibited by the product of the final reaction. This avoids an excess of the product.

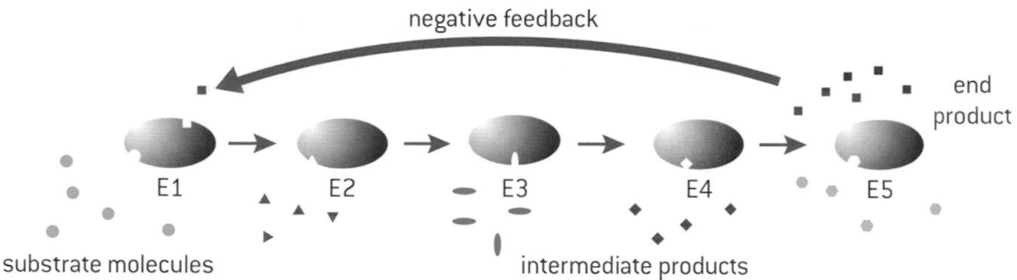

▲ **Figure 8.1.2.** Negative feedback by end-product inhibition

Example 8.1.2.

Explain how isoleucine controls its own production.

Solution

The production of isoleucine is controlled by end-product inhibition. It inhibits the pathway that converts threonine to isoleucine. Through a series of five enzyme-catalysed reactions, the amino acid threonine is converted to isoleucine. As the reaction takes place, the concentration of isoleucine increases. Isoleucine then binds to the allosteric site of the first enzyme in the chain, threonine deaminase. This enzyme is therefore inhibited. Isoleucine acts as a non-competitive inhibitor. When the level of isoleucine is low again, the enzyme will start breaking down threonine again.

8.2 CELL RESPIRATION

You should know:

✔ energy is converted to a usable form in cell respiration.

✔ cell respiration involves the oxidation and reduction of electron carriers.

✔ phosphorylation of molecules makes them less stable.

✔ in glycolysis, glucose is converted to pyruvate in the cytoplasm.

✔ glycolysis gives a small net gain of ATP without the use of oxygen.

✔ in aerobic cell respiration pyruvate is decarboxylated and oxidized, converted into an acetyl compound and attached to coenzyme A to form acetyl coenzyme A in the link reaction.

✔ in the Krebs cycle, the oxidation of acetyl groups is coupled to the reduction of hydrogen carriers, liberating carbon dioxide.

✔ energy released by oxidation reactions is carried to the cristae of the mitochondria by reduced NAD and FAD.

✔ transfer of electrons between carriers in the electron transport chain in the membrane of the cristae is coupled to proton pumping.

✔ in chemiosmosis protons diffuse through ATP synthase to generate ATP.

✔ oxygen is needed to bind with the free protons to maintain the hydrogen ion gradient and to accept electrons from the end of the electron transport pathway, resulting in the formation of water.

✔ the structure of a mitochondrion is adapted to the functions it performs.

You should be able to:

- ✔ explain how energy is made available through respiration.
- ✔ outline the reactions in glycolysis, the link reaction and the Krebs cycle.
- ✔ explain the electron transport chain and the involvement of NAD and FAD in respiration.
- ✔ explain chemiosmosis and the function of ATP synthase in the production of ATP.
- ✔ analyse electron tomography images of active mitochondria.
- ✔ analyse diagrams of the pathways of aerobic respiration to deduce where decarboxylation and oxidation reactions occur.
- ✔ annotate a diagram of a mitochondrion to show how it is adapted to its function.

In Topic 2.8 you studied cell respiration and in Topic 1.2 the ultrastructure of cells, including mitochondria, was covered.

Cell respiration involves the oxidation and reduction of electron carriers. Oxidation and reduction are chemical processes that always occur together in what is called redox reactions. Oxidation is when electrons are lost from a substance and reduction is when electrons are gained. Phosphorylation is the addition of phosphate to a molecule, for example, when an inorganic phosphate ion is added to ADP to form ATP. In this process, electrons transferred are part of a hydrogen atom, so redox reactions happen by transfer of hydrogen atoms in both respiration and photosynthesis.

Example 8.2.1.

Which of the following processes includes oxidation?

A. Production of $FADH_2$

B. Production of NADH from NAD

C. Production of ADP from ATP

D. Production of acetyl CoA from pyruvate

Solution

The correct answer is **D**. In the link reaction, pyruvate is decarboxylated and oxidized, converted into an acetyl compound and attached to coenzyme A to form acetyl coenzyme A.

In glycolysis, glucose is converted to pyruvate in the cytoplasm. Glycolysis gives a small net gain of ATP without the use of oxygen. In anaerobic respiration the pyruvate can be used in alcoholic or lactic fermentation. Alcoholic fermentation produces carbon dioxide and alcohol, while lactic fermentation produces lactic acid. Lactic acid and lactate are effectively the same thing. When a proton is given up from an organic acid the name changes from '-ic acid' to '-ate'.

Example 8.2.2.

What is the total net number of ATP molecules produced during glycolysis?

A. 0 **B.** 1 **C.** 2 **D.** 4

Solution

The net amount is what is left at the end of the process. The correct answer is **C**, as two ATP molecules are used in the phosphorylation of glucose to glucose 6-phosphate and two ATP molecules are produced when triose phosphate is transformed into pyruvate. As there are two triose phosphate molecules, this produces a total of four molecules of ATP. Therefore, two are used and four are produced, leaving a net total of two molecules of ATP produced.

After glycolysis, pyruvate enters the mitochondrion. In aerobic cell respiration pyruvate is decarboxylated and oxidized, and converted into an acetyl compound. It is then attached to coenzyme A to form acetyl coenzyme A in the link reaction. In the Krebs cycle (sometimes called the citric acid cycle), the oxidation of acetyl groups is coupled to the reduction of hydrogen carriers, liberating carbon dioxide. Energy released by oxidation reactions is carried to the cristae of the mitochondria by reduced NAD and FAD.

The reduced NAD and FAD (NADH and $FADH_2$, respectively) produced in glycolysis, the link reaction and the Krebs cycle are carried to the cristae of the mitochondria. Transfer of electrons between carriers in the electron transport chain in the membrane of the cristae is linked to proton pumping. The molecules of the electron transport chain include protein complexes such as cytochromes. Each donor will pass the electron to a more electronegative acceptor through redox reactions. This liberates energy that will be used to build up a proton (H^+) gradient across the inner mitochondrial membrane. Oxygen is the final acceptor of electrons in the transport chain. Oxygen is needed to bind with the free protons to maintain the hydrogen ion gradient, resulting in the formation of water.

Energy can be released in the mitochondrion from the breakdown of glucose, amino acids or lipids and stored as ATP.

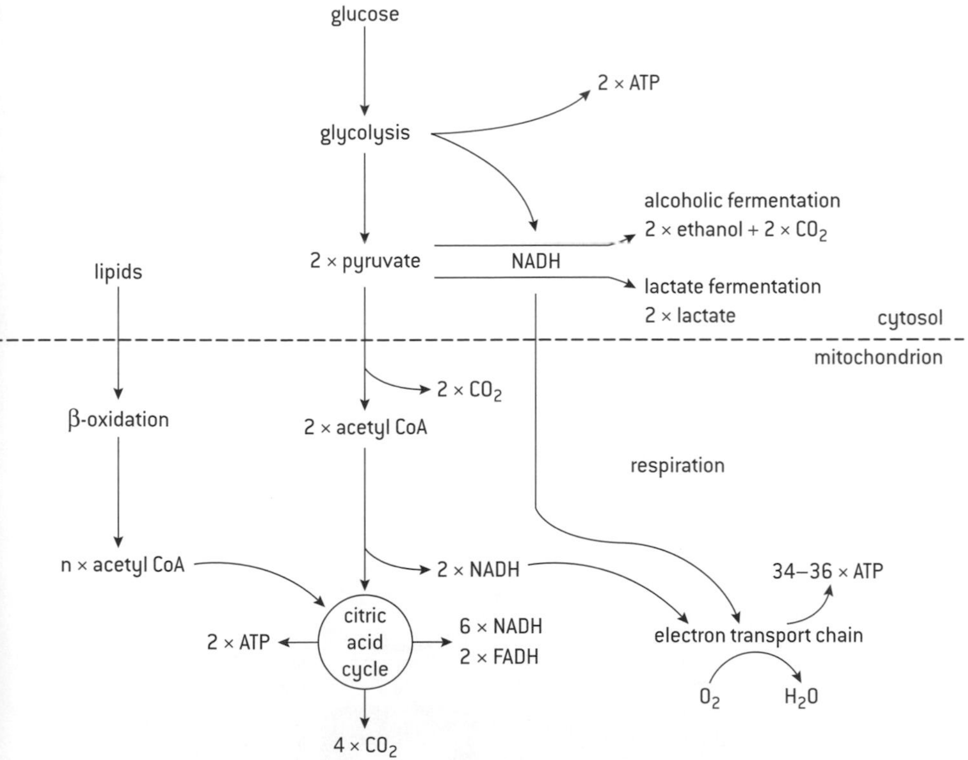

▲ **Figure 8.2.1.** Summary of the processes producing energy in a cell

Chemiosmosis is the movement of ions across a partially permeable membrane, down their electrochemical gradient. In the mitochondria, chemiosmosis occurs when protons diffuse through ATP synthase to generate ATP. The energy to maintain the hydrogen ion gradient comes from the redox reactions of the electron transport chain. Protons are pumped out of the membrane into the intermembrane space where they build up. When they come back through the ATP synthase enzyme they enable the phosphorylation of ATP.

The chemiosmotic theory led to a paradigm shift in the field of bioenergetics. Peter Mitchell's chemiosmotic theory encountered years of opposition before it was finally accepted. Previously it was believed that the proteins in the electron transport chain flowed freely, but Mitchell showed they were found in the mitochondrial cristae and that the difference in pH required to drive phosphorylation of ATP is obtained thanks to the small gap between mitochondrial membranes.

- **Oxidative phosphorylation** is the phosphorylation of ADP to ATP using the oxidation of molecules such as reduced NAD (NADH) and reduced FAD ($FADH_2$).
- The **electron transport chain** is the process by which electrons are passed from one carrier to another liberating energy that is used to transfer protons across the inner membrane from the matrix to the intermembrane space.
- **Chemiosmosis** is the process by which protons move across the inner mitochondrial membrane, down the concentration gradient, releasing the energy needed for the enzyme ATP synthase to make ATP.
- The **Krebs cycle** is a series of chemical reactions to reduce NAD and FAD, by oxidation of two-carbon acetyl groups. It occurs in the matrix of the mitochondrion.

Example 8.2.3.

What occurs during chemiosmosis in cell respiration?

A. NADH and $FADH_2$ provide energy through their phosphorylation.

B. Protons move to the intermembrane space along a concentration gradient.

C. ATP synthase transports protons to the matrix of the mitochondrion.

D. A six carbon molecule is broken down into a four carbon molecule.

Solution

The correct answer is **C**. In chemiosmosis, protons are pumped out of the membrane into the intermembrane space where they build up. When they come back to the matrix through the ATP synthase enzyme they enable the phosphorylation of ATP. Answer **A**. is incorrect because NADH and $FADH_2$ (which are reduced) become oxidized to supply energy, not phosphorylated. Answer **B** is incorrect because the movement to the intermembrane space requires energy provided by the redox reactions in the electron transport chain. Answer **D** relates to the Krebs cycle.

The structure of a mitochondrion is adapted to the functions it performs. The mitochondrial matrix contains enzymes for the link reaction and Krebs cycle. It is also the site of oxidation of fats (*β* oxidation) and has many ribosomes and DNA for protein synthesis. It has a double membrane with a small intermembrane gap between the inner and outer membranes where a gradient of protons can develop. The cristae are folds in the inner membrane that give it a large surface area for ATP synthesis by chemiosmosis, as this is where the proton pumping and electron transport chains occur. ATP synthase (also called ATP synthetase) has stalked particles that generate ATP from ADP and inorganic phosphate.

Stereogram of the mitochondrion

outer membrane
inner membrane
mitochondrial envelope
cristae
ATP synthase complex

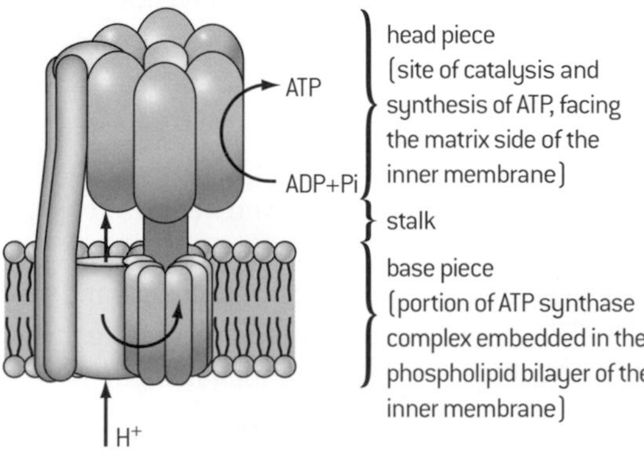

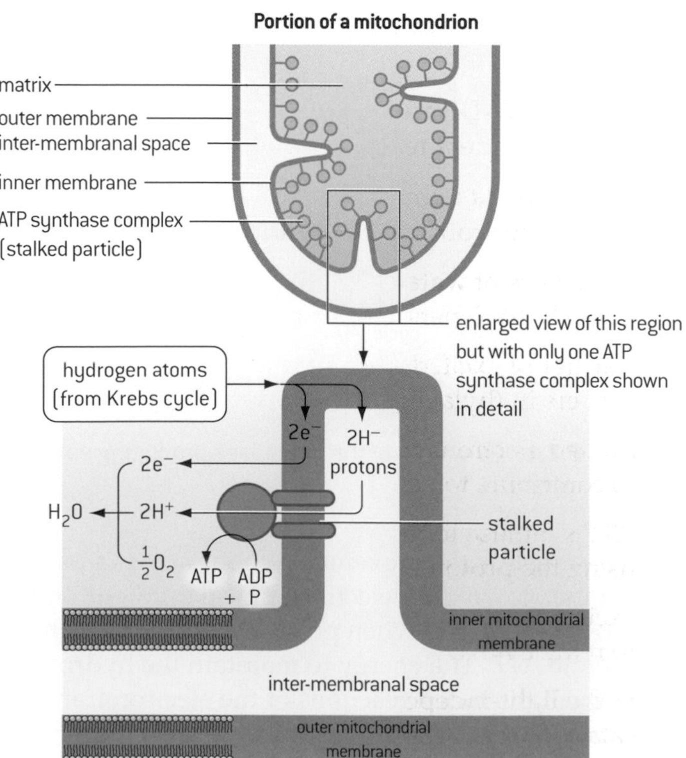

▲ **Figure 8.2.2.** The mitochondrion is adapted to its function

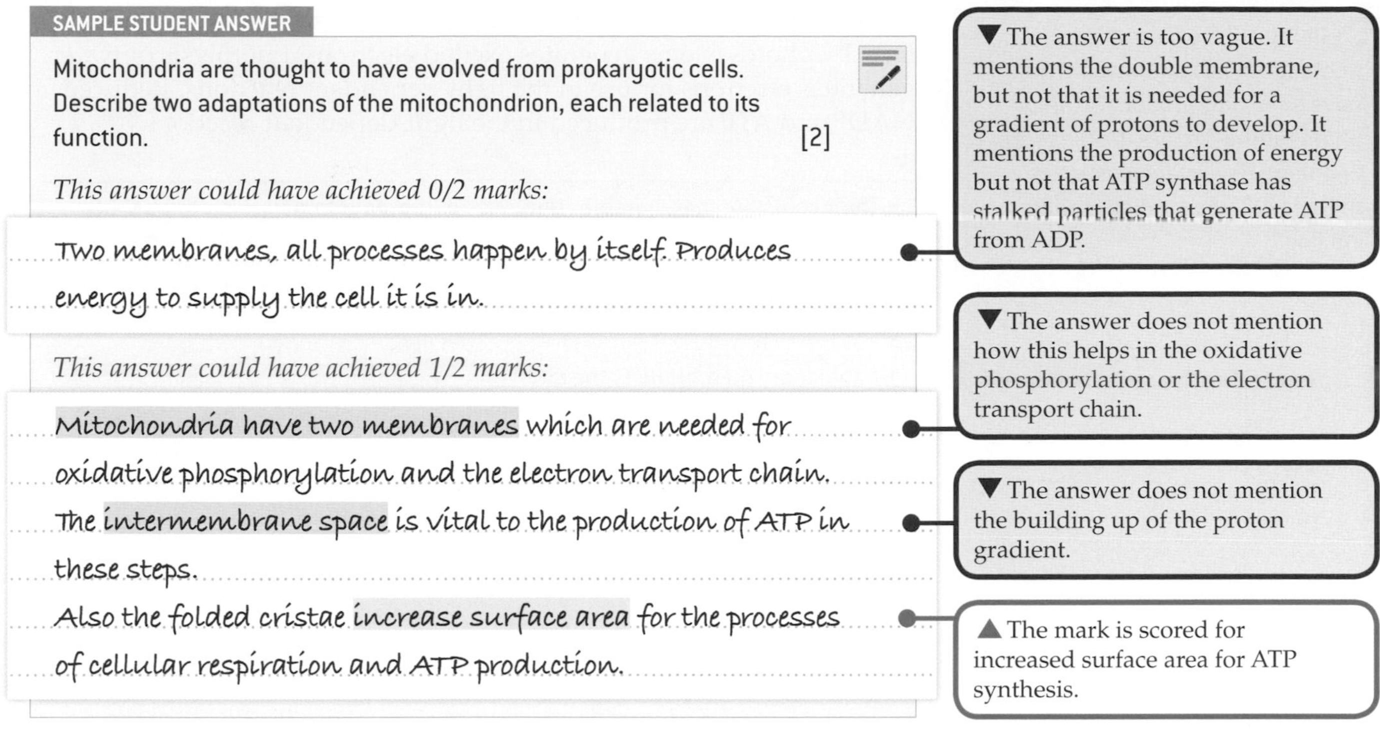

SAMPLE STUDENT ANSWER
Mitochondria are thought to have evolved from prokaryotic cells. Describe two adaptations of the mitochondrion, each related to its function. [2]
This answer could have achieved 0/2 marks:
Two membranes, all processes happen by itself. Produces energy to supply the cell it is in.
This answer could have achieved 1/2 marks:
Mitochondria have two membranes which are needed for oxidative phosphorylation and the electron transport chain.
The intermembrane space is vital to the production of ATP in these steps.
Also the folded cristae increase surface area for the processes of cellular respiration and ATP production.
▼ The answer is too vague. It mentions the double membrane, but not that it is needed for a gradient of protons to develop. It mentions the production of energy but not that ATP synthase has stalked particles that generate ATP from ADP.
▼ The answer does not mention how this helps in the oxidative phosphorylation or the electron transport chain.
▼ The answer does not mention the building up of the proton gradient.
▲ The mark is scored for increased surface area for ATP synthesis.

8.3 PHOTOSYNTHESIS

You should know:

- ✔ light energy is converted into chemical energy.
- ✔ light-dependent reactions take place in the intermembrane space of the thylakoids.
- ✔ light-independent reactions take place in the stroma.
- ✔ reduced NADP and ATP are produced in the light-dependent reactions.
- ✔ absorption of light by photosystems generates excited electrons.
- ✔ photolysis of water generates electrons for use in the light-dependent reactions.
- ✔ transfer of excited electrons occurs between carriers in thylakoid membranes.
- ✔ excited electrons from photosystem II are used to contribute to generating a proton gradient.
- ✔ ATP synthase in thylakoids generates ATP using the proton gradient.
- ✔ excited electrons from photosystem I are used to reduce NADP.
- ✔ in the light-independent reactions a carboxylase catalyses the carboxylation of ribulose bisphosphate.
- ✔ glycerate 3-phosphate is reduced to triose phosphate using reduced NADP and ATP.
- ✔ triose phosphate is used to regenerate RuBP and produce carbohydrates.
- ✔ ribulose bisphosphate is reformed using ATP.

You should be able to:

- ✔ annotate diagrams to show how a chloroplast is adapted to its function.
- ✔ describe the steps involved in the light-dependent reactions of photosynthesis.
- ✔ explain how photosystems in the chloroplasts absorb lights, generating excited electrons.
- ✔ describe the process of photolysis.
- ✔ explain the transfer of excited electrons between carriers in thylakoid membranes.
- ✔ explain how NADP and ATP are produced in the light-dependent reactions.
- ✔ explain Calvin's experiment to elucidate the carboxylation of RuBP.
- ✔ analyse diagrams of the pathways of photosynthesis to deduce where decarboxylation and oxidation reactions occur.

In Topic 2.9 you studied photosynthesis. In Topic 4.2 you have seen how producers that photosynthesize are the first step of energy flow in an ecosystem. In Topic 4.3 you studied carbon cycling.

In the light-dependent reactions of photosynthesis, the absorption of light by photosystems generates excited electrons. Photolysis of water generates electrons for use in the light-dependent reactions. Reduced NADP and ATP are produced in the light-dependent reactions.

- **Thylakoids** are membrane-bound compartments inside chloroplasts, formed by a membrane surrounding a thylakoid space. They form stacks of disks called grana.
- A **thylakoid membrane** is the site of light absorption and the electron transport chain.
- A **thylakoid space** is the site of photolysis and proton build-up.
- **Photosystems** are proteins that absorb light and transfer electrons.
- **Photolysis** is the breaking down of water molecules by photons present in light.
- **Reduced NADP** (NADPH) is the molecule produced in the last step of the electron transport chain. It is used as reducing power for the biosynthetic reactions in the Calvin cycle.

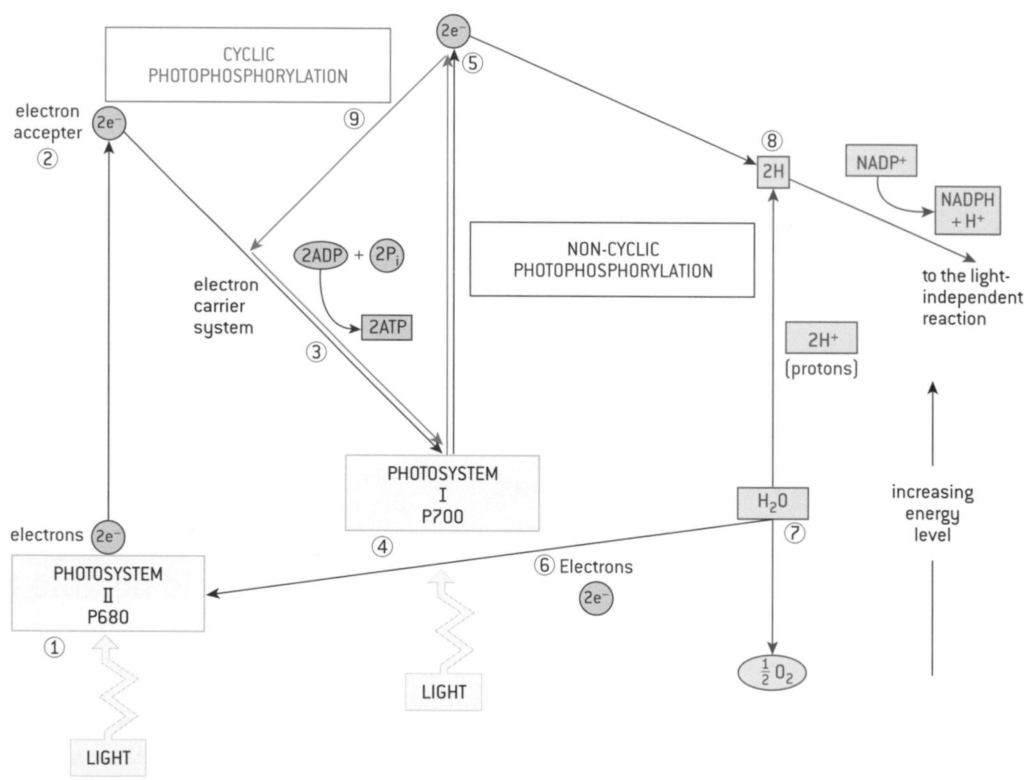

▲ **Figure 8.3.1.** Light-dependent reactions in photosynthesis

SAMPLE STUDENT ANSWER

The diagram represents a detail of the chloroplast showing light-dependent reactions linked to chemiosmosis.

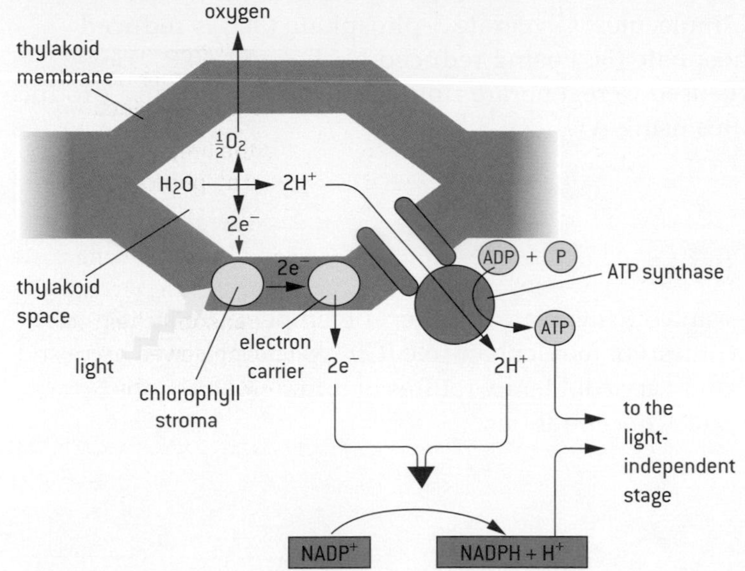

a) State the reaction occurring in the ATP synthase. [1]

This answer could have achieved 1/1 marks:

Phosphorylation of ADP to ATP.

b) Using the diagram, suggest how the chloroplast is adapted to perform the light-independent process of photosynthesis. [8]

This answer could have achieved 7/8 marks:

The chloroplasts are adapted to perform the process of photosynthesis. They have the inner membrane called the thylakoid membrane that stacks into grana to provide a large surface area. Absorption of light occurs in photosystems embedded in the thylakoid membrane and generates excited electrons which are carried by carriers such as cytochrome in the thylakoid membranes. The membrane allows these proteins to be one next to the other. Light-dependent reactions take place in the intermembrane space of the thylakoids. The electrons lost from photosystem II get replaced by the oxidation of water, which is lysed into protons and oxygen. The electron transfer in the thylakoid membrane causes these protons to be actively pumped across the thylakoid membrane into the thylakoid space. The protons then flow down their electrochemical potential gradient through ATP synthase, creating ATP by phosphorylation of ADP in the process of chemiosmosis. Oxygen diffuses out of the chloroplasts. Excited electrons from photosystem I are used to reduce NADP forming NADPH to be used in the light-independent reactions in the stroma of the chloroplast.

▲ This answer explains how the chloroplast is adapted to perform its function.

▼ Although the answer was good enough to score many marks, it has some faults. For example, water is not oxidized and it is not lysed into just protons and oxygen.

- The **Calvin cycle** is the light-independent stage of photosynthesis consisting of carbon fixation, reduction reactions, and ribulose 1,5-bisphosphate (RuBP) regeneration.
- **Carbon fixation** is the addition of carbon dioxide to organic molecules.
- **Ribulose bisphosphate carboxylase oxidase** (RubisCo) is the enzyme that catalyses the incorporation of carbon dioxide.
- The **stroma** is a fluid part of the chloroplast and is the site of the Calvin cycle.

Sources of ^{14}C and autoradiography enabled Calvin to elucidate the pathways of carbon fixation. The lollipop experiment used to work out the biochemical details of the Calvin cycle shows considerable creativity. At the start of the experiment, in the lollipop vessel, radioactive carbon dioxide with ^{14}C was supplied to algae. Samples of the algae were removed at time intervals and carbon compounds in the algae containing radioactive ^{14}C were recorded.

The light-independent reactions occur in the stroma of the chloroplast. A carboxylase catalyses the carboxylation of ribulose bisphosphate (5C sugar). In a carboxylation, carbon dioxide is added to the molecule. The 6C sugar formed is broken down into two 3C molecules. Glycerate 3-phosphate (3C) is reduced to triose phosphate (3C) using reduced NADP and ATP. Triose phosphate is used to regenerate ribulose bisphosphate and produce carbohydrates using ATP.

Example 8.3.1.

Scientists wanted to evaluate the effect of temperature on the rate of photosynthesis of maidenhair tree (*Ginkgo biloba*) leaves, exposed to normal and very cold temperatures at different atmospheric carbon dioxide concentrations.

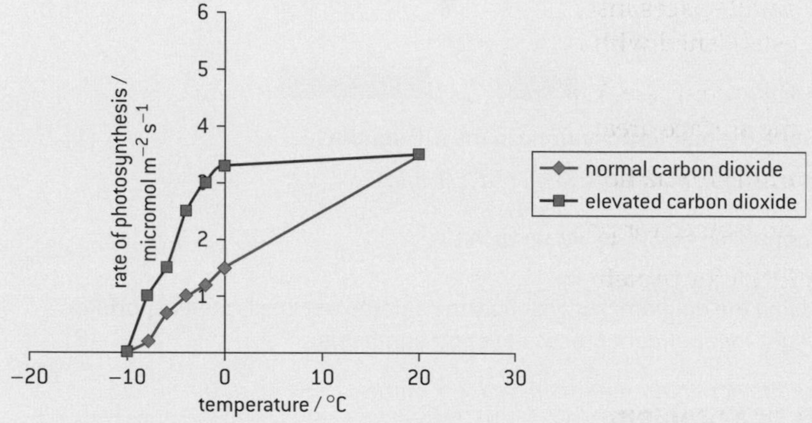

a) Calculate the difference in rate of photosynthesis between normal and elevated carbon dioxide levels at 0°C.

b) (i) Outline the effect of temperature on rate of photosynthesis at different CO_2 levels.

(ii) Suggest how the change in photosynthetic rate will affect plant growth.

c) Suggest which reactions of photosynthesis could be affected by cold temperatures.

Solution

a) 3.3–1.5 = 1.8 micromol m^{-2} s^{-1}

b) (i) At 20°C, with either normal or elevated CO_2 levels, the plants have the same rate of photosynthesis. As the temperature decreases the rate decreases, but it decreases much faster in normal carbon dioxide levels. At −10°C both rates are 0.

(ii) The lower the rate of photosynthesis the less the plant will grow, as photosynthesis produces glucose that is stored as starch. This is used to produce energy and for structural growth.

c) Temperature affects enzymes; therefore the enzyme-catalysed reactions will be affected. In the Calvin cycle, ribulose bisphosphate carboxylase oxidase, for example, could be affected. In chemiosmosis, ATP synthase is the affected enzyme.

Assessment tip

Remember to write the units.

Example 8.3.2.

Outline adaptations of the chloroplast to its function.

Solution

Adaptations of the chloroplast are:

- a double membrane enclosing and protecting the chloroplast
- internal thylakoid membranes exposing photosystems to capture light energy
- presence of large amounts of light-absorbing pigments in the photosystems
- internal thylakoid membranes to increase the surface area to hold electron carriers
- spaces between thylakoid membranes to allow a build-up of protons for chemiosmosis
- small spaces inside thylakoids, so the proton gradient is established with fewer protons
- thylakoids stacked into grana to save space while increasing the surface area:volume ratio
- fluid stroma holding enzymes for the Calvin cycle
- ribosomes for protein synthesis
- DNA for protein synthesis.

Assessment tip

Three answers from this list are expected in a three-mark question.

Practice problems for Topic 8

Problem 1

The enzyme α-1,4-glucosidase catalyses the breakdown of maltose to two glucose molecules. Its activation by maltose and inhibition by glucose was studied. The activity of α-glucosidase was measured in bacteria which were grown in a medium containing 2% maltose only, 2% glucose only or a mixture of 1.5% maltose and 0.5% glucose.

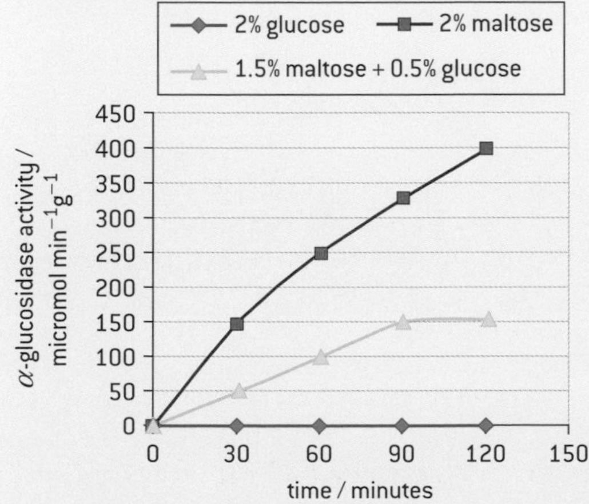

a) Calculate the percentage difference in α-glucosidase activity after 120 minutes in bacteria grown with only maltose compared with bacteria grown with a mixture of both maltose and glucose. Show your working.

b) Analyse the effect of incubating the bacteria with either maltose or glucose in terms of metabolic pathways.

c) Acarbose, an α-glucosidase inhibitor, competitively and reversibly inhibits α-glucosidase in the intestines. Discuss whether acarbose may be used to prevent the development of diabetic symptoms.

Problem 2

Explain how the rate of reaction of an enzyme can be controlled inside a cell.

Problem 3

a) Outline the light-dependent reactions of photosynthesis.

b) Explain the effect of light intensity and temperature on the rate of photosynthesis.

Problem 4

Fomepizole is a drug used to treat antifreeze (methanol) ingestion. Methanol is toxic because enzymes break it down to formic acid, which is very toxic. Fomepizole acts to inhibit the breakdown of methanol into active toxic metabolites. It is a competitive inhibitor of the enzyme alcohol dehydrogenase found in the liver. This enzyme plays a key role in the metabolism of methanol.

Patients who had accidentally ingested antifreeze were treated with fomepizole and their blood plasma methanol and formic acid levels were measured after several hours. The line graph shows the changes in levels of methanol and formic acid with time.

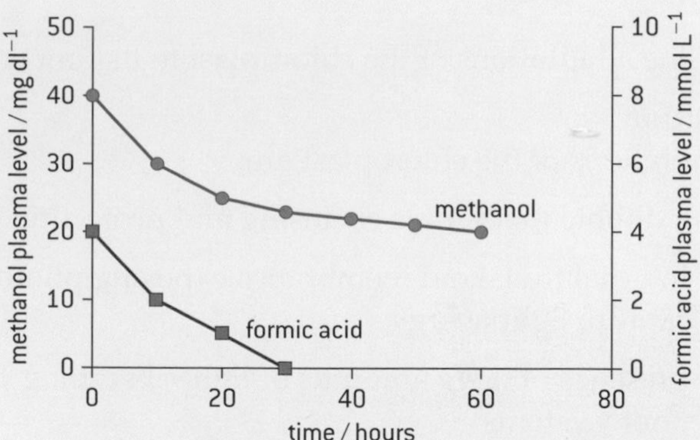

a) State the reason the breaking down of methanol is considered a metabolic pathway.

b) Compare and contrast the trends in methanol and formic acid metabolism with time.

c) Suggest whether the treatment with fomepizole is efficient for antifreeze intoxication.

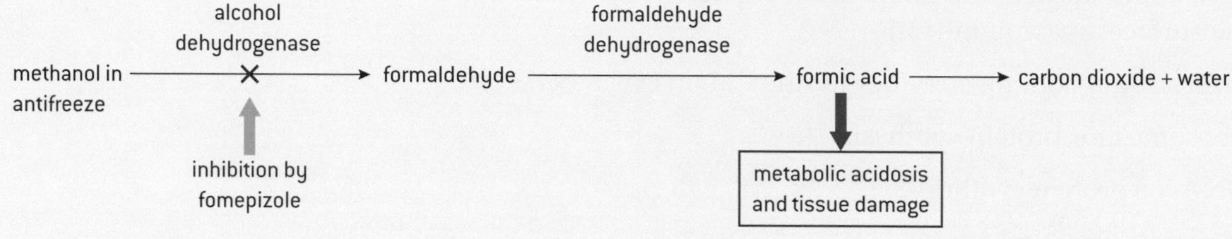

9 PLANT BIOLOGY (AHL)

9.1 TRANSPORT IN THE XYLEM OF PLANTS

You should know:

- ✔ transpiration is the inevitable consequence of gas exchange in the leaf.
- ✔ plants transport water from the roots to the leaves to replace losses from transpiration.
- ✔ the cohesive property of water and the structure of the xylem vessels allow transport under tension.
- ✔ the adhesive property of water and evaporation generate tension forces in leaf cell walls.
- ✔ active uptake of mineral ions in the roots causes absorption of water by osmosis.

You should be able to:

- ✔ explain how structure and function are correlated in the xylem of plants.
- ✔ explain how plants in deserts and in saline soils are adapted for water conservation.
- ✔ design models of water transport in xylem using simple apparatus, including blotting or filter paper, porous pots and capillary tubing.
- ✔ draw the structure of primary xylem vessels in sections of stems based on microscope images.
- ✔ measure transpiration rates using potometers.
- ✔ design experiments to test hypotheses about the effect of temperature or humidity on transpiration rates.
- ✔ calculate stomatal apertures and distribution of stomata from light micrographs or leaf casts.

Water molecules are polar and the partial negative charge on the oxygen atom in one water molecule attracts the hydrogen atom in another water molecule, causing cohesion. Water is also attracted to hydrophilic parts of the cell walls of xylem. This is adhesion. As a result of the cohesion between the molecules, water can be pulled up through the xylem in what is called the transpiration pull.

Water is transported in the plant through xylem vessels. It enters through the root hair cells in the epidermis of the roots and reaches the xylem in the centre of the root. It then travels up the stem through the vascular bundles till it reaches the leaves, where it exits by transpiration as water vapour through the stomata.

In Topic 2.2 you studied the properties of water and in Topics 2.9 and 8.3 you studied the process of photosynthesis.

- **Cohesion** is the attraction between water molecules due to their hydrogen bonds.
- **Adhesion** is the attraction of water to other substances, for example the xylem walls.
- **Xylem** is the tissue that carries water from root to leaf.
- A **meniscus** is the concave surface of a liquid resulting from surface tension.

Example 9.1.1.

A student set up the following apparatus and measured the movement of the air–water meniscus along the scale in the capillary tube.

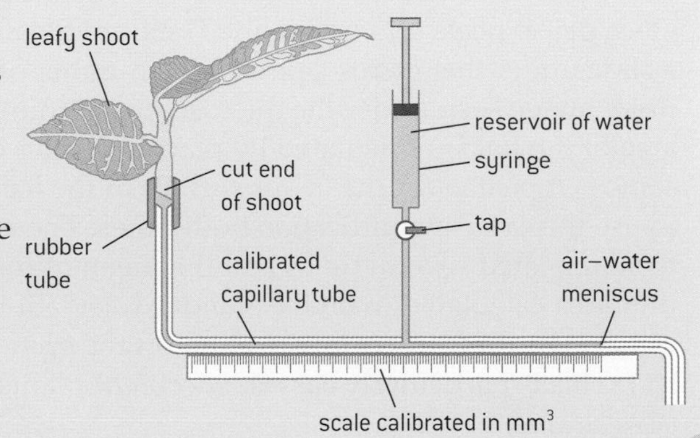

a) State the name of this apparatus.

b) Describe the function of this apparatus.

c) Explain how changing the environmental temperature would affect the results of this experiment.

d) Suggest an experimental way to modify the environmental humidity in this experiment.

Solution

a) Potometer.

b) This apparatus is used to measure the rate of water uptake in a twig. It is a model to measure the rate of transpiration in plants. The capillary tube represents the xylem vessels. The water reservoir is used to top up the capillary tube to start a new measurement.

c) If the temperature is increased, the stomata of the leaves open and water vapour is eliminated by transpiration at a faster rate. This means that the air–water meniscus in the capillary tube will move at a faster pace. The distance moved can be measured in a unit of time (for example minutes). The warmer the temperature, the faster the movement, therefore increasing the rate of water uptake (measured in mm^3 per minute). Warmer temperatures increase the rate of photosynthesis. This process requires water, so more water will be taken up by the leaves.

d) A way of increasing the atmospheric humidity is by placing the apparatus under a bell jar or in a plastic bag. This will maintain the water vapour eliminated by transpiration in the environment. A method of decreasing the environmental humidity is using a fan. The fan blows away the transpired water vapour, drying the environment. By placing a fan at different distances from the potometer, one can model different environmental humidities.

Use models as representations of the real world—mechanisms involved in water transport in the xylem can be investigated using apparatus and materials that show similarities in structure to plant tissues. Experiments using the potometer have allowed scientists to work out how different conditions affect transpiration rates. The introduction of image processing software and digital microscopes has further increased the capacity to gather even more data to ensure reliability.

Transpiration is the inevitable consequence of gas exchange in the leaf. During photosynthesis in the leaves, carbon dioxide is required and oxygen is produced. These gases are exchanged in structures found in the leaf epidermis called stomata. As the stomata open for gaseous exchange, water is also lost by transpiration.

Example 9.1.2.

Design an experiment to compare the number of stomata per unit area in the lower and upper surfaces of a leaf.

Solution

A leaf cast can be prepared with nail varnish. The surface of the leaf is painted with two or three layers of transparent nail varnish. Once dry, it peels off very easily. The upper leaf cast is placed on a slide under the microscope and the number of stomata per field (total image seen under the microscope) is counted. The surface area of the field is calculated by placing a ruler on the stage. The same is repeated for the lower surface of the leaf. It is important to use the same magnification both times. The experiment needs to be repeated several times and the mean number of stomata per unit area calculated, with the standard deviation. The experiment can also be done by simply peeling off the epidermis from the leaf. This is particularly easy in succulent plants such as *Kalanchoe blossfeldiana*.

The light micrograph shows part of a leaf surface of 1 mm by 1 mm. The number of stomata in this case would be 4 per mm². Both upper and lower epidermis must be investigated to compare the stomatal distribution.

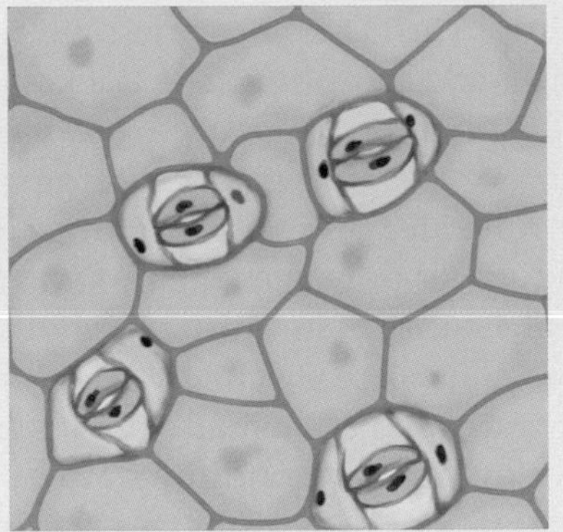

>> **Assessment tip**

There are different possible answers to this question. This is a reasonable answer.

>> **Assessment tip**

In experimental design questions it is useful to sketch how you would perform the experiment. In this case it is complicated, but you could sketch what the results would look like.

Evaporation of water in the mesophyll of the leaf creates a tension or negative pressure potential, which creates a pulling force called the transpiration pull. Water is drawn through cell walls out of the xylem in the vascular bundle of the leaf by capillary action due to adhesion to cellulose. Low pressure causes a suction force in the xylem. Cohesion allows water to be pulled up under tension in the transpiration pull. Xylem vessels resist tension and do not collapse because they have thickened and lignified walls. Plants transport water from the roots to the leaves in the xylem to replace water lost from the leaves by transpiration. Active transport of ions into the roots enables osmosis of water into the roots from the soil.

Example 9.1.3.

The diagram below shows a vascular bundle in the stem of a plant.

a) Label the xylem.

b) Structure and function are correlated in the xylem of plants. Explain adaptations of xylem vessels to their function.

Solution

a)

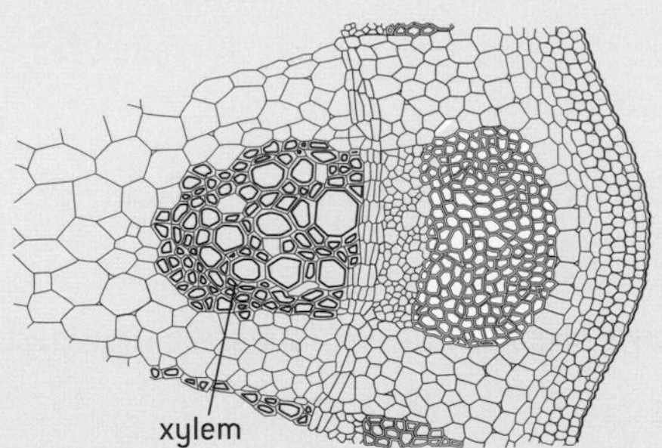

b) Xylem vessels are continuous tubes of piled cells arranged one after the other. These cells have thickened walls that become lignified. This strengthens the walls, so that they can withstand low pressures without collapsing. The cells in xylem vessels die after some time, leaving only the walls of the non-living cells. This provides a larger space for water transport, and also determines that the water movement in these vessels must be passive, as there are no mitochondria to supply energy. The divisions between cells also disappear, making a continuous tube. Small holes or pits allow substances to pass in and out of the vessels.

>> **Assessment tip**

You can sketch your answer in an annotated diagram, but remember it must be very clear.

Plants that live in deserts adapted for water conservation are called xerophytes. Deep or wide-ranging root systems increase the chances of water absorption. The thick, waxy cuticle of the leaf surface reduces transpiration. Small or no leaves or thorns instead of leaves reduce the surface area of leaves. Few stomata or the presence of stomata in pits or rolled leaves will also reduce evaporation. Hairs on the leaf surface reduce air flow near the leaf and reflect sunlight. The stomata open at night rather than during the hot daytime conditions, to reduce water loss. Stems are succulent as they are capable of storing large amounts of water and are usually photosynthetic. Plants that are adapted to live in high salt conditions are called halophytes. They have similar adaptations to xerophytes and some have structures for removing salt.

SAMPLE STUDENT ANSWER

a) Outline the properties of water molecules that permit water to move upwards in plants. [2]

This answer could have achieved 2/2 marks:

They are polar: this allows for hydrogen bonding between water molecules creating cohesion (when one molecule of water pulls upwards, they are all pulled). Polarity also allows adhesion to xylem walls which enables the transpiration pull to work by dragging water up in a continuous mass, hence the transport can be passive.

▼ Although the answer did score full marks, it includes a misconception—adhesion of water to the xylem vessel wall can only draw water up in an air-filled vessel. If the vessel is already filled with water or sap, adhesion cannot draw water up higher. It can draw water through the cell walls of leaf cells, though, if water has been lost from them by evaporation.

This answer could have achieved 1/2 marks:

Water evaporates from the surface of the leaf, creating negative pressure. Water flows under tension from the roots through xylem to the leaves. Water forms cohesive forces with other water molecules and adhesive forces between water and xylem vessels which enable them to move upwards.

▲ This answer scored one mark for the adhesion of water allowing water to move upwards.

▼ This answer is incomplete. It does not mention that water evaporates by transpiration. It does not mention that cohesion is due to the polarity of the water allowing for hydrogen bonding, thus forming a continuous water column.

b) In hot, dry conditions plants lose water rapidly due to transpiration. Explain how the structures and processes of the plant allow this water to be replaced. [8]

This answer could have achieved 0/8 marks:

- Mass flow of water allows plants to replace water being lost through transpiration.
- The xylem takes up water from the roots and takes it upwards to the rest of the plant.
- Cohesion allows the water to stick to each other while being taken up.
- Adhesion allows water to stick to the walls of the xylem.

▼ This answer does not score marks, as it does not answer the question. There is no explanation of how the structure is adapted to function.

This answer could have achieved 4/8 marks:

Plants in hot, dry places develop many structures and functions to preserve water. One structure is the presence of rolled up leaves to prevent loss of water by way of the stomata. Also, these plant leaves typically have less stomata

▲ The marks are for less stomata in rolled leaves and long roots to absorb water. A third mark was given for stomata open at night to reduce water loss. The fourth mark was for saying that the waxy cuticle reduces water loss by transpiration.

for this reason as well. These plants also have long roots in order to reach water and minerals that are below the ground. These photosynthetic processes take advantage of night and day by opening their stomata for transpiration during the night because it is cooler and there is no sun to evaporate the water quickly. These plants also have a waxy cuticle that prevents water loss by way of transpiration. These plants also compartmentalize or store water for long periods of time.

▼ The answer did not mention that evaporation of water lowers the pressure in leaves that at the same time creates a pressure potential or pulling force called the transpiration pull. The water is drawn through cell walls and up the stem by a pulling force in xylem due to hydrogen bonds making water molecules cohesive and also adhesive to xylem walls. The xylem does not collapse because of thickened walls allowing water to travel from the roots (where it is absorbed by osmosis) to the leaves to replace losses by transpiration.

9.2 TRANSPORT IN THE PHLOEM OF PLANTS

You should know:

- ✔ plants transport organic compounds from sources to sinks.
- ✔ incompressibility of water allows transport along hydrostatic pressure gradients.
- ✔ active transport is used to load organic compounds into phloem sieve tubes at the source.
- ✔ high concentrations of solutes in the phloem at the source lead to water uptake by osmosis.
- ✔ raised hydrostatic pressure causes the contents of the phloem to flow towards sinks.

You should be able to:

- ✔ explain how the structure of phloem sieve tubes is related to their function.
- ✔ explain how plants load and transport organic compounds from sources to sinks.
- ✔ explain water uptake in plant tissues.
- ✔ identify xylem and phloem in microscope images of stem and root.
- ✔ analyse data from experiments measuring phloem transport rates using aphid stylets and radioactively labelled carbon dioxide.

Plants transport organic compounds from sources (photosynthetic tissues and storage organs) to sinks (roots and parts of the plant requiring energy). A concentration gradient of sucrose is required for transport in the phloem. Active transport is used to load organic compounds into phloem sieve tubes at the source. Hydrogen ions are actively transported out of the companion cells using ATP. The build-up of hydrogen ions makes them flow down a concentration gradient back into the companion cells through a protein that co-transports sucrose into the cells. The incompressibility of water allows transport along hydrostatic pressure gradients. High concentrations of solutes in the phloem at the source lead to water uptake by osmosis. Raised hydrostatic pressure causes the contents of the phloem to flow towards sinks.

You studied membrane transport in Topic 1.4 and water transport in xylem in Topic 9.1.

- **Organic compounds** are formed by carbon joined to atoms of other elements, most commonly hydrogen, oxygen, or nitrogen.
- **Translocation** is the transport of organic compounds in plant phloem. It requires energy.
- A **source** is a part of the plant where carbohydrates and amino acids are loaded into the phloem.
- A **sink** is where the carbohydrates and amino acids are unloaded to be used.

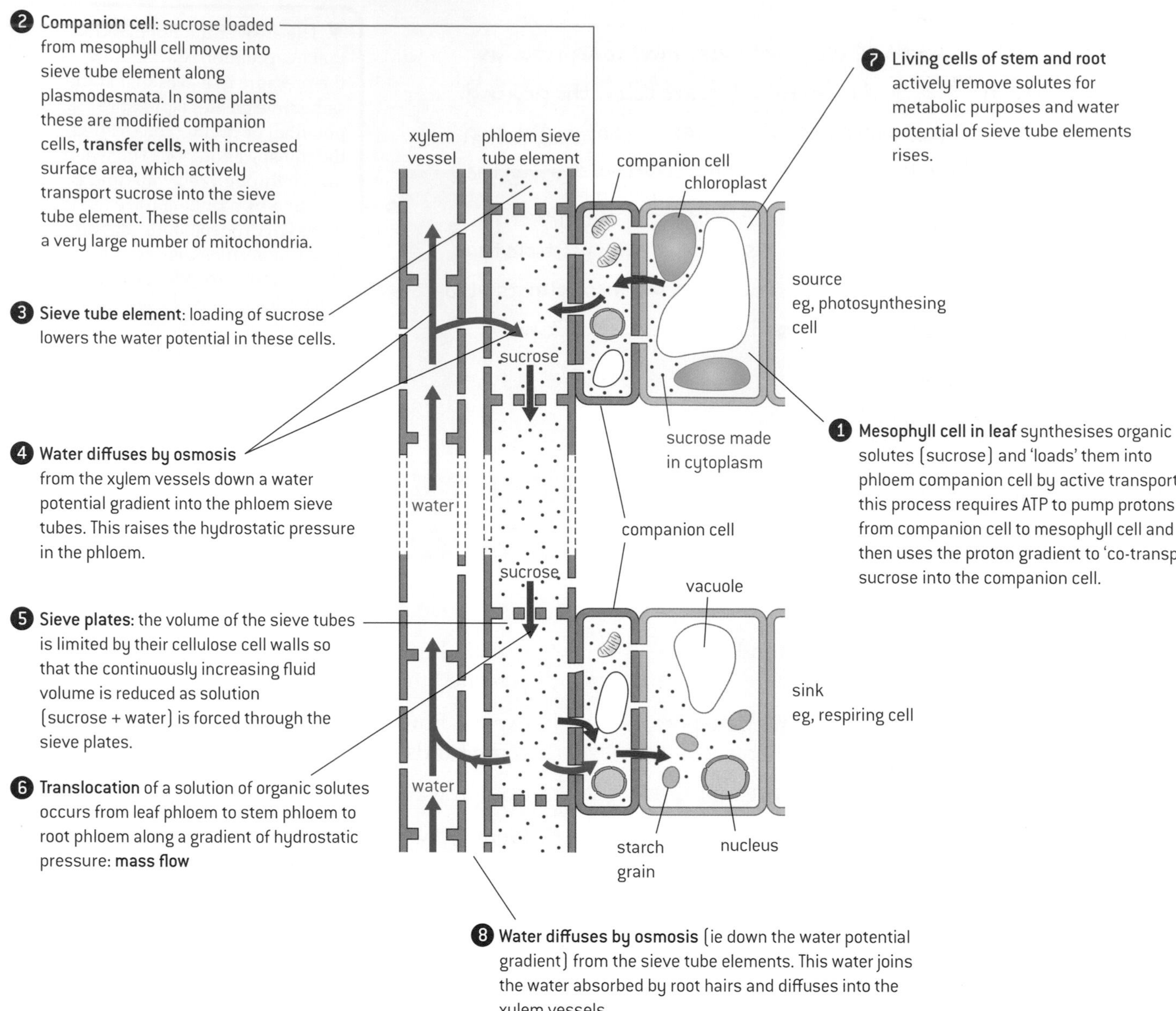

2 Companion cell: sucrose loaded from mesophyll cell moves into sieve tube element along plasmodesmata. In some plants these are modified companion cells, transfer cells, with increased surface area, which actively transport sucrose into the sieve tube element. These cells contain a very large number of mitochondria.
3 Sieve tube element: loading of sucrose lowers the water potential in these cells.
4 Water diffuses by osmosis from the xylem vessels down a water potential gradient into the phloem sieve tubes. This raises the hydrostatic pressure in the phloem.
5 Sieve plates: the volume of the sieve tubes is limited by their cellulose cell walls so that the continuously increasing fluid volume is reduced as solution (sucrose + water) is forced through the sieve plates.
6 Translocation of a solution of organic solutes occurs from leaf phloem to stem phloem to root phloem along a gradient of hydrostatic pressure: mass flow
xylem vessel
phloem sieve tube element
companion cell
chloroplast
7 Living cells of stem and root actively remove solutes for metabolic purposes and water potential of sieve tube elements rises.
source eg, photosynthesing cell
sucrose
sucrose made in cytoplasm
1 Mesophyll cell in leaf synthesises organic solutes (sucrose) and 'loads' them into phloem companion cell by active transport – this process requires ATP to pump protons from companion cell to mesophyll cell and then uses the proton gradient to 'co-transport' sucrose into the companion cell.
water
companion cell
sucrose
vacuole
sink eg, respiring cell
water
starch grain
nucleus
8 Water diffuses by osmosis (ie down the water potential gradient) from the sieve tube elements. This water joins the water absorbed by root hairs and diffuses into the xylem vessels.

▲ Figure 9.2.1. Transport in plant cells

Example 9.2.1.

In a dicotyledonous plant, the products of photosynthesis are transported from the leaves to the rest of the plant through which tissues?

I Xylem vessels II Companion cells III Sieve tube cells

A. I only **C.** II and III only

B. II only **D.** I, II and III

Solution

The correct answer is **C**, as although most of the carbohydrates are transported in the sieve tube cells, some are transported in the companion cells.

The phloem tissue is composed of sieve tubes. Sieve tubes are composed of sieve tube elements which are cells that are no longer living and have no nucleus. Sieve tube elements are closely associated with companion cells. These cells are connected to the sieve tube elements by plasmodesmata.

Sieve tube elements maintain the sucrose and organic molecule concentration that has been established by active transport. Companion cells are rich in mitochondria and have a large surface area of cell membrane to support the active transport of sucrose and aid transport in the phloem. The cell walls of the sieve tube elements are rigid to allow for the establishment of the pressure necessary to achieve the flow of sap in the phloem. Individual sieve tube elements are separated by perforated walls called sieve plates, allowing quick flow from cell to cell.

longitudinal view
sieve plate with pores
cellulose cell wall of companion cell
nucleus of companion cell
cytoplasm of companion cell
cellulose cell wall of sieve tube element
cytoplasm of sieve tube element

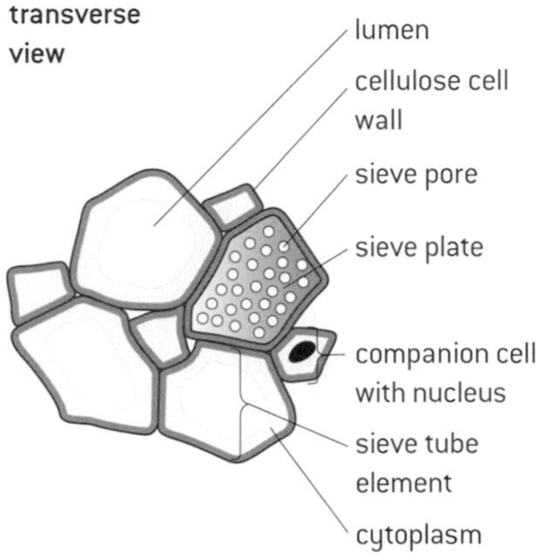

▲ **Figure 9.2.2.** Structure of the sieve tubes

SAMPLE STUDENT ANSWER

The image shows a severed aphid stylet embedded in plant tissue.

a) Identify the tissue labelled II. [1]

This answer could have achieved 1/1 marks:

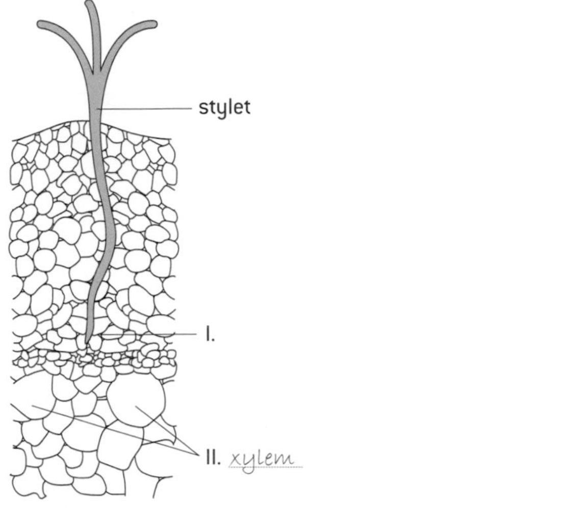

b) Outline **one** piece of evidence that the tissue labelled I is phloem tissue. [1]

This answer could have achieved 1/1 marks:

The stylet is located in the tissue labelled I and not in the tissue labelled II. Aphid stylets only go into the phloem tissue to obtain the content of phloem sap; therefore tissue I is phloem tissue.

▲ The answer recognizes the tissue as the place aphids go for food. Another answer could have been that it is closer to the surface than the xylem. Also that its cells are smaller than the xylem tissue and it has smaller companion cells adjacent to the larger sieve tube cells.

c) Explain how aphid stylets can be used to study the movement of solutes in plant tissues. [3]

This answer could have achieved 3/3 marks:

Aphids take up solutes from the phloem of plant tissues using the stylet. As an aphid is taking up phloem sap, its stylet can be severed. Once this occurs, the phloem sap (which contains solutes) continues to flow out, and the closer the stylet was to the sink, the greater the flow. The flow of phloem sap can be analysed for content and rate of transportation. It can be analysed for transportation of solutes in phloem, which is a tissue that transports solutes in plants.

▲ The answer gained full marks, but could also have mentioned that if plants are grown in radioactive CO_2 ($^{14}CO_2$) which is incorporated into carbohydrates, the radioactively labelled carbon can be detected later in the phloem sap. Stylets inserted at different parts of the plant can show different rates of movement.

9.3 GROWTH IN PLANTS

You should know:

- ✔ plants adapt their growth to environmental conditions.
- ✔ undifferentiated cells in the meristems of plants allow indeterminate growth.
- ✔ mitosis and cell division in the shoot apex provide cells needed for extension of the stem and development of leaves.
- ✔ plant hormones control growth in the shoot apex.
- ✔ plant shoots respond to the environment by tropisms.
- ✔ auxin efflux pumps can set up concentration gradients of auxin in plant tissue.
- ✔ auxin influences cell growth rates by changing the pattern of gene expression.

You should be able to:

- ✔ explain how undifferentiated cells are used for growth and micropropagation.
- ✔ identify meristems in plant micrographs.
- ✔ design micropropagation experiments of plants using tissue from the shoot apex, nutrient agar gels and growth hormones.
- ✔ explain uses of micropropagation such as rapid bulking up of new varieties, production of virus-free strains of existing varieties and propagation of orchids and other rare species.
- ✔ explain the action of auxin in plants.
- ✔ design experiments to show tropisms.

In Topic 3.5 you studied genetic modification and biotechnology in plants.

- **Totipotent cells** are able to develop into any differentiated cell of the organism.
- **Meristems** are totipotent undifferentiated dividing cells located at the tips of a plant's shoots and at the ends of its roots.
- **Auxins** are plant hormones with important roles in the coordination of many growth processes.
- **Micropropagation** is the multiplication of a selected plant using small pieces of plant tissue. It is done in sterile conditions *in vitro* using agar gel containing nutrients and hormones.

Plants adapt their growth to environmental conditions. Plant cells have the ability to generate a whole plant; this means their cells are totipotent. Plants grow mainly at meristems where undifferentiated cells allow indeterminate growth; this means that there is no fixed final size or fixed number of parts such as branches. Mitosis and cell division in the shoot apex provide cells needed for extension of the stem and development of leaves. Meristems can be used for micropropagation, which is used for rapid bulking up of new varieties, production of virus-free strains of existing varieties and propagation of orchids and other rare species.

SAMPLE STUDENT ANSWER

Achieving successful rooting of cuttings is difficult in some shrub species. An experiment was undertaken to determine whether juvenile shoots of trees root more successfully than mature shoots.

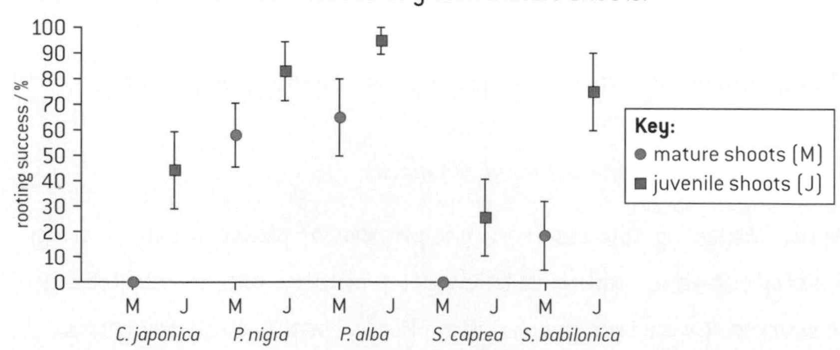

a) Distinguish between the rooting success of the juvenile shoots and that of the mature shoots. [1]

This answer could have achieved 1/1 marks:

Overall the juvenile shoots had greater rooting success than the mature shoots.

Assessment tip

Remember to use a comparative word in a "distinguish" question. An answer saying that juvenile shoots have a high rooting success would not have scored a mark.

b) Suggest **one** reason for the differences in rooting success in the juvenile shoots and the mature shoots. [1]

This answer could have achieved 1/1 marks:

Juvenile shoots contain more undifferentiated cells in the meristems, allowing for more successful rooting.

▲ The answer shows understanding of undifferentiated tissues. A more correct answer would have been that juvenile shoots have a faster response to auxin.

c) Outline **one** variable that would need to be controlled in this experiment. [1]

This answer could have achieved 0/1 marks:

Equal amounts of hormones should be maintained because hormones can impact the development and growth of shoots.

▼ Leaf area, size, mass and length of cutting should be kept similar in all groups. Light, temperature, nutrients, rooting mixture or moisture should be the same for all plants.

d) Auxin is a hormone that can be applied to improve the percentage success of rooting in those study plants with poor rooting success. Explain the effects of auxin on plant cells. [3]

This answer could have achieved 1/3 marks:

- auxin is a hormone produced by the meristem.
- auxin promotes fruit and seed development in plants.
- plant cells have auxin receptors on their membrane to which auxin can bind and either inhibit or promote cell growth.

▼ Auxin has an effect on rate of mitosis. It changes the pattern of gene expression, promoting transcription of some genes. Auxin also changes the pH of the extracellular environment and cell wall, increasing activity of proton pumps. It breaks cross-links between cellulose fibres in cell walls, increasing cell wall plasticity. Varying auxin concentrations have different effects in different parts of the plant.

▲ The answer scored the mark for the last point, that auxin increases cell growth.

Plant hormones control growth in the shoot apex. Auxins, especially indole 3-acetic acid (IAA), are involved in plant growth behaviours. Plant shoots respond to the environment by tropisms. Shoots have positive phototropism and negative gravitropism. Auxin efflux pumps can set up concentration gradients of auxin in plant tissue. Auxin influences cell growth rates by changing the pattern of gene expression.

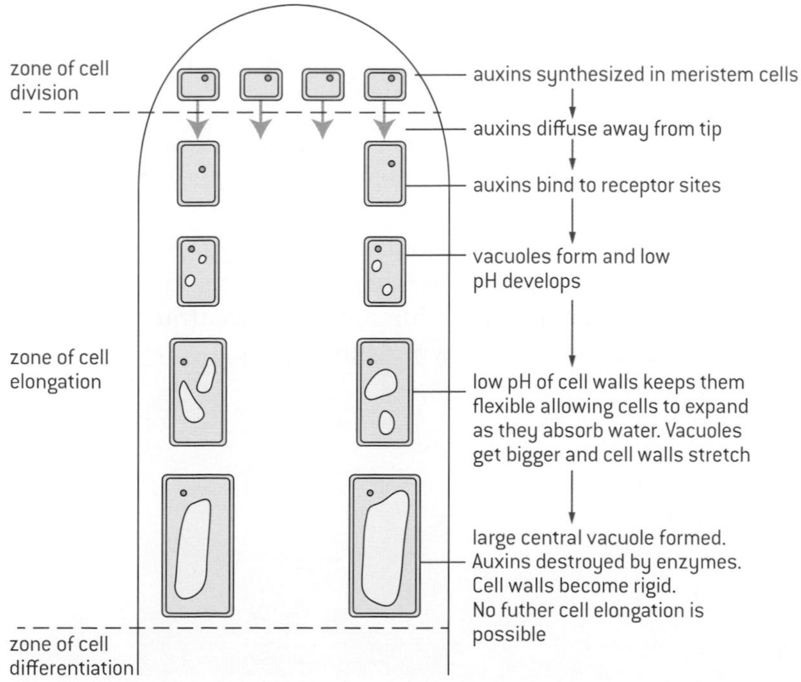
zone of cell division
zone of cell elongation
zone of cell differentiation
auxins synthesized in meristem cells
auxins diffuse away from tip
auxins bind to receptor sites
vacuoles form and low pH develops
low pH of cell walls keeps them flexible allowing cells to expand as they absorb water. Vacuoles get bigger and cell walls stretch
large central vacuole formed. Auxins destroyed by enzymes. Cell walls become rigid. No futher cell elongation is possible

▲ **Figure 9.3.1.** The effect of auxin on apical shoot growth

Improvements in analytical techniques allowing the detection of trace amounts of substances has led to advances in the understanding of plant hormones and their effect on gene expression. The demand for increased crop productivity and the predicted challenges related to plant survival under adverse environmental conditions have renewed the interest in research in root biology. Research in the areas of hormone signalling, understanding metabolism involved in hormone action and the transport of plant hormones can be used for investigating the impact of abiotic stress factors on molecular mechanisms.

Example 9.3.1.

The brown planthopper (*Nilaparvata lugens*) is a destructive piercing-sucking insect pest of rice (*Oryza sativa*). The plant hormone jasmonic acid (JA) plays an important role in plant–pest interactions. JA induces the rice gene JAmyb that encodes a transcription factor necessary for plant defence. Total RNAs were isolated from rice seedlings following infestation with different numbers of brown planthoppers. The levels of expression of mRNA of JAmyb in rice seedlings were determined at different numbers of infesting insects per plant. Ribosomic RNA is a housekeeping gene; it is always expressed, therefore it is used as a control.

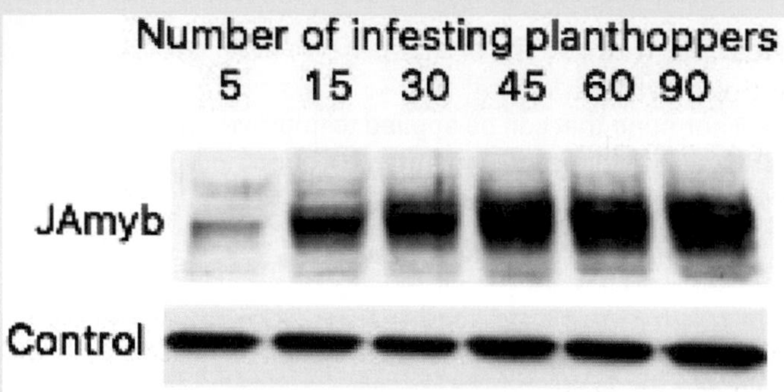

a) Describe the effect of the number of planthoppers on the production of JAmyb in rice.

b) Suggest **one** reason ribosomic RNA is used as a control.

c) Suggest whether the data supports the hypothesis that JA is involved in the defence of rice plants in a planthopper infestation by affecting gene expression of JAmyb.

Solution

a) The greater the number of planthoppers, the greater the amount of JAmyb produced till approximately 45 where it is maintained constant (values cannot be precise in this gel).

b) Ribosomic RNA is a good control to make sure that the RNA has been extracted correctly (it is a positive control).

c) The amount of JAmyb increases with the number of infesting planthoppers, therefore one could conclude that it is expressed in response to infestation (thus supporting the hypothesis). Nevertheless, the data does not really give any information on the effect of JA on the expression of JAmyb. One could have modified the experiment to add different JA concentrations to rice instead of numbers of brown planthoppers.

9.4 REPRODUCTION IN PLANTS

You should know:

- ✔ reproduction in flowering plants is influenced by the biotic and abiotic environments.
- ✔ flowering involves a change in gene expression in the shoot apex.
- ✔ the switch to flowering is a response to the length of light and dark periods in many plants.
- ✔ success in plant reproduction depends on pollination, fertilization and seed dispersal.
- ✔ most flowering plants use mutualistic relationships with pollinators in sexual reproduction.

You should be able to:

- ✔ distinguish between pollination, fertilization and seed dispersal.
- ✔ design experimental methods to induce short-day plants to flower out of season.
- ✔ draw the internal structure of seeds.
- ✔ draw half-views of animal-pollinated flowers.
- ✔ design experiments to test hypotheses about factors affecting germination.

Flowering plants reproduce by sexual reproduction. The flowers are the reproductive structures produced by the shoot apical meristem that contain the gametes. Female gametes or ovules are found in the ovary. Male gametes or pollen are found in the anther.

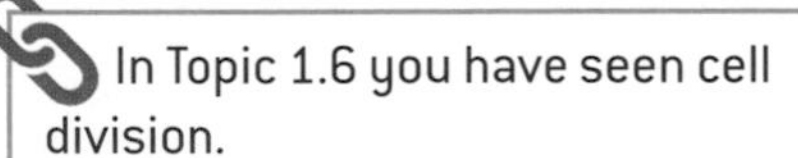

In Topic 1.6 you have seen cell division.

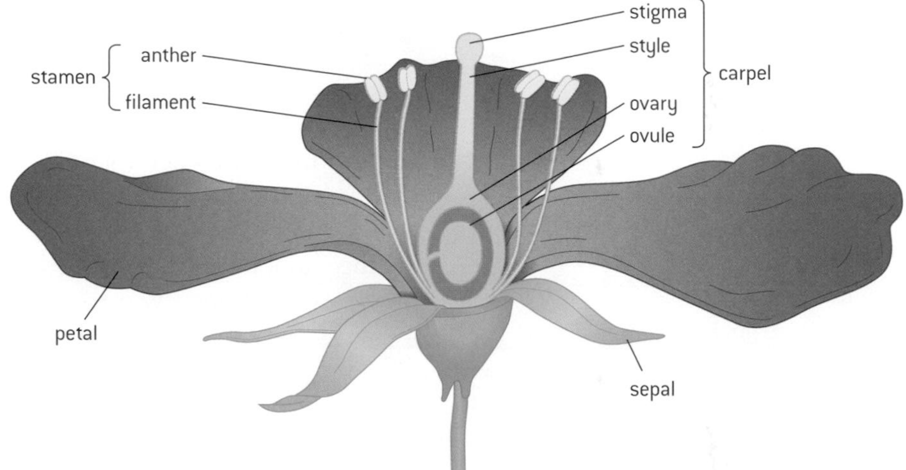

▲ **Figure 9.4.1.** Half-view of an animal-pollinated flower

Flowering involves a change in gene expression in the shoot apex. In many plants, the switch to flowering is a response to the length of light and dark periods in many plants that affects phytochromes. Phytochromes are red (Pr) or far-red (Pfr) light photoreceptors. Sunlight contains mainly red light (not far red), therefore during the day, the Pr is activated to Pfr. At night, Pfr slowly returns to the inactive Pr form. In long-day plants Pfr promotes the transcription of genes for flowering, while in short-day plants Pfr inhibits this transcription.

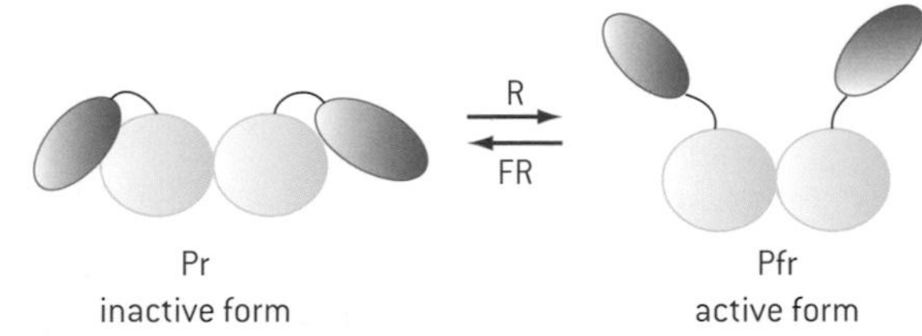

▲ **Figure 9.4.2.** Phytochromes

Success in plant reproduction depends on pollination, fertilization and seed dispersal. Most flowering plants use mutualistic relationships with pollinators for sexual reproduction.

- **Long-day plants** flower with short nights (summer).
- **Short-day plants** flower only if nights are long (autumn).
- **Phytochromes** are pigments that absorb light and can detect the length of night.

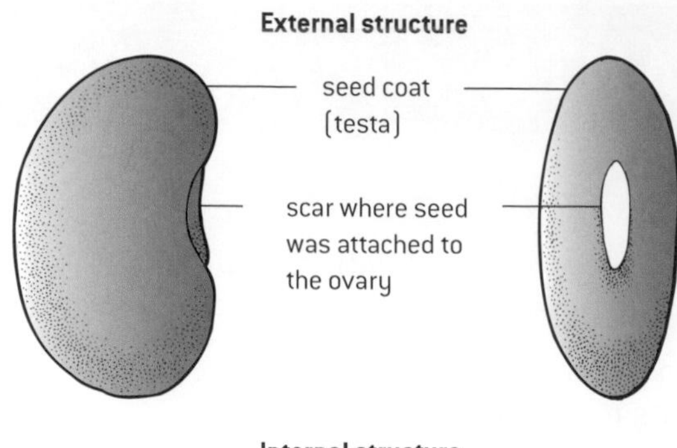

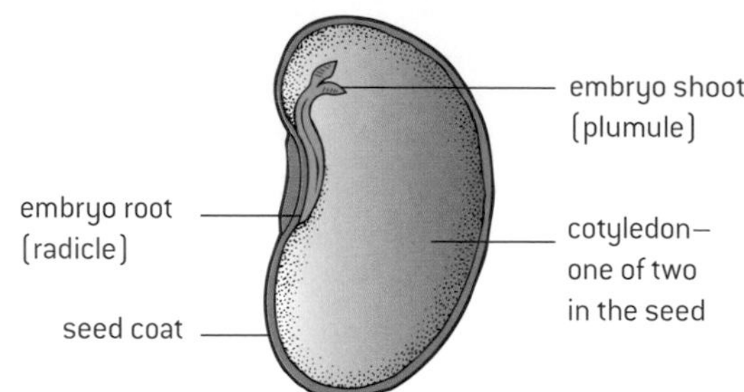

▲ **Figure 9.4.3.** Internal structure of a seed

As the seed germinates, the radicle emerges from the testa with positive gravitropism (geotropism). The shoot (plumule) grows from the cotyledon.

Example 9.4.1.

What does a long-day plant require to flower?

A. Days with a lot of sunlight

B. Long dark nights

C. Days with more light hours than dark hours

D. Long days and long nights

Solution

Although they are called long-day plants, what they really require are short nights to avoid too much Pfr changing back to Pr, therefore the correct answer is **C**.

- **Pollination** is the transfer of pollen from the anther to the stigma.
- **Fertilization** is the union of the female and male gametes.
- **Seed dispersal** is the transport of the seed away from the parent plant.

Example 9.4.2.

A series of experiments was conducted with the seeds of cocklebur (*Xanthium pennsylvanicum*) to determine the conditions necessary for germination. Each lot of seeds was put on wet absorbent cotton in a jar and was subjected to certain conditions of atmospheric pressure at recorded room temperature for 10 days. The elongation of the shoot below the cotyledon (hypocotyl), followed by the geotropic response, was used as a criterion of germination.

Jar	Atmospheric pressure / mm	Oxygen pressure / mm	Temperature range / °C	Germinated seeds / %	Average length of hypocotyl / mm
A	99	20	19–22	75	15
B	90	19	21–22	80	23
C	72	15	20–28	45	11
D	72	15	20–22	30	6
E	28	6	21–24	0	0

a) State the jar with the best conditions for germination in this experiment.

b) Describe the effect of oxygen pressure on germination and shoot elongation.

c) Suggest with a reason which jars could be used to see the effect of temperature on germination.

d) Evaluate the experimental design that allowed the researcher to obtain the results shown.

Solution

a) Jar B.

b) The lower the pressure, the lower the germination rate. The lower the pressure, the shorter the hypocotyl. At pressures of 6 mm the seeds did not germinate.

Paradigm shift—more than 85% of the world's 250,000 species of flowering plant depend on pollinators for reproduction. The University of Göttingen, in Germany, found that 87 of 115 of the leading global crop plants depend to some degree upon animal pollination, including bees. This knowledge has led to protecting entire ecosystems rather than individual species. The European Commission (EC) and the United States Environmental Protection Agency (EPA) have prohibited the use of certain neonicotinoid pesticides when bees are present.

c) All temperatures are similar, so it is difficult to see the effect of temperature. The only ones that could be used are jars C and D, as the rest of the conditions are the same—only the temperature changes. In C the temperature had a wider range and reached a higher value. At higher temperatures there was a higher germination rate and the shoot elongation was greater.

d) The experimental design has many flaws. There should have been many replications of each jar. One environmental condition should have been changed at a time, for example, the atmospheric pressure and temperature kept fixed and then the oxygen pressure changed. Temperature should have been kept constant, not varied through a wide range.

Practice problems for Topic 9

Problem 1

The drawing shows the rolled leaf of marram grass (*Ammophila arenaria*).

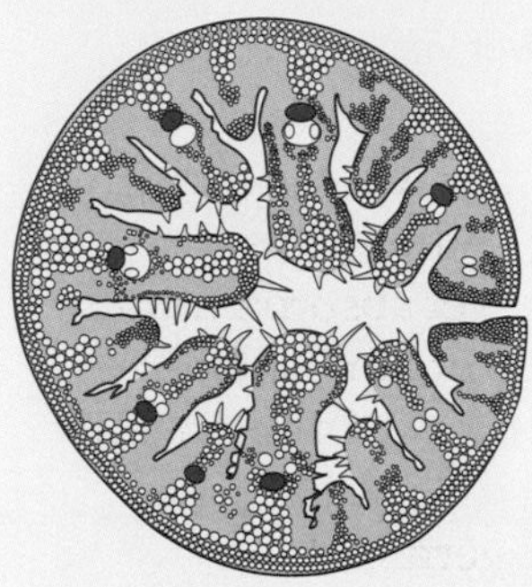

Explain how the marram grass is adapted to live in a dry environment.

Problem 2

In an experiment, the pressure inside the xylem and the rate of water flow were measured at different times of the day in common oak trees (*Quercus robur*). The graph shows the results obtained during one day.

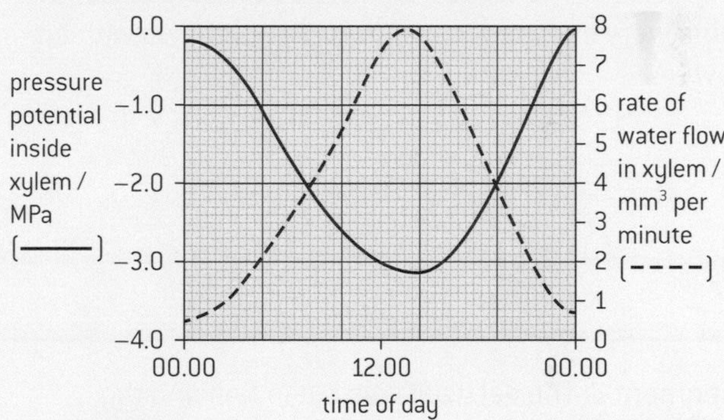

a) State the rate of water flow in the xylem at noon (12.00).

b) Calculate the change in pressure potential in the xylem from midnight (00.00) to noon (12.00).

c) Explain the changes in water flow at different times of the day.

d) Suggest a reason for the change in water pressure in the xylem.

Problem 3

Compare and contrast the structure of xylem and phloem vessels.

Problem 4

In three experiments, seedlings were kept in the dark for two days and then two shoots were cut and placed on a gel for a few hours. One shoot was kept intact and the other was split in half. The concentration of auxin in the gel was measured in nmol g^{-1} fresh weight in all three experiments.

Experiment 1: Seedlings kept in the dark and the auxin concentration measured.

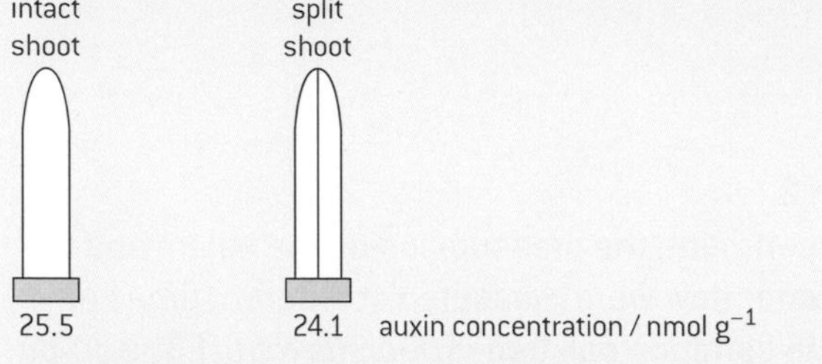

Experiment 2: The measurement was repeated after the shoots were exposed to unilateral light (shown with an arrow).

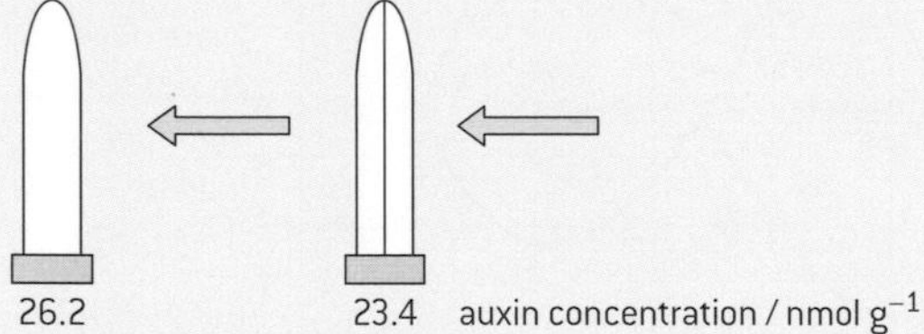

Experiment 3: The gel itself was cut in half and the concentrations on the left and right sides of the gel were recorded after unilateral light exposure.

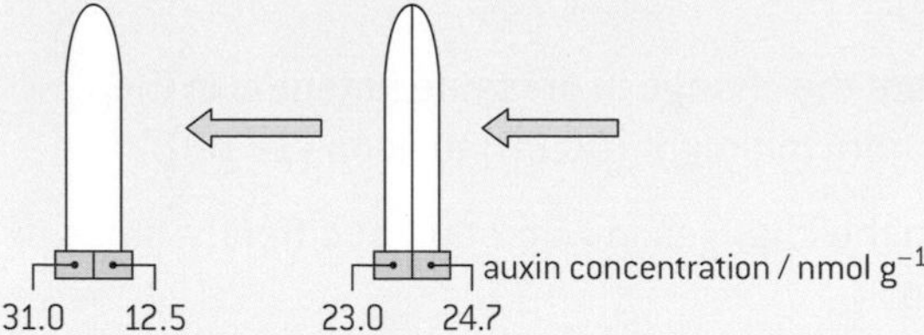

a) Calculate the concentration difference between the left and the right sides of the gel in the intact shoot and in the split shoot in experiment 3.

b) Compare and contrast the results for intact and split shoots on the concentration of auxin measured in experiments 1 and 2.

c) Explain the results obtained in experiment 3.

d) In reality, the intact shoot in experiment 3 will look different to how it is shown here. Draw the resulting shape of this shoot.

Problem 5

In a study, the percentage of insect visitors to flowers and the crop yields were recorded across small and large farms in different continents over 5 years. The percentage of flower visitors was calculated as the number of bees visiting 100 flowers in determined quadrats.

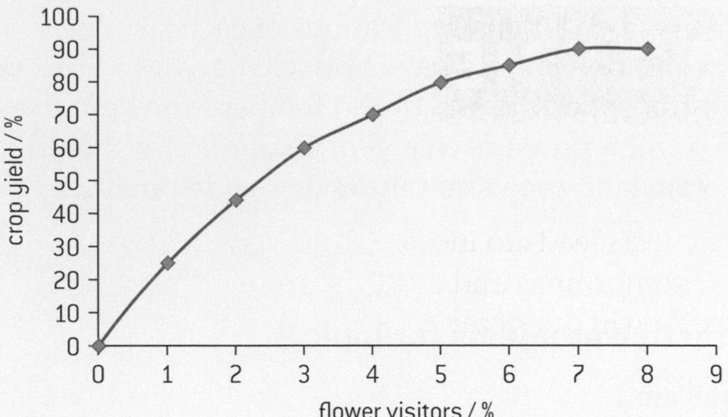

a) State the crop yield at 3% flower visitors.

b) **(i)** Describe the effect of flower visitors on crop yield.

(ii) Suggest **one** reason for the results obtained in (i).

10 GENETICS AND EVOLUTION (AHL)

10.1 MEIOSIS

You should know:

- ✔ meiosis leads to independent assortment of chromosomes and a unique composition of alleles in daughter cells.
- ✔ chromosomes replicate in interphase before meiosis.
- ✔ crossing over is the exchange of DNA material between non-sister homologous chromatids.
- ✔ crossing over produces new combinations of alleles on the chromosomes of the haploid cells.
- ✔ chiasmata formation between non-sister chromatids can result in an exchange of alleles.
- ✔ homologous chromosomes separate in meiosis I.
- ✔ sister chromatids separate in meiosis II.
- ✔ independent assortment of genes is due to the random orientation of pairs of homologous chromosomes in meiosis I.

You should be able to:

- ✔ describe the processes in meiosis.
- ✔ explain the consequences of crossing over.
- ✔ draw diagrams to show chiasmata formed by crossing over.
- ✔ explain the outcome of meiosis.
- ✔ analyse micrographs to identify cells in meiosis.

Meiosis produces gametes with haploid cells. Gametes are haploid so that the chromosome number after fertilization is diploid. Meiosis occurs in two divisions, meiosis I and meiosis II. Chromosomes replicate in interphase before meiosis. In meiosis I the number of chromosomes is halved as the homologous chromosomes migrate to each pole and the cells divide into two. In meiosis II the sister chromatids of each chromosome are separated into two new cells. The total outcome is four different cells with half the number of chromosomes of the parent cell.

In Topic 1.6 you studied cell division and in Topic 3.3 you looked at meiosis. In Topic 11.4 you will study sexual reproduction.

Example 10.1.1.

In which stage of meiosis is the number of chromosomes halved?

A. Prophase I **B.** Anaphase I **C.** Metaphase II **D.** Telophase II

Solution
The correct answer is **B** as one from each pair of chromosomes migrates to the opposite pole, making two cells with half the number of chromosomes in telophase I.

Interphase
Immediately before meiosis DNA replicates so that the cell now contains four, rather than the original two, copies of each chromatid.

Prophase I
The chromosomes shorten and fatten and come together in their homologous pairs to form a bivalent. The chromatids wrap around one another and attach at points called chiasmata. The chromosomes may break at these points and swap similar sections of chromatids with one another in a process called crossing over. Finally the nucleolus disappears and the nuclear envelope disintegrates.

Metaphase I
Centromeres attach to the spindle and the bivalents arrange themselves randomly on the equator of the cell with each of a pair of homologous chromosomes facing opposite poles.

Anaphase I
One of each pair of homologous chromosomes is pulled by spindle fibres to opposite poles.

Telophase I
Microtubules pull two sides of the cell surface membrane together so that the cell becomes narrower towards its centre until the opposite parts of the membrane fuse to give two separate cells. In most animal cells a nuclear envelope re-forms around the chromosomes at each pole, but in most plant cells there is no telophase I and the cell goes directly into metaphase II.

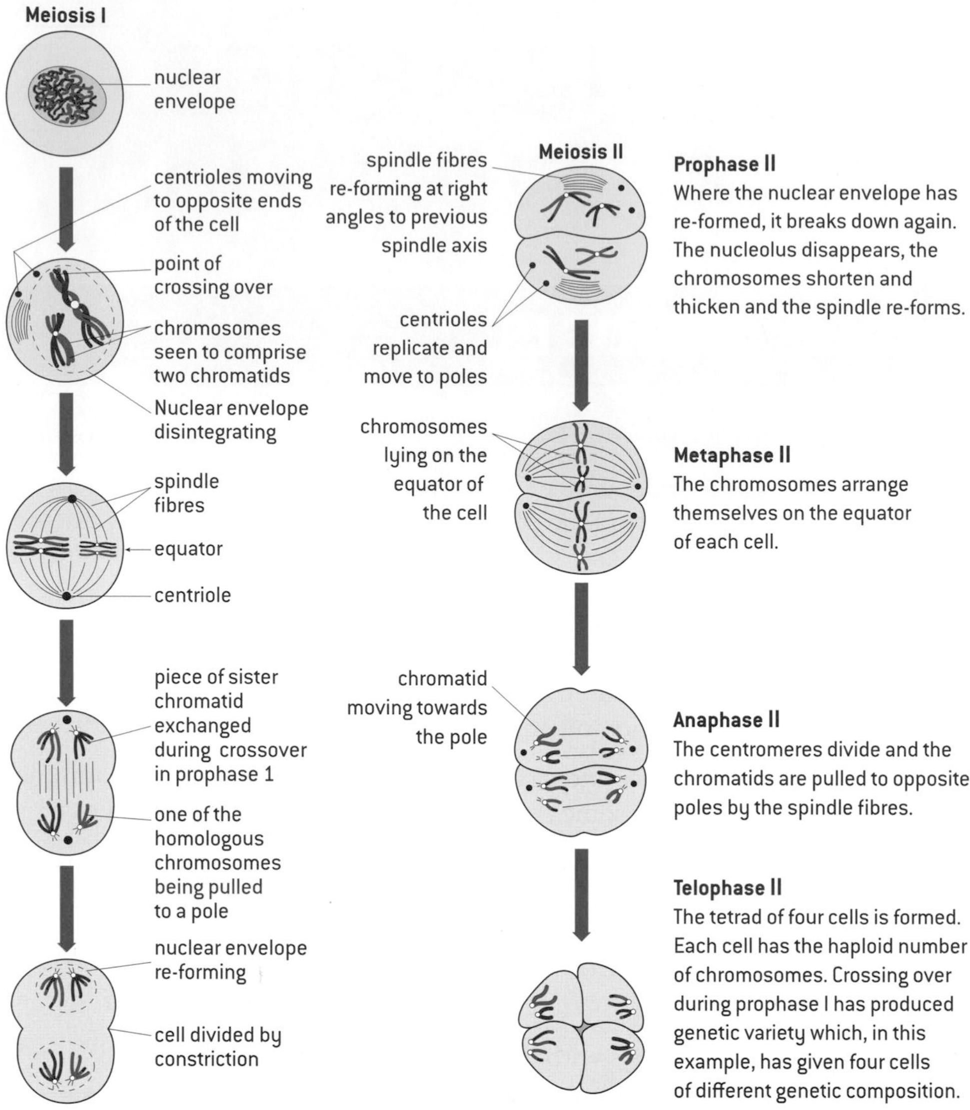

Prophase II
Where the nuclear envelope has re-formed, it breaks down again. The nucleolus disappears, the chromosomes shorten and thicken and the spindle re-forms.

Metaphase II
The chromosomes arrange themselves on the equator of each cell.

Anaphase II
The centromeres divide and the chromatids are pulled to opposite poles by the spindle fibres.

Telophase II
The tetrad of four cells is formed. Each cell has the haploid number of chromosomes. Crossing over during prophase I has produced genetic variety which, in this example, has given four cells of different genetic composition.

▲ Figure 10.1.1. Stages in meiosis

- **Homologous chromosomes** are one set of maternal and one set of paternal chromosomes that have the same genes in the same loci. They pair up during meiosis.
- **Crossing over** is the process in which DNA is exchanged between non-sister chromatids of homologous pairs.
- The **chiasma** is the point where genetic material is exchanged in crossing over.

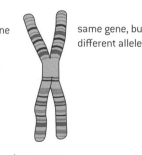

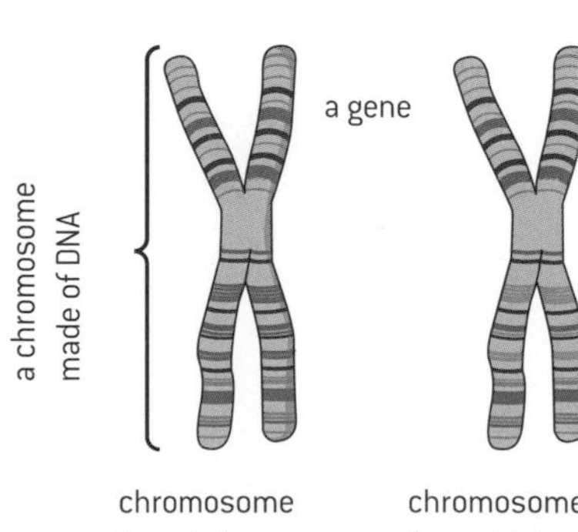

▲ Figure 10.1.2. A homologous pair of chromosomes

Meiosis leads to an independent assortment of chromosomes and a unique composition of alleles in daughter cells. Crossing over occurs in prophase I. It is the exchange of DNA material between non-sister homologous chromatids, producing new combinations of alleles on the chromosomes of the haploid cells. The independent assortment of genes is due to the random orientation of pairs of homologous chromosomes in meiosis I.

Example 10.1.2.

The diagram shows six chromosomes of a diploid cell in early metaphase I (although not shown as bivalents).

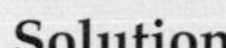

Which is one pair of homologous chromosomes?

A. A and B

B. A and C

C. B, D, E and F

D. There is no pair of homologous chromosomes.

Solution

The correct answer is **B.** The homologous chromosomes have the centromere in the same place and the chromatids are of the same length. **C** is incorrect, as although the chromosomes all look the same, they are not a pair—there are four not two chromosomes.

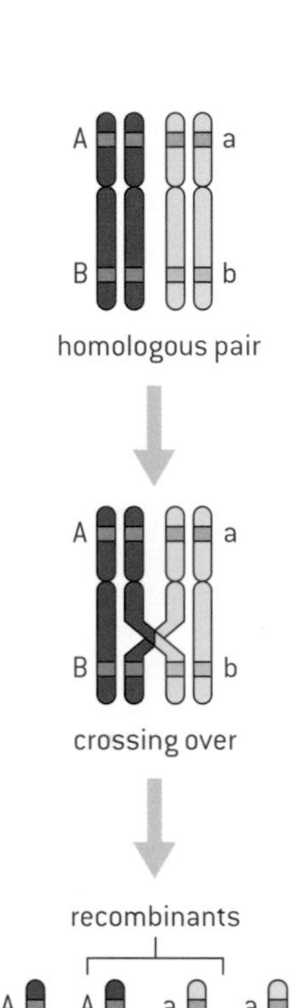

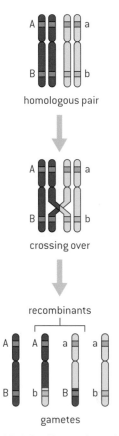

▲ Figure 10.1.3. Formation of recombinants

Example 10.1.3.

The light micrograph shows a thin section through an anther of a lily (*Lilium sp.*), showing pollen cells dividing. The magnification is ×70.

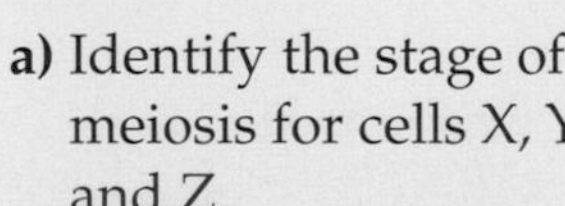

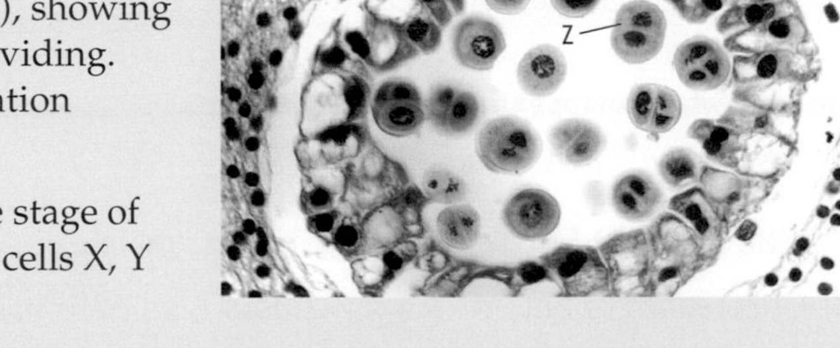

a) Identify the stage of meiosis for cells X, Y and Z.

b) Explain the reason why cells in the anthers of *Lilium* must undergo meiosis.

Solution

a) X: prophase I; Y: telophase II; Z: telophase I.

b) Cells in the anther of *Lilium* must undergo meiosis because they will form the pollen which will fertilize the egg cell in the ovary of the plant. The number of chromosomes must, therefore, be reduced to half. This occurs by the migration of homologous chromosomes during anaphase I to opposite poles and the separation into two cells in telophase I, forming haploid cells.

Careful observation and record keeping revealed anomalous data that Mendel's law of independent assortment could not account for. Thomas Hunt Morgan developed the notion of linked genes to account for these anomalies.

Crossing over is the exchange of DNA material between non-sister homologous chromatids occurring during prophase I. A single strand break (SSB) in the DNA is a cut between the non-sister chromatids of homologous chromosomes. Chiasmata (or chiasma) formation between non-sister chromatids can result in an exchange of alleles. This exchange of alleles can produce recombinants.

Crossing over produces new combinations of alleles on the chromosomes of the haploid cells.

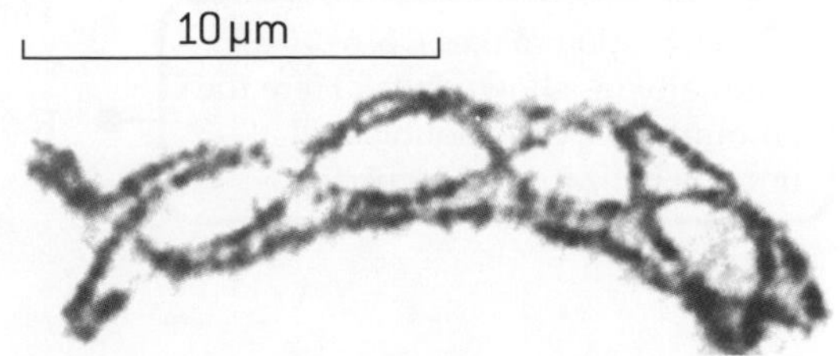

▲ Figure 10.1.4. Micrograph image of crossing over during meiosis

When crossing over occurs, homologous chromosomes form structures with strange shapes called tetrads (named after the presence of four chromatids) or bivalents (named after two chromosomes). The chromatids pull away from each other but are held together at the chiasmata.

The diagram shows the changes in the amount of DNA per cell in the anther of a plant.

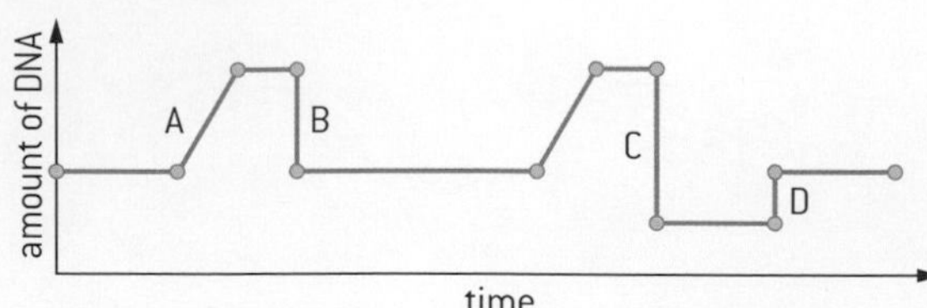

a) State the process occurring at A. [1]

This answer could have achieved 1/1 marks:

DNA replication.

b) Explain the reason the mass of DNA decreases at B. [2]

This answer could have achieved 2/2 marks:

At B the cell is going through mitosis. The sister chromatids of each chromosome have migrated to the opposite pole in anaphase, in telophase the nuclear membrane reappears and the cell has divided into two cells in cytokinesis.

▲ The answer identifies the process and stages of DNA reduction.

c) Suggest what is happening at C. [2]

This answer could have achieved 2/2 marks:

At C the process of meiosis is occurring. The amount of DNA decreases to half during meiosis I, as each homologous chromosome migrates to a different pole in anaphase I, producing two different cells with half the number of chromosomes in cytokinesis. In the second division, in meiosis II, each sister chromatid of each chromosome migrates to the opposite pole in anaphase II, and new cell nuclear membranes form in telophase II. In cytokinesis four cells are formed. These cells have half the number of chromosomes and 1/4 the amount of DNA compared with the original cell.

▲ This answer is complete.

d) State the name of the process occurring at D. [1]

This answer could have achieved 0/1 marks:

Replication.

▼ Replication makes DNA gradually (as shown in A). Here the amount of DNA is doubled all at once. Fertilization is occurring.

10.2 INHERITANCE

You should know:

- ✔ gene loci are said to be linked if on the same chromosome.
- ✔ unlinked genes segregate independently as a result of meiosis.
- ✔ variation can be discrete or continuous.
- ✔ the phenotypes of polygenic characteristics tend to show continuous variation.
- ✔ Chi-squared tests are used to determine whether the difference between an observed and an expected frequency distribution is statistically significant.

You should be able to:

- ✔ construct Punnett squares for dihybrid traits.
- ✔ describe Morgan's discovery of non-Mendelian ratios in Drosophila.
- ✔ calculate the predicted genotypic and phenotypic ratio of offspring of dihybrid crosses involving unlinked autosomal genes.
- ✔ identify recombinants in crosses involving two linked genes.
- ✔ deduce results of a chi-squared test on data from dihybrid crosses.
- ✔ explain how polygenic traits such as human height may also be influenced by environmental factors.

Genes may be linked or unlinked and are inherited accordingly. Gene loci are said to be linked if they are on the same chromosome. Unlinked genes segregate independently as a result of meiosis. Linkage can be autosomal or sex-linked; in sex-linked inheritance the genes are usually located on the X chromosome.

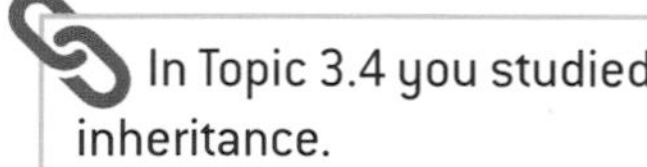
In Topic 3.4 you studied inheritance.

Alleles are usually shown side by side in dihybrid crosses, for example, AaBb. In representing crosses involving linkage, it is more common to show them as vertical pairs, for example:

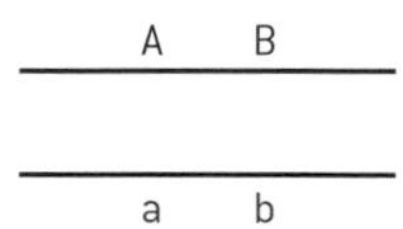

Assessment tip

The lines symbolize chromosomes in the notation used for linked genes.

- **Linked genes** are those on the same chromosome.
- A **monohybrid cross** is a genetic cross involving one gene locus.
- A **dihybrid cross** is a genetic cross involving two gene loci.
- A **chi-squared test** is a statistical test to determine whether the difference between an observed and an expected frequency distribution is significant.

Example 10.2.1.

The sketch shows two pairs of human chromosomes, 1 and 12. On chromosome 1, the gene for CHD5 is represented by the alleles R and r, and the gene for CLIC4 by S and s. On chromosome 12, the gene for ZNF26 is represented by T and t.

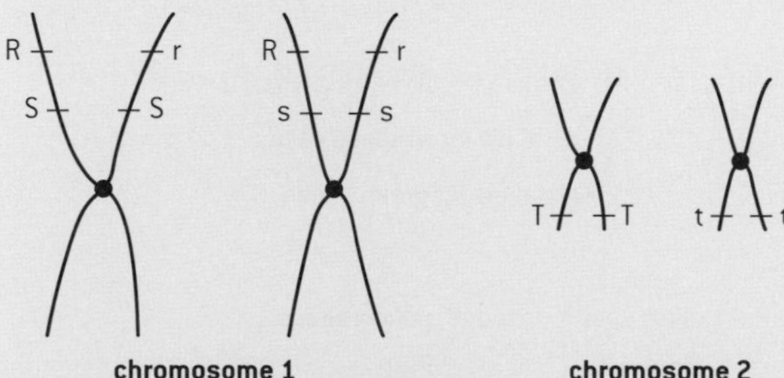

a) State two genes that are linked.

b) Suggest with a reason the process that resulted in this combination of alleles R and r on the homologous pair of chromosome 1.

c) Draw the possible genotypes of gametes formed with these chromosomes using a Punnett square.

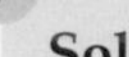

Solution

a) The linked genes are on the same chromosome, therefore the answer is CHD5 and CLIC4 (R and S), as they are on chromosome 1.

b) The alleles on one chromosome should be the same, as one sister chromatid is formed by replication of the other. In this case through the process of crossing over, the R from one chromosome was exchanged with the r of one of the non-sister chromatids of the homologous pair.

c)

	RS	rS	Rs	rs
T	RST	rST	RsT	rsT
t	RSt	rSt	Rst	rst

A dihybrid cross is a genetic cross involving two gene loci, each with two different alleles.

Data from dihybrid crosses can be tested with a chi-squared test. These tests are used to determine whether the difference between an observed and an expected frequency distribution is statistically significant.

Example 10.2.2.

The seeds of a plant can be round-shaped (R) or wrinkled (r) and can have yellow (G) or green (g) colour. Round shape is dominant over wrinkled and yellow is dominant over green. In a cross between two plants with round and yellow seeds the following offspring were produced:

947 round and yellow seeds

284 round and green seeds

322 wrinkled and yellow seeds

109 wrinkled and green seeds.

a) Calculate the expected ratio of phenotypes assuming these are unlinked characteristics. Show your results in a Punnett square.

b) Use the chi-squared formula and table shown in example 4.1.3 to determine whether these characteristics are linked. Show your working.

Solution

a) The Mendelian crossing for unlinked traits is as follows:

R = allele for round seeds r = allele for wrinkled seeds

G = allele for yellow seeds g = allele for green seeds

Parental phenotypes	round, yellow seeds	round, yellow seeds
Parental genotypes	RrGg	RrGg
	meiosis	**meiosis**
Gametes	RG Rg rG rg	RG Rg rG rg

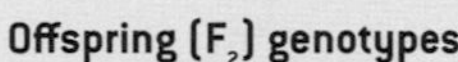

Offspring (F_2) genotypes

	♂ Gametes			
♀ Gametes	RG	Rg	rG	rg
RG	RRGG	RRGg	RrGG	RrGg
Rg	RRGg	RRgg	RrGg	Rrgg
rG	RrGG	RrGg	rrGG	rrGg
rg	RrGg	Rrgg	rrGg	rrgg

Offspring (F_2) phenotypes

9 round, yellow seeds
3 round, green seeds
3 wrinkled, yellow seeds
1 wrinkled, green seed

b) There are two hypotheses: the null hypothesis (H_0) and the alternative hypothesis (H_1).

H_0 = The characteristics assort independently (are not linked).

H_1 = The characteristics do not assort independently (thus are linked).

The total number of seeds is 1662.

To calculate the expected frequencies we use the data obtained in (a) for the expected ratios of round yellow:round green:wrinkled yellow:wrinkled green of 9:3:3:1.

First add the expected frequencies: $9 + 3 + 3 + 1 = 16$

The expected frequency for round yellow is 9/16 so it is calculated as: $947 \times 9/16 = 935$

This is done for all possible characteristics. The results are shown in the table:

Phenotypes	Observed (O)	Expected (E)	(O – E)	$(O - E)^2/E$
round yellow	947	935	−12	0.2
round wrinkled	284	312	28	2.4
wrinkled yellow	322	312	−10	0.3
wrinkled green	109	104	-5	0.3
Total	**1662**	**1662**		**3.2**

The value for chi squared is = **3.2.**

The degrees of freedom for each characteristic is calculated as: (number of rows – 1) = (4 – 1) = **3**

Then use the table to find the critical value at $p = \mathbf{0.05}$ (significant at 5%) and 3 degrees of freedom. The value is **7.815**.

If your results give a value above the critical value in the table, you reject the null hypothesis and accept the alternative hypothesis. If it is below the critical value, you accept the null hypothesis.

In this example, the value for chi-squared is below the critical value (ie, statistically insignificant), therefore the null hypothesis is accepted, confirming that the shape of the seed and the colour of the seed are not linked.

Assessment tip

It is a convention to use the same letter for alleles of the same gene, using a capital letter for the dominant characteristic.

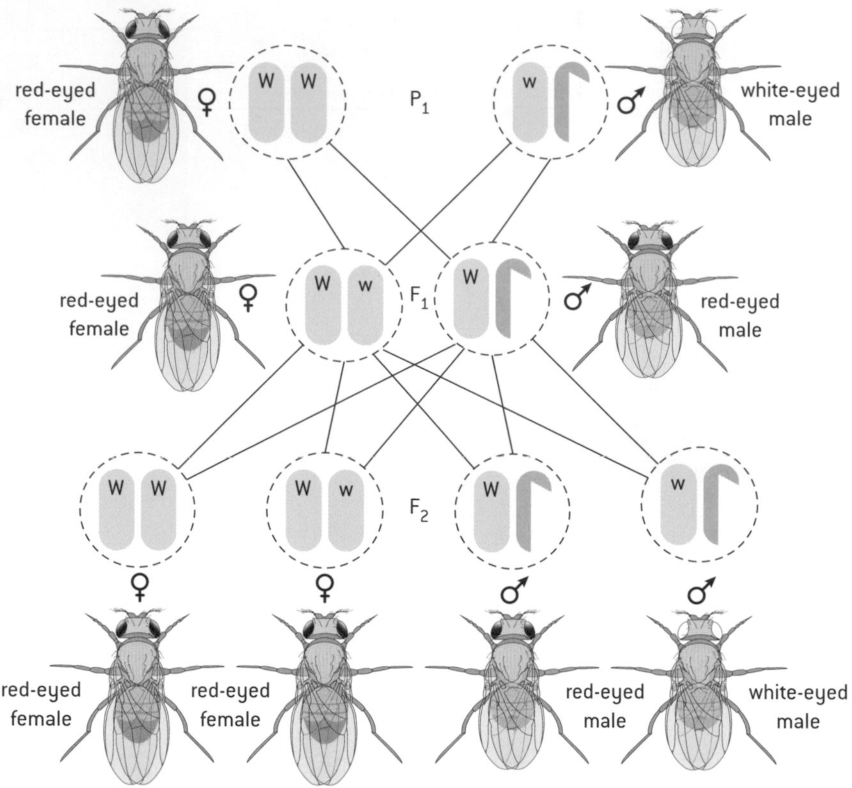

▲ Figure 10.2.1. Linked genes inheritance

Morgan discovered non-Mendelian ratios in Mendel's experiments with Drosophila. Morgan crossed a red-eyed female fruit fly *Drosophila melanogaster* with a white-eyed male. When he crossed heterozygous F1 red-eyed flies, the traits of the F2 progeny did not assort independently. The expected Mendelian 1:1:1:1 ratio of red-eyed females, red-eyed males, white-eyed males, and white-eyed females did not occur. Instead, the observed phenotypes in his F2 generation were a ratio of 2:1:1:0 of red-eyed females, red-eyed males, white-eyed males, and white-eyed females. Morgan therefore hypothesized that the eye-colour trait was connected with the sex factor.

- **Discrete variation** is a discontinuous variation found in a number of distinct categories.
- **Continuous variation** is the combined effect of many genes and/or the effect of the environment on genes. It can be plotted as a normal distribution curve.

Variation can be discrete (discontinuous) or continuous. In discrete variation a characteristic is found in two or more distinct forms. An example of discrete variation is the blood groups. The phenotypes of polygenic characteristics tend to show continuous variation. Polygenic traits such as human height may be influenced by environmental factors such as diet.

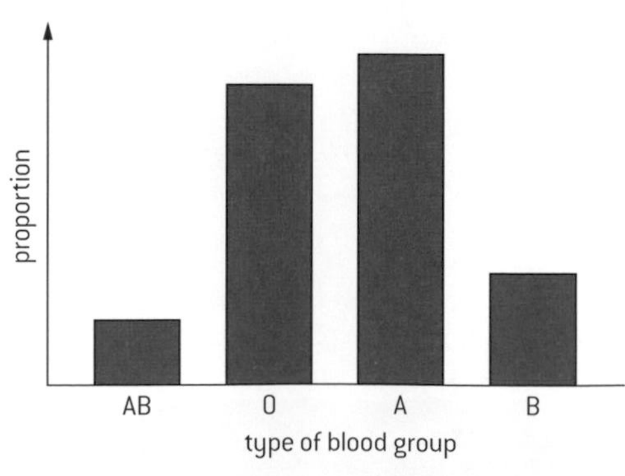

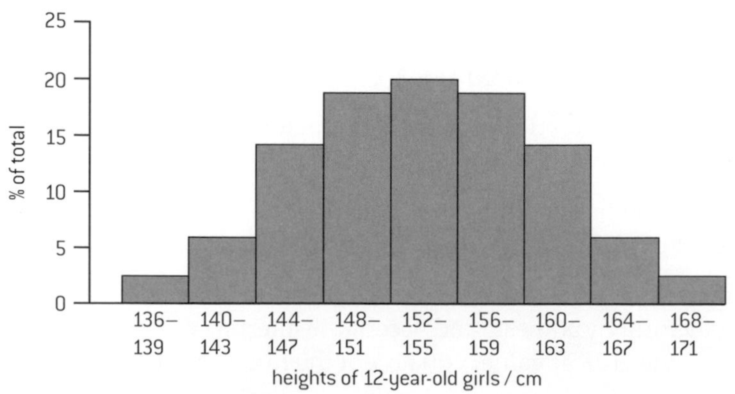

▲ Figure 10.2.2. Discrete variation (left) and continuous variation (right)

10.3 GENE POOLS AND SPECIATION

You should know:

- ✓ gene pools change over time.
- ✓ a gene pool consists of all the genes and their different alleles, present in an interbreeding population.
- ✓ allele frequencies change with time in populations.
- ✓ reproductive isolation of populations can be temporal, behavioural or geographic.
- ✓ speciation due to divergence of isolated populations can be gradual.
- ✓ speciation can occur abruptly.

You should be able to:

- ✓ explain evolution in terms of changing of allele frequencies.
- ✓ explain speciation due to isolation.
- ✓ compare and contrast temporal, behavioural and geographic reproductive isolation.
- ✓ compare allele frequencies of geographically isolated populations.
- ✓ identify examples of directional, stabilizing and disruptive selection.
- ✓ explain speciation in the genus *Allium* by polyploidy.

A gene pool is the sum of all the population's genetic material at a given time. Gene pools change over time through evolution. For example, in sickle cell anemia natural selection tends to maintain a mutated allele and a normal allele in the genetic pool, as heterozygous organisms do not develop sickle cell anemia and are resistant to malaria.

In Topic 5.1 you studied evidence for evolution.

- A **gene pool** is all the genes and their different alleles present in an interbreeding population.
- **Reproductive** isolation is a mechanism that prevents production of offspring. Isolation of populations can be temporal, behavioural or geographic.
- **Polyploidy** is the multiplication of sets of chromosomes. It can lead to speciation as it produces hybrids.

In speciation, the gene pool is split if populations are reproductively isolated and therefore do not interbreed. Reproductive isolation of populations can be sympatric or allopatric. Sympatric isolation occurs in the same place and includes temporal and behavioural isolation. Allopatric isolation is geographic isolation. In different environments there are different selection pressures as there are different habitats or niches to exploit. These changes cause natural selection. The allele frequencies change, causing the populations to diverge.

SAMPLE STUDENT ANSWER

Outline how reproductive isolation can occur in an animal population. [3]

This answer could have achieved 2/3 marks:

Reproductive isolation occurs when organisms from a species are unable to mate with another species. Reproductive isolation can be temporal, behavioural or geographic. Temporal - the species mate at different times in the year (seasonally), hence resulting in reproductive isolation. Behavioural - the species have different courtship patterns. Geographical - the species are in geographically different places hence cannot reproduce.

▲ Although the candidate mentions different species, it is implied that they were from the same species before speciation. This answer scored a mark for saying that there was temporal isolation by members of different populations reproducing at different times and behavioural isolation by differences in courtship behaviours.

▼ The answer did not score a mark for geographic isolation by a population being separated because it did not mention the example of barrier, for example, a river, mountain or road.

Assessment tip

This is a three mark-question, therefore three isolations are expected.

Speciation due to divergence of isolated populations can be gradual or can occur abruptly.

In gradualism, the evolution occurs uniformly at a steady rate with a gradual and continuous transformation of the species, with a long

In some plants, the duplication of the genetic material has led to speciation. In species such as those of the genus *Allium* (for example onions) the number of chromosomes has duplicated by polyploidy. When a diploid organism (two sets of chromosomes) duplicates its sets it becomes a tetraploid (four sets of chromosomes). Polyploidy is a type of sympatric speciation, because polyploids cannot breed with the original diploids. Many crop species have been created to be polyploid. Polyploidy increases allelic diversity and permits novel phenotypes to be generated, leading to hybrid vigour. The domestication of wheat involved multiple polyploidization events between several species to obtain hexaploid bread wheat (*Triticum aestivum*). This new species has larger grains which produce more flour than the wild ancestral species.

sequence of intermediate forms. In punctuated equilibrium, speciation occurs abruptly and the population becomes stable immediately. Punctuated equilibrium implies long periods without appreciable change and short periods of rapid evolution.

- **Gradualism** is the process by which speciation occurs at a steady rate with a long sequence of intermediate forms.
- A **punctuated equilibrium** is the speciation that occurs abruptly without intermediate forms.
- **Directional selection** favours one extreme phenotype over the other.
- **Stabilizing selection** favours middle phenotypes.
- **Disruptive selection** favours both extreme phenotypes over middle phenotypes.

Example 10.3.1.

Describe the pace of evolution in the theory of punctuated equilibrium.

Solution

There are long stable periods with little change and sudden short periods of rapid evolution. An example of punctuated evolution occurs in the case of a volcanic eruption or a meteor impact, causing sudden climatic or environmental changes.

Evolution requires that allele frequencies change with time in populations. Natural selection can occur due to directional, stabilizing or disruptive selection.

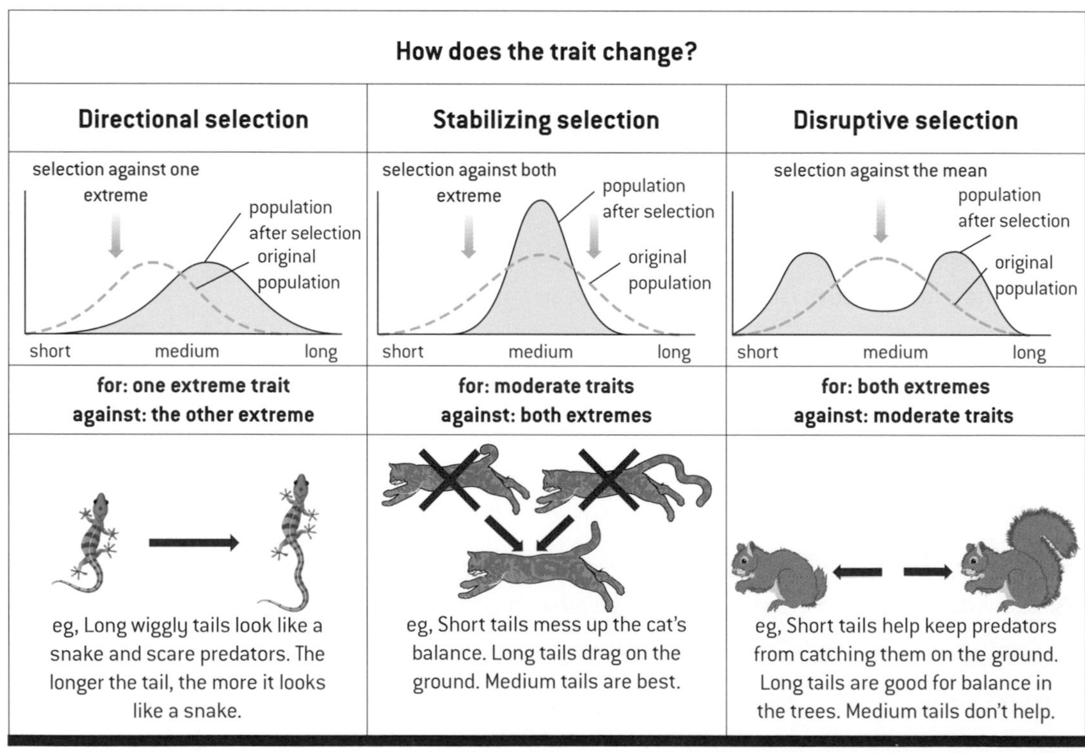

▲ Figure 10.3.1. Types of selection

Practice problems for Topic 10

Problem 1

The diagram shows a pair of homologous chromosomes during meiosis in a cell in the human testis. The positions of the alleles of some genes are indicated.

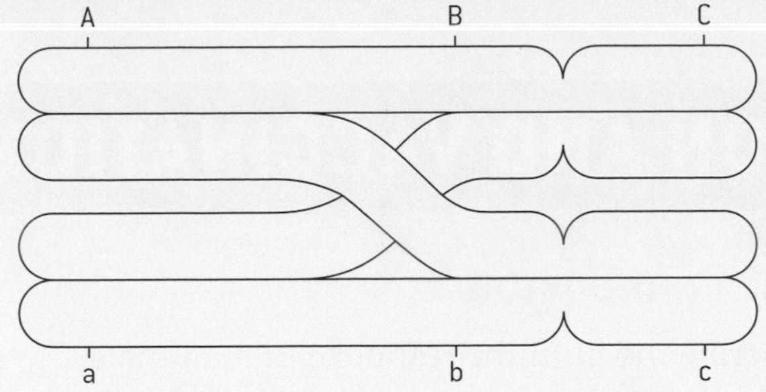

a) (i) Draw the possible chromosomes that could be found in the gametes formed.

(ii) Identify in your diagrams in (i) which chromosomes are recombinant.

b) Explain the consequence of crossing over on variability.

Problem 2

Two genes determine the coat colour in mice. One gene controls whether the coat is coloured and the other controls the colour. The gene for coloured (C) is dominant over non-coloured or albino (c). Agouti (A) is for grey-coloured hair and is dominant over black (a). Albino mice (cc) do not have coat colour because these mice are unable to produce any colour.

a) Using a Punnett square, predict the possible genotypes for colour of coat of the offspring of a crossing of two agouti mice heterozygous for both genes.

b) Identify the possible phenotypes for the offspring in (a).

Problem 3

Starting from the concept of a gene pool, explain briefly how populations of early vertebrates could have evolved into different groups.

11 ANIMAL PHYSIOLOGY (AHL)

11.1 ANTIBODY PRODUCTION AND VACCINATION

You should know:

- ✔ every organism has unique molecules on the surface of its cells.
- ✔ pathogens can be species-specific although others can cross species barriers.
- ✔ B lymphocytes are activated by T lymphocytes in mammals.
- ✔ activated B cells multiply to form clones of plasma cells and memory cells.
- ✔ plasma cells secrete antibodies.
- ✔ antibodies aid the destruction of pathogens.
- ✔ immunity depends upon the persistence of memory cells.
- ✔ vaccines contain antigens that trigger immunity but do not cause the disease.
- ✔ white cells release histamine in response to allergens which cause allergic symptoms.
- ✔ fusion of a tumour cell with an antibody-producing plasma cell creates a hybridoma cell.
- ✔ monoclonal antibodies are produced by hybridoma cells.

You should be able to:

- ✔ describe the immune response in mammals, including antibody production.
- ✔ explain the reason antigens on the surface of red blood cells stimulate antibody production in a person with a different blood group.
- ✔ explain the effects of vaccination on the immune system.
- ✔ describe how smallpox was the first infectious disease of humans to have been eradicated by vaccination.
- ✔ analyse epidemiological data.
- ✔ explain the use of monoclonal antibodies to HCG in pregnancy test kits.

In Topic 6.3 you studied defence against infectious diseases.

Immunity is based on the recognition of self and the destruction of foreign material. Every organism has unique molecules on the surface of its cells called antigens, which are recognized by specific antibodies. Antibodies are proteins that attach to the antigens. Antibodies are produced when the white blood cells encounter the antigen. Lymphocytes are a type of white blood cell. In mammals, B lymphocytes are activated by T lymphocytes. Activated B cells multiply to form clones of plasma cells and memory cells. Plasma cells secrete antibodies that aid in the destruction of pathogens. Memory cells are mature B lymphocytes that remember how to produce the antibodies in the event of a second encounter with the antigen. Immunity depends upon the persistence of memory cells.

Once marked by antibodies, foreign matter such as bacteria is destroyed by phagocytes. Phagocytes are non-specific, as they destroy any foreign matter.

- An **antigen** is a molecule (usually a peptide or glycoprotein) found on the surface of the cells of pathogens.
- An **antibody** is a protein of the immune system produced by plasma cells that attacks antigens.
- **B lymphocytes** are white blood cells that mature into either plasma cells or memory cells.
- **T lymphocytes** are white blood cells that activate B lymphocytes.
- **Plasma cells** are mature B lymphocytes that produce antibodies.
- **Memory cells** are mature B lymphocytes that remember how to produce the antibodies.

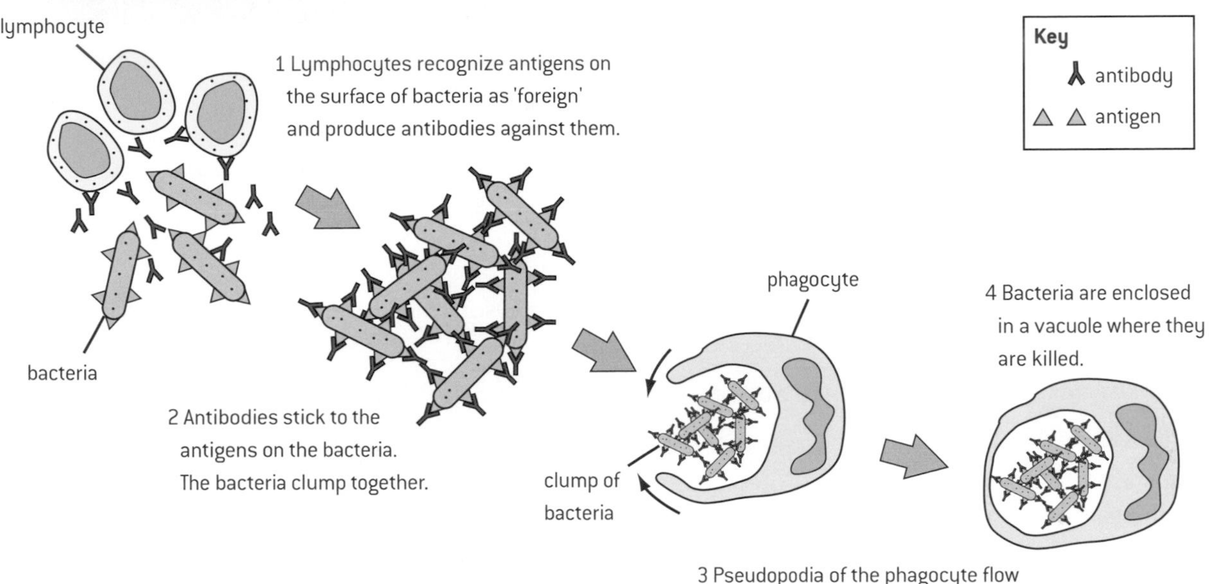

lymphocyte
1 Lymphocytes recognize antigens on the surface of bacteria as 'foreign' and produce antibodies against them.
Key
antibody
antigen
bacteria
2 Antibodies stick to the antigens on the bacteria. The bacteria clump together.
phagocyte
clump of bacteria
4 Bacteria are enclosed in a vacuole where they are killed.
3 Pseudopodia of the phagocyte flow around the bacteria.

▲ Figure 11.1.1. Defence against disease

Example 11.1.1.

Which cells produce antibodies?

A. erythrocytes

B. immature B lymphocytes

C. plasma cells

D. mature T lymphocytes

Solution

The correct answer is **C.** Erythrocytes are red blood cells; they have no nucleus so they cannot produce proteins. B lymphocytes need chemicals produced by T lymphocytes, such as cytokines, which make them mature into plasma cells and only then can they produce antibodies.

Some pathogens are species-specific but others can affect more than one species. Vaccines contain antigens that trigger immunity but do not cause the disease. *Salmonella enterica* subspecies *enterica* is a bacterium containing different types (or serovars) that can act either as non-species-specific or as species-specific pathogens. The type *Salmonella* Typhimurium is associated with gastroenteritis in humans and has animal reservoirs in a broad range of reptilian, avian and/or mammalian species. However, the type *Salmonella* typhi is a specialist; it is associated with disseminated septicemic infections such as typhoid fever which affect only humans.

Example 11.1.2.

The graph shows the concentration of antibodies in blood after vaccination (on day 0) and after a second encounter with the antigen.

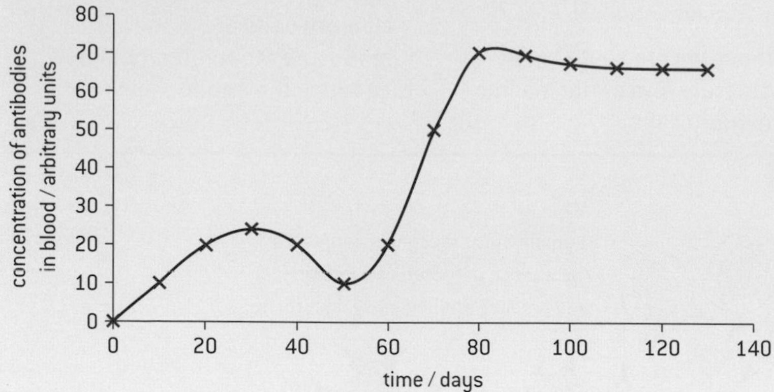

a) State the concentration of antibodies on day 30.

b) (i) Identify the day the person encountered the antigen for a second time.

(ii) Suggest a possible source of this encounter.

c) Explain the body's response to a second encounter with the antigen.

> **Assessment tip**
>
> You must always write the units. In this case as they are arbitrary units it is not really necessary, but it is always better to include them.

Solution

a) 24 (arbitrary units)

b) (i) day 50

(ii) The person could have had a second shot of the vaccine or could have been infected with the pathogen.

c) Following vaccination, the B lymphocytes have matured into plasma cells with the help of the T lymphocytes. Plasma cells start producing antibodies against the antigen. After 30 days the antibodies start decreasing. Some B lymphocytes mature into memory cells that remember how to make the antibodies. These cells are activated as soon as they encounter the antigen the second time, therefore antibodies are produced immediately and in a greater quantity.

Epidemiological data usually shows the distribution, causes and risk factors of diseases in a defined population. It is usually based on a collection of data and statistical analysis. It helps provide information to determine policies to eradicate and prevent diseases.

Example 11.1.3.

Cholera is an infection caused by the water-borne bacterium *Vibrio cholerae*. The symptoms of cholera are watery diarrhoea, stomach pain and cramps. It can lead to dehydration and death.

John Snow's completed map of cholera cases in Soho, London, in 1854, shows each case as one dot. The black squares show the communal water pumps. After analysing the data, Snow persuaded the local council to disable one of the water pumps by removing its handle. The cholera outbreak soon ended.

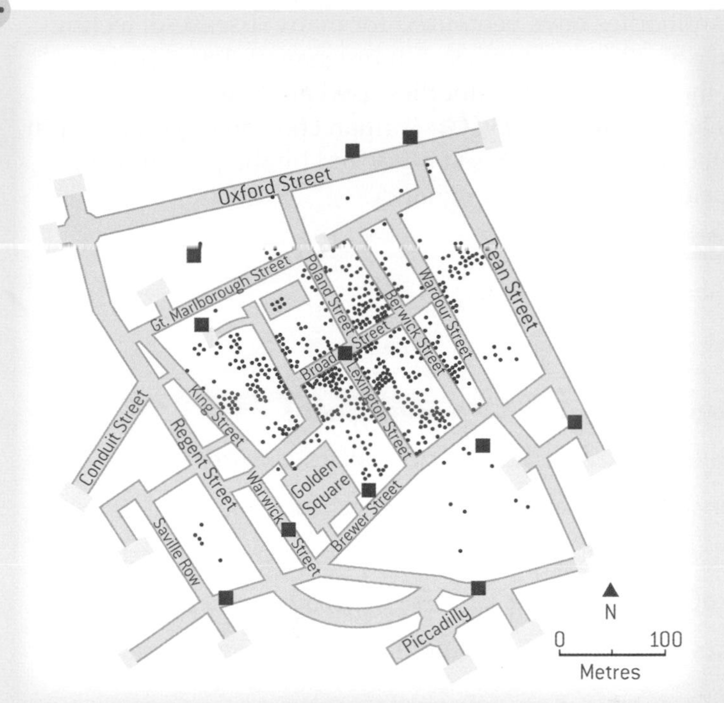

a) State **two** streets with relatively higher numbers of cholera cases.

b) Outline how the information in this map might have helped to stop the cholera outbreak.

Solution

a) Broad Street has the most cases. In second place it could be Lexington Street.

b) The map was a good way to represent the epidemiological information gathered by Snow. It clearly identifies the areas with more cholera cases. It shows that most cases were close to the water pump in Broad Street and Lexington Street. When Snow inactivated this water pump, the source of contamination was stopped, therefore eradicating the outbreak.

There are ethical implications of research. Jenner tested his vaccine for smallpox on a child. Nowadays it would be regarded as completely unethical to do so without testing it first on many animals. The World Health Organization initiated the campaign for the global eradication of smallpox in 1967. Vaccination policies that led to immunization at a global level and the fact that humans are the only reservoir of the smallpox virus (there are no other carriers) played an important role in the decline of the disease. Inadequate reporting of cases and the impossibility of funding the vaccination or reaching inaccessible areas of some countries slowed down the process. The campaign was deemed a success and smallpox was globally eradicated in 1977, only 10 years later.

- **Polyclonal antibodies** are a group of antibodies against many antigens.
- **Monoclonal antibodies** are specific against only one determined antigen.

Fusion of a tumour cell with an antibody-producing plasma cell creates a hybridoma cell which produces monoclonal antibodies.

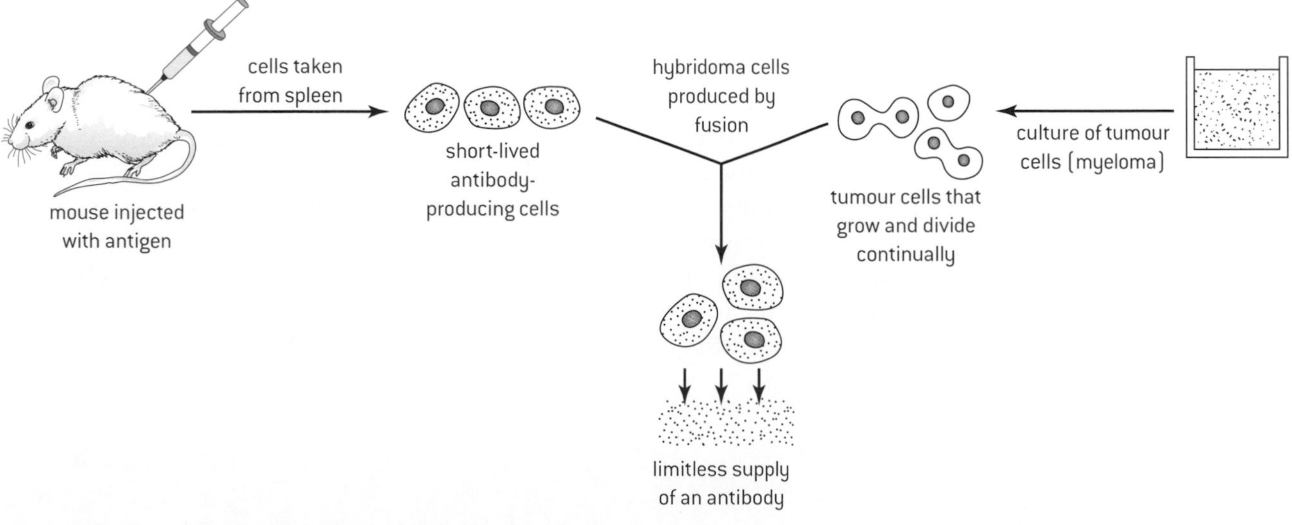

▲ Figure 11.1.2. Production of monoclonal antibodies

Monoclonal antibodies have been used for many disease-detecting kits. They have also been very useful in pregnancy-detection kits. In these kits, the monoclonal antibodies used are anti-human chorionic gonadotropin (anti-HCG). Human chorionic gonadotropin is a glycoprotein hormone that is synthesized by the placenta, and therefore appears only during pregnancy.

▲ This answer scored a mark for the following. In step 1, macrophage phagocytes engulf the pathogen. In step 2, macrophages then display the antigen from the pathogen and antigens of a pathogen correspond to a specific T helper cell. In step 3, each antibody corresponds to a specific antigen and lymphocytes are activated by antigen binding. In step 5, some activated B cells act as memory cells.

▼ The answer failed to say that antibodies are necessary for immunity or resistance to infectious disease. In step 2, it does not say that the macrophage digests the pathogen. In step 3, although it is correct that T cells activate B cells, it is not through hormones that this is done, but through peptides such as interferon and cytokines. The answer should have said that B cells divide by mitosis to form clones of plasma cells and it is these plasma cells that secrete specific antibodies.

SAMPLE STUDENT ANSWER

Explain the production of antibodies. [7]

This answer could have achieved 6/7 marks:

1. A pathogen is engulfed by a macrophage.
2. Macrophage presents the antigen on its surface and releases hormones to activate T helper cells.
3. The T helper cells bind to the antigen and thereby obtain the information as to what specific antibody is needed. It then releases hormones to activate B cells and passes on the information.
4. B cells start to divide by mitosis and produce and secrete antibodies.
5. Some B cells remain in the body as memory cells to be able to produce antibodies the next time this pathogen invades the body.

Assessment tip

It can be useful to use annotated diagrams to explain an answer.

Allergies occur when the immune system over-reacts to a specific allergy-causing antigen called an allergen. The most common allergens are pollen, dust and foods such as peanuts. In response to allergens, white blood cells release histamine which causes the allergic symptoms.

Histamines released by white blood cells (mast cells) that are found in the connective tissues increase the permeability of blood capillaries. This causes itching, rash and swelling that can eventually lead to death. Some foods are naturally high in histamines and should be avoided. These include fermented foods and alcohol (especially red wine).

Antigens on the surface of red blood cells stimulate antibody production in a person with a different blood group. This will determine the different blood donor possibilities. Blood is tested before a transfusion to avoid an incompatibility reaction. This reaction includes symptoms such as fever, chills, breathing difficulty, muscle ache, blood in urine due to kidney damage and jaundice due to liver damage. It can lead to death as blood clotting may occur throughout the body, shutting off the blood supply to the main organs.

Example 11.1.4.

Compare and contrast antigens and antibodies in different blood groups by completing the table.

Solution

	Blood type			
Feature	**A**	**B**	**AB**	**O**
Membrane antigens	A	B	AB	O
Plasma antibodies anti-blood group produced	anti-B	anti-A	none	anti-A and anti-B
Blood type that can be received in a transfusion	B and O	A and O	A, B, AB and O	only O

11.2 MOVEMENT

You should know:

- bones and exoskeletons provide anchorage for muscles and act as levers.
- synovial joints allow certain movements but not others.
- movement of the body requires muscles to work in antagonistic pairs.
- skeletal muscle fibres are multinucleate and contain specialized endoplasmic reticulum.
- muscle fibres contain many myofibrils, each made up of contractile sarcomeres.
- the contraction of the skeletal muscle is achieved by the sliding of actin and myosin filaments.
- ATP hydrolysis and cross bridge formation are necessary for the filaments to slide.
- calcium ions and the proteins tropomyosin and troponin control muscle contractions.

You should be able to:

- describe antagonistic pairs of muscles in an insect leg.
- annotate a diagram of the human elbow.
- draw labelled diagrams of the structure of a sarcomere.
- explain muscle contraction.
- analyse electron micrographs to find the state of contraction of muscle fibres.
- design experiments using grip strength data loggers to assess muscle fatigue.

The roles of the musculoskeletal system are movement, support and protection. Movement is achieved mainly by muscles pulling on bones, while bones provide most support and protection. Bones and exoskeletons provide anchorage for muscles and act as levers. Movement of the body requires muscles to work in antagonistic pairs.

In Topic 2.8 you studied cell respiration and ATP production and in Topic 1.2, the ultrastructure of cells.

- **Abduction** is the movement of a limb away from the midline of the body.
- **Adduction** is the movement of a limb towards the midline of the body.
- **Ligaments** attach bone to bone.
- **Tendons** attach muscle to bone.
- **Antagonistic muscles** produce opposing movement at a joint—when one contracts, the other relaxes.
- **Cartilage** prevents the friction of bones.
- **Synovial fluid** provides lubrication preventing friction.
- The **joint capsule** seals the joint and helps to prevent dislocation.

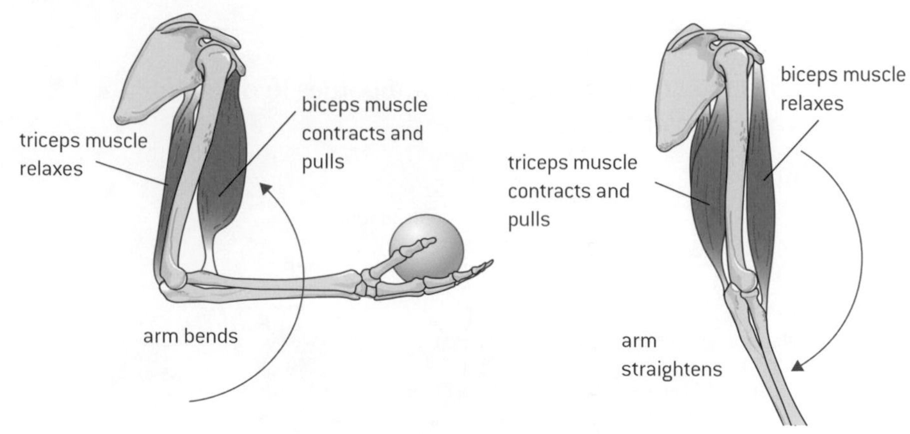

▲ Figure 11.2.1. Movement of the human arm

Synovial joints allow certain movements but not others. The hip joint, where the femur joins the pelvis, is a ball and socket joint. This type of joint allows movement in all directions including rotation, flexion (bending) and extension, abduction and adduction. The elbow allows less movement because it is a hinge joint, allowing flexion and extension only.

SAMPLE STUDENT ANSWER

The diagram shows a human elbow.

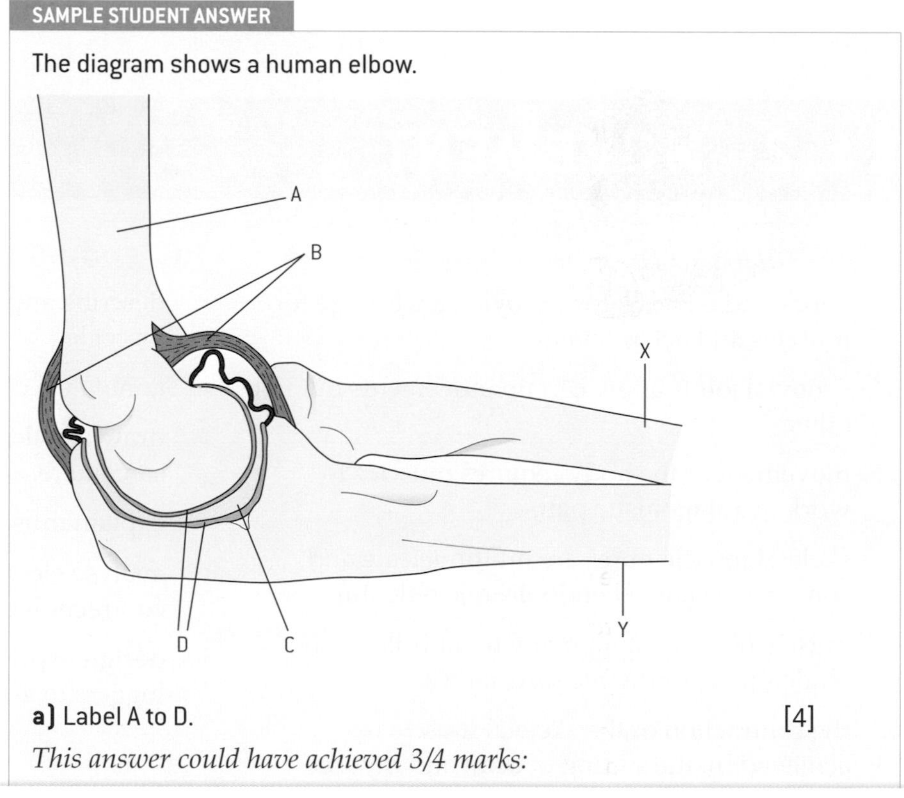

a) Label A to D. [4]

This answer could have achieved 3/4 marks:

A: bone, B: ligament, C: synovial fluid and D: cartilage.

▼ Label A is correct, but too vague. The correct answer was the humerus.

b) State the function of C in this elbow. [1]

This answer could have achieved 1/1 marks:

C lubricates the joint and prevents friction.

c) Explain the function of bones X and Y. [4]

This answer could have achieved 4/4 marks:

X is the radius and Y the ulna. The radius is attached to the biceps muscle through tendons while the ulna is attached to the triceps. These muscles are antagonistic. When the biceps contracts, the lower arm lifts; when the triceps contracts and the biceps relaxes the lower arm moves down. These bones act together with the humerus as levers. This is a third class lever—using this lever, less force is required to lift objects.

▲ The answer explains the function and anchorage of muscles and the function of the system as a lever.

Muscles contain many fibres, which are formed by many myofibrils. Myofibrils are organelles composed of units called sarcomeres containing actin and myosin bound together. The myofibrils are secured at either end to the cell membrane.

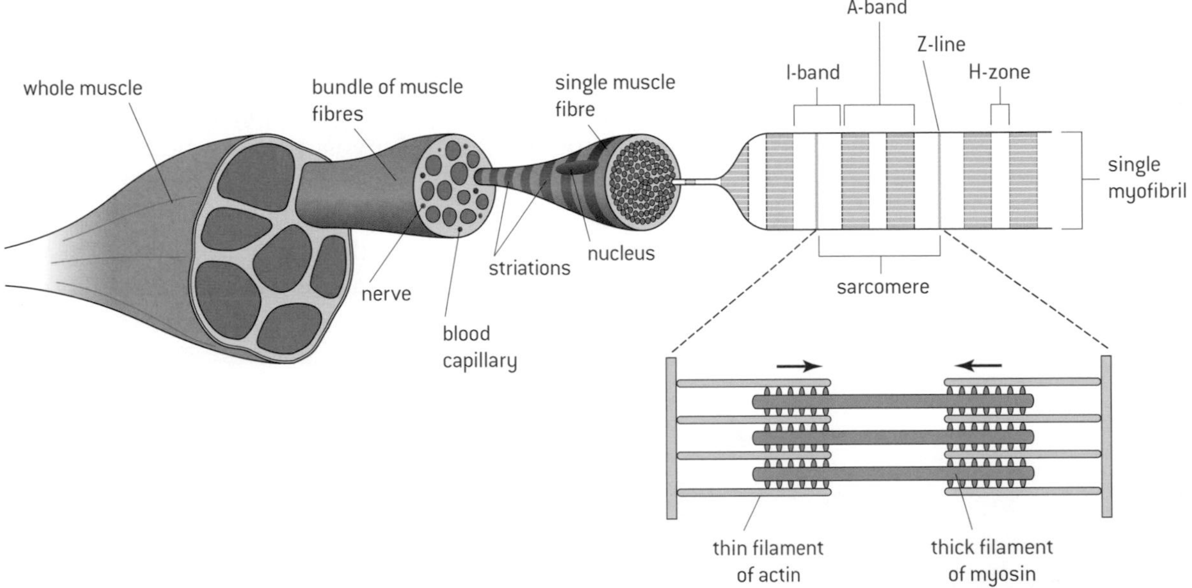

▲ Figure 11.2.2. Structure of muscles

Muscle cells are suited for their function. Skeletal muscle fibres are multinucleate with the nuclei displaced to the periphery of the muscle cell. They also contain a specialized endoplasmic reticulum, which stores and releases calcium ions to regulate muscle contraction. Muscle fibres contain many myofibrils, each made up of contractile sarcomeres. The light micrograph in Figure 11.2.3 shows a longitudinal section through a myofibril (×900). A number of repeating units that alternate between light and dark bands are visible. These are the sarcomeres.

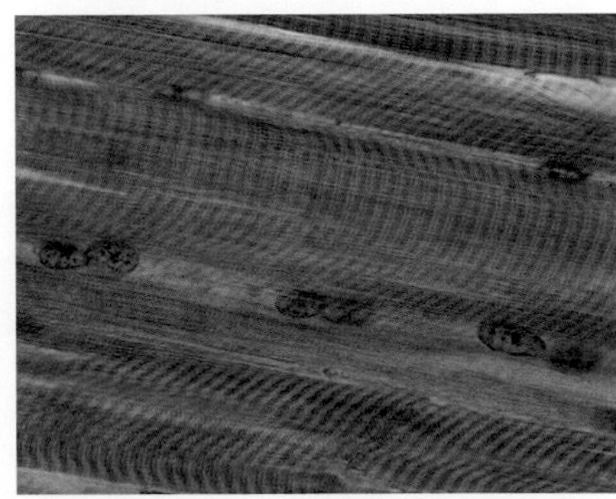

▲ Figure 11.2.3. Light microscope image of a myofibril

Example 11.2.1.

Annotate diagrams of the structure of a sarcomere showing the differences when relaxed and contracted.

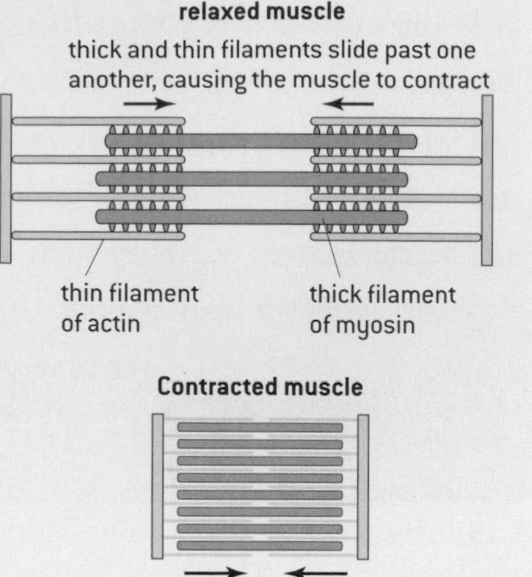

Solution

In a relaxed muscle, the Z-lines of the sarcomere are further apart, the light bands wider and the sarcomere longer. When the muscle contracts, the myosin filaments pull the actin filaments inwards, towards the centre of the sarcomere, which becomes shorter. The dark bands in A become closer and the light band H becomes much smaller.

The diagram below shows that troponin (T) is bound to tropomyosin, preventing the myosin head from forming a cross bridge with the actin filament. The propagation of a nerve action potential releases calcium ions (purple) into the muscle cell. The calcium ion binds to troponin, changing its shape and exposing the myosin binding site on the actin filament. The binding of actin and myosin causes the muscle to contract. The contraction of skeletal muscle is achieved by the sliding of actin and myosin filaments. The myosin then binds ATP, which allows it to release the actin and return to the starting position. This movement requires the hydrolysis of ATP and cross bridge formation for the filaments to slide.

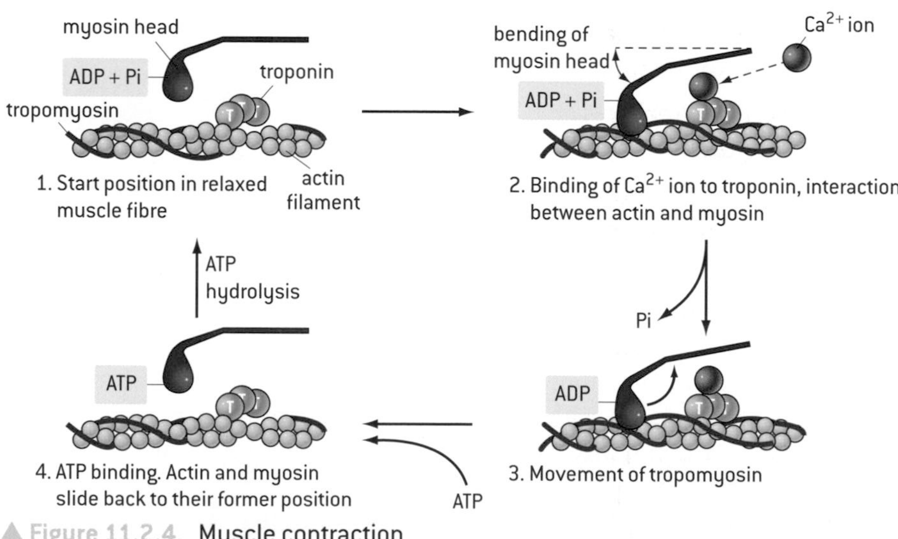

▲ Figure 11.2.4. Muscle contraction

Example 11.2.2.

Fluorescent calcium ions have been used to study muscle contraction. The fluorescence decomposition micrograph shows the calcium levels in a muscle cell (red being the highest) before and while contracting.

Explain how the increased level of calcium ions triggers the contraction of muscle cells.

Solution

When the nerve action potential reaches the muscle end plate, it induces the release of calcium ions into the muscle cell. The calcium ion binds to troponin, changing its shape and exposing the myosin binding site on the actin filament. The binding of actin and myosin causes the muscle to contract.

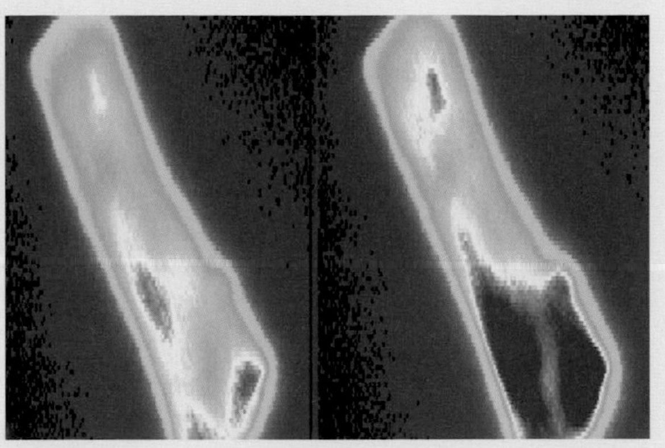

11.3 THE KIDNEY AND OSMOREGULATION

You should know:

- animals are either osmoregulators or osmoconformers.
- the Malpighian tubule system in insects and the kidney carry out osmoregulation and removal of nitrogenous wastes.
- the composition of blood in the renal artery is different from that in the renal vein.
- the ultrastructure of the glomerulus and Bowman's capsule facilitate ultrafiltration.
- the proximal convoluted tubule selectively reabsorbs useful substances by active transport.
- the loop of Henlé maintains hypertonic conditions in the medulla.
- ADH controls reabsorption of water in the collecting duct.
- the length of the loop of Henlé is positively correlated with the need for water conservation in animals.
- the type of nitrogenous waste in animals is correlated with evolutionary history and habitat.

You should be able to:

- compare and contrast osmoregulator and osmoconformer animals.
- describe the structures of kidneys and Malpighian tubes.
- draw an annotated diagram of the human kidney.
- state the functions of the different parts of the kidney.
- annotate diagrams of the nephron.
- explain osmoregulation in humans and describe the consequences of dehydration and overhydration.
- explain the treatment of kidney failure by hemodialysis or kidney transplant.
- state that blood cells, glucose, proteins and drugs are detected in urinary tests.

All animals excrete nitrogenous waste products and some animals also balance water and solute concentrations. Nitrogenous wastes are produced from the breakdown of amino acids and nucleic acids. The Malpighian tubule system in insects and the kidneys in humans carry out osmoregulation and removal of nitrogenous wastes. Not all animals are osmoregulators; some are osmoconformers.

In Topic 1.3 you studied the membrane structure and in Topic 1.4, membrane transport.

- **Osmoconformers** are organisms that change according to the external environment.
- **Osmoregulators** maintain their internal environment regardless of the external environment.
- **Kidneys** are the organs in humans used to carry out osmoregulation and removal of nitrogenous wastes.
- **Malpighian tubules** are excretory organs in arthropods (insects).

The Malpighian tubules are a series of branched tubes that reach the alimentary canal in most arthropods. They absorb water and wastes from the body fluids by diffusion and mineral ions by active transport. These substances are released from the Malpighian tubules into the alimentary canal. Mineral ions are reabsorbed by the rectum by active transport and water is reabsorbed by osmosis. The wastes are excreted as nitrogenous wastes together with undigested food.

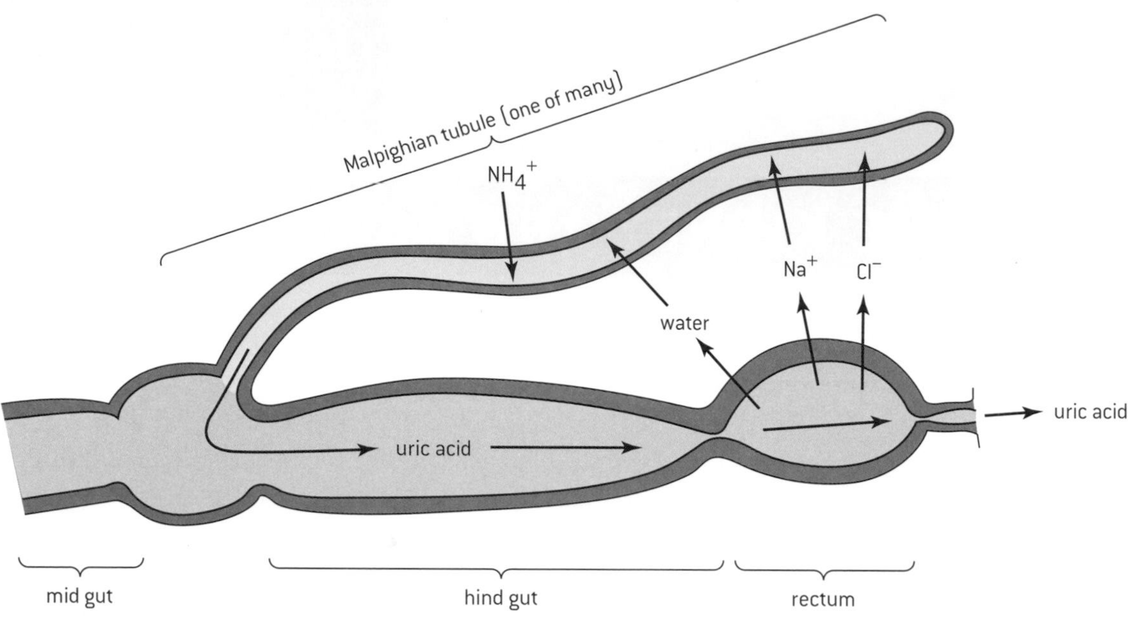

▲ Figure 11.3.1. Malpighian tubule

- The **glomerulus** is the high-pressure capillary causing ultrafiltration in the Bowman's capsule.
- The **Bowman's capsule** is where glomerular filtrate collects to be further processed.
- The **proximal convoluted tubule** selectively reabsorbs useful substances by active transport. It contains many mitochondria and microvilli for absorption.
- The **loop of Henlé** maintains hypertonic conditions in the medulla. Animals adapted to dry habitats have long loops of Henlé.
- The **distal convoluted tubule** regulates mineral ion levels and maintains pH.
- The **collecting duct** is the site where the amount of water reabsorption can be varied as a part of osmoregulation.
- **ADH** is the antidiuretic hormone that controls water reabsorption.

Example 11.3.1.

The diagram shows a human kidney.

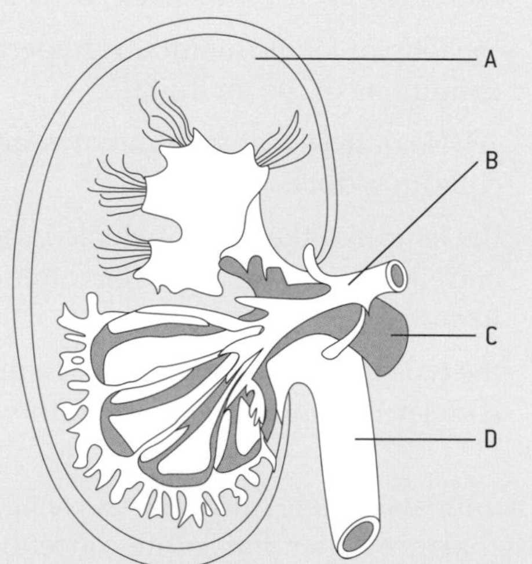

a) Which label shows the cortex?

b) Which label shows the blood vessel carrying blood into the glomerulus?

c) What is the pelvis of the kidney?

Solution

a) The cortex is **A**.

b) The blood vessel carrying blood to the glomerulus is **B**, the renal artery. It can be distinguished from the renal vein (**C**) by the width of the vessel. **D** is the ureter, where all the collecting ducts join.

c) The pelvis is the funnel-like structure leading to the ureter.

SAMPLE STUDENT ANSWER

Explain how the structure of the nephron and its associated blood vessels enable the kidney to carry out its functions. [8]

This answer could have achieved 7/8 marks:

In the glomerulus of the nephron, the afferent arteriole is much larger than the efferent arteriole. This creates large amounts of pressure, forcing out the urea, glucose, water and mineral ions from the blood but leaving the proteins in the blood vessels. The rest enters the Bowman's capsule, which is lined by podocytes, which wrap capillaries. This is ultrafiltration. The filtrate moves to the proximal convoluted tubule for reabsorption. Here, the glucose and ions are reabsorbed by active transport. The cells have mitochondria to help with this. Water flows by osmosis. The large surface area increases the reabsorption in the proximal convoluted tubule. The fluid then moves to the loop of Henlé.

-descending loop is permeable to water but not to ions.

-ascending loop is permeable to ions but not to water.

This structure allows for a high concentration of salt ions in the medulla. The longer the loop of Henlé, the more water is conserved.

The fluid then moves to the distal convoluted tubule, where more is reabsorbed and to the collecting duct, where water is reabsorbed if ADH is present.

Blood vessels surround the nephron to allow reabsorption to the bloodstream. Glucose is reabsorbed into the blood capillaries as well as ions and water, but urea is removed as urine.

The nephron is in the kidney and functions to filter the blood. Desert animals have long loops of Henlé to conserve water.

▲ The answer scored marks for saying that high blood pressure in the glomerulus is due to a larger arteriole bringing blood than the arteriole taking away blood. One mark was given for the (selective) reabsorption of glucose and useful substances such as mineral ions in the proximal convoluted tubule. Explaining that the selective reabsorption of glucose and useful substances occurs in the proximal convoluted tubule, which has a large surface area, scored two marks. Another mark was given for water reabsorbed in the descending limb of loop of Henlé and one for the ascending limb being impermeable to water. One mark was for the loop of Henlé creating a high solute concentration in the medulla. Two marks were for water reabsorbed in collecting duct, which has microvilli to give a large surface area.

▼ Marks could have been obtained for saying that a function of the kidney is osmoregulation or excretion of nitrogenous waste such as urea. Although the candidate does mention ultrafiltration, the answer does not mention that capillary walls in the glomerulus are permeable to smaller molecules and that the basement membrane with podocytes of the Bowman's capsule act as a filter thus preventing the loss of large molecules, such as protein, and blood cells. The answer failed to mention the active transport or active pumping of sodium ions out of the ascending limb from the filtrate to the medulla. The fact that the distal convoluted tubule adjusts the pH and concentration of Na^+, K^+ and H^+ is not mentioned. That the collecting duct permeability to water varies with the number of aquaporins and amount of ADH or osmoregulation by varying the amount of water reabsorbed in the collecting duct could have been mentioned for a mark.

Assessment tip

No marks are given for ultrafiltration in the Bowman's capsule because this answer is too vague.

Assessment tip

Although you may find vasopressin mentioned in books, you should use the term ADH in the exam rather than vasopressin.

ADH or antidiuretic hormone is a peptide hormone produced by the hypothalamus and stored in the pituitary gland. It controls the reabsorption of water by changing the permeability of the collecting ducts. When ADH is present, the collecting ducts become permeable to water and more is reabsorbed into blood, therefore the urine becomes less concentrated and less urine is produced. Aquaporins are integral membrane proteins that serve as channels in the transfer of water across the membrane. They are expressed in the kidney collecting ducts, where they move from intracellular sites to the plasma membrane under the control of ADH.

The length of the loop of Henlé positively correlates with the need for water conservation in animals, as does the amount of ADH produced. The type of nitrogenous waste animals excrete is correlated with their evolutionary history and habitat. The main nitrogenous wastes excreted by animals are ammonia, uric acid and urea. Ammonia is highly toxic and soluble in water. Uric acid is less toxic but insoluble in water, while urea is also less toxic but it is soluble in water. Aquatic animals excrete ammonia directly into the water, where it is safely diluted as soon as it is excreted. Birds excrete uric acid in the form of a paste, as this allows them to retain water. Terrestrial animals and marine mammals convert ammonia into urea which they excrete dissolved in water, as urine.

Desert animals prevent water loss in a number of ways. Some live in burrows, some have thick outer coverings to reduce moisture loss and some have lower resting metabolic rates or lower body masses to avoid heat production and water loss. Some produce small amounts of highly concentrated urine by modulating the amount of ADH.

- **Ammonia** is a toxic nitrogenous waste that is soluble in water.
- **Uric acid** is a less toxic nitrogenous waste that is insoluble in water.
- **Urea** is a less toxic nitrogenous waste that is soluble in water.
- **Dehydration** is loss of total body water. It can be caused by lack of drinking or by excess diarrhoea or vomiting, fever, excessive sweating or increased urination. It can lead to heart injury, kidney problems, seizures and low blood volume.
- **Overhydration** is excess of water in the body, usually caused by over drinking. The symptoms are swelling of cells by excess osmosis.
- **Hemodialysis** is the extracorporeal purification of blood.

Blood cells, glucose, proteins and drugs are detected in urinary tests. Doctors can monitor how the kidneys are working according to the substances that are present in the urine. In the event of kidney failure, hemodialysis is used to maintain osmoregulation. In hemodialysis, the blood is taken out of the patient and in the dialysis machine it is filtered through a selectively permeable membrane that takes out waste matter, keeping useful substances in the blood. The purified blood is then returned to the patient. Kidneys can be transplanted from a living or deceased donor to replace a malfunctioning kidney. There are risks with the procedure that the transplanted kidney could be rejected by the patient's immune system, but the advantages are not having to go through hours of inconvenient dialysis every few days and also that it can greatly increase life expectancy.

11.4 SEXUAL REPRODUCTION

You should know:

- ✔ spermatogenesis and oogenesis both involve mitosis, cell growth, two divisions of meiosis and differentiation.
- ✔ processes in spermatogenesis and oogenesis result in different numbers of gametes with different amounts of cytoplasm.
- ✔ fertilization in animals can be internal or external.
- ✔ fertilization involves mechanisms that prevent polyspermy.
- ✔ implantation of the blastocyst in the endometrium is essential for the continuation of pregnancy.
- ✔ HCG stimulates the ovaries to secrete progesterone during early pregnancy.
- ✔ the placenta facilitates the exchange of materials between the mother and fetus.
- ✔ estrogen and progesterone are secreted by the placenta once it has formed.
- ✔ birth is mediated by positive feedback involving estrogen and oxytocin.

You should be able to:

- ✔ explain the formation of gametes in spermatogenesis and oogenesis.
- ✔ annotate diagrams of the seminiferous tubule and ovary to show the stages of gametogenesis.
- ✔ annotate diagrams of a mature sperm and egg to indicate their functions.
- ✔ describe internal and external fertilization.
- ✔ explain implantation in mammals.
- ✔ explain pregnancy and the hormones involved.
- ✔ explain birth and the hormones involved.
- ✔ analyse graphs showing the correlation between animal size and the development of the young at birth in mammals.

Sexual reproduction involves the development and fusion of haploid gametes. Gametes therefore need to undergo a reduction in the number of chromosomes through meiosis. Spermatogenesis is the development of the male gametes (spermatozoa or sperms) and occurs in the seminiferous tubules of the testes. Oogenesis is the formation of the female gamete (ovum) and occurs in the ovaries. Spermatogenesis and oogenesis both involve mitosis, cell growth, two divisions of meiosis and differentiation.

In Topic 3.3 you studied meiosis and in Topic 6.6, hormones, homeostasis and reproduction.

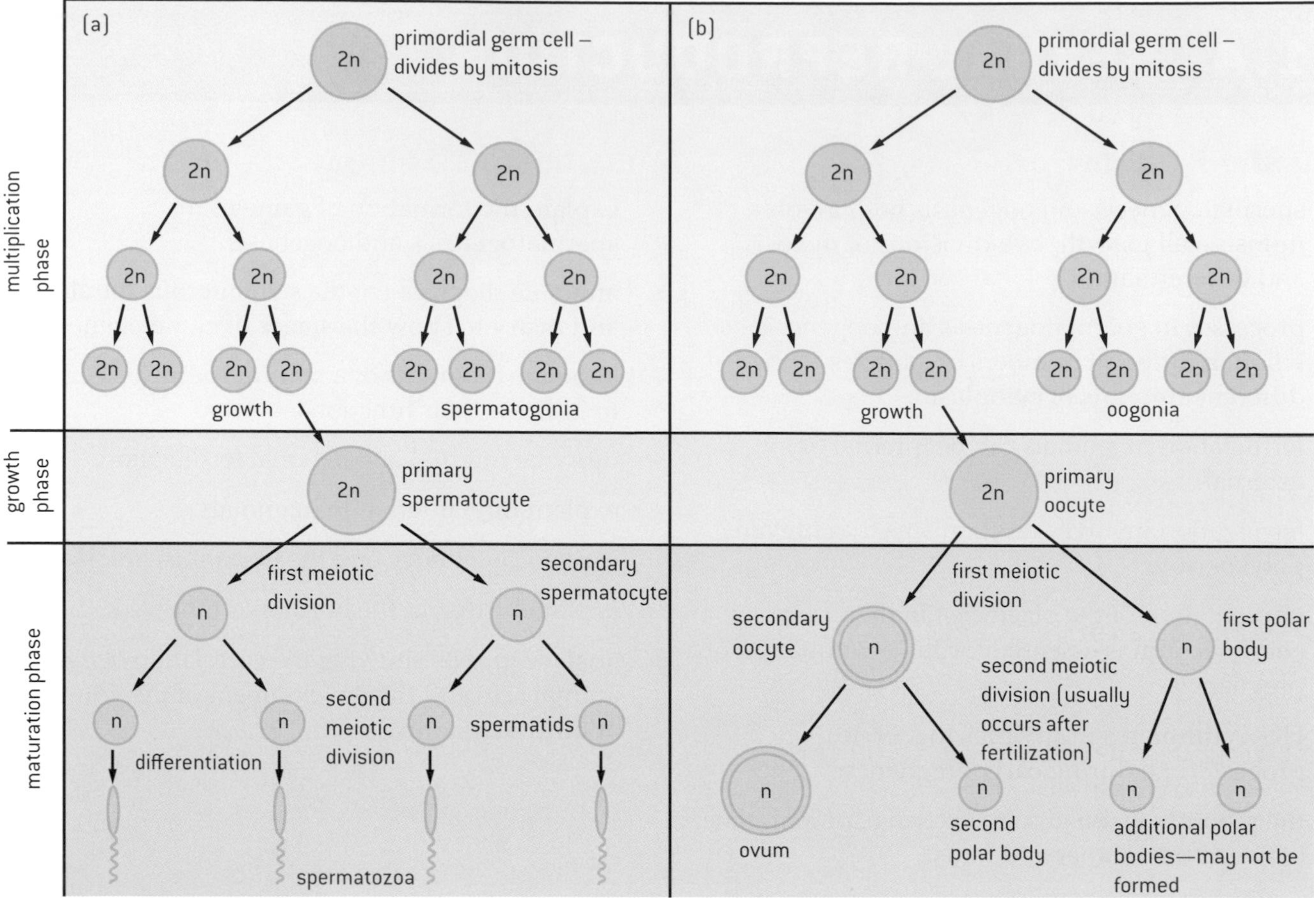

▲ Figure 11.4.1. (a) Spermatogenesis and (b) oogenesis

Assessment tip

It is always good to use a diagram to explain an answer, but in this case it did not get more marks as it was not annotated to add more information than that given in the text, therefore it was a waste of precious time.

SAMPLE STUDENT ANSWER

Describe the different cell types in the seminiferous tubules that are involved in the process of spermatogenesis. [4]

This answer could have achieved 2/4 marks:

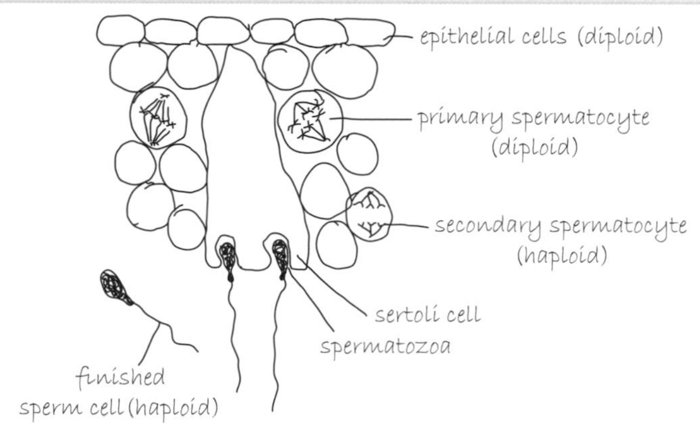

The cells which are furthest inside are the most mature.
A primary spermatocyte undergoes the first cell division in meiosis I.
Secondary spermatocytes are undergoing the second cell division to become haploid.
Sertoli cells nurse the spermatozoa till they develop flagella.
They can detach and swim into the epididymis, where they are stored.

▲ The answer scored a mark for primary spermatocytes dividing by meiosis I into secondary spermatocytes. The second mark is for Sertoli or nurse cells providing nourishment to these developing cells.

▼ The answer could have mentioned that spermatogonia (2n) are undifferentiated germ cells. The spermatogonia mature and divide by mitosis into primary spermatocytes. The answer did not score a mark for secondary spermatocytes dividing into spermatids because it does not mention that it is by meiosis II. The answer does not mention that spermatids differentiate or mature into sperm nor that Leydig or interstitial cells produce testosterone.

Fertilization in animals can be internal or external. Aquatic animals often use external fertilization. Usually many eggs are fertilized because there is a great risk of loss by predation. In terrestrial animals and marine mammals, fertilization is usually internal. This increases the likelihood of success, so fewer organisms need to be produced. In humans, fertilization involves the acrosome reaction, fusion of the plasma membrane of the egg and sperm and the cortical reaction. The process of fertilization in humans occurs in the oviduct (fallopian tube).

- **Fertilization** is the joining of the ovum and the nucleus of the sperm.
- An **acrosome reaction** is the release of antigens and enzymes for binding to the membrane and digesting through the ovum's zona pellucida.
- A **cortical reaction** is the release of cortical granules to prevent any additional sperm from fertilizing the egg, avoiding polyspermy.

Example 11.4.1.

The diagram shows an ovum.

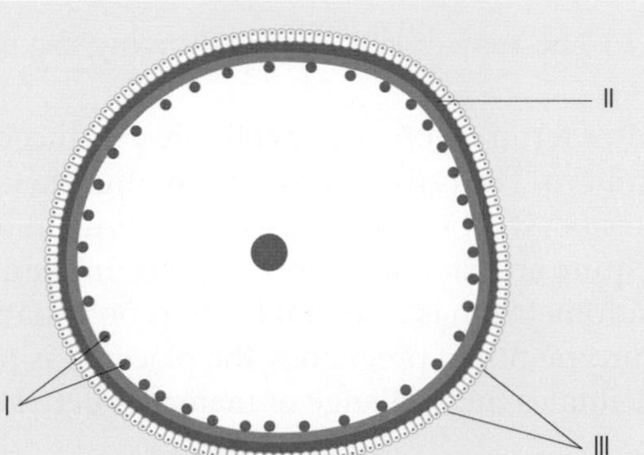

a) Label I to III.

b) Outline one function of I.

c) State the organ in which the ovum is formed.

d) Explain the process by which the diploid germ cell becomes a haploid ovum.

Solution

a) I are cortical granules, II is the zona pellucida (glycoprotein layer) and III are the follicle cells (corona radiata).

b) When fertilization of the ovum by the sperm occurs, the cortical granules are activated. These vesicles release their contents by exocytosis, producing the cortical reaction. In this reaction, the enzymes of the cortical granules produce changes that harden the zona pellucida, not allowing more sperm to enter the ovum, preventing polyspermy.

c) Ovary.

d) The process by which the diploid germ cell becomes a haploid ovum is called oogenesis. This process occurs in the ovary. The germ cells in the ovary of the fetus divide by mitosis (multiplication phase). Around the fourth month of development, the cells start dividing by meiosis. Just before or shortly after birth, the oogonia differentiate into primary oocytes (growth phase). These primary oocytes enter into prophase of meiosis I and primary follicles are formed. Each primary follicle consists of one ovum and follicle cells around them. No further primary oocytes are produced. At the start of each menstrual cycle, a primary follicle is stimulated by FSH to mature into a secondary oocyte by completing meiosis I (maturation phase). Of the two cells produced, only one transforms into a secondary oocyte and survives; the other cell turns into a smaller cell called a polar body which degenerates. The secondary oocyte halts meiosis II in metaphase II. Upon fertilization, the secondary oocyte will finish the second meiotic division. When the secondary oocyte undergoes meiosis II, one cell becomes the ovum and the other another polar body. The result is one ovum which is much larger than the male sperm.

The risks to human male fertility were not adequately assessed before steroids related to progesterone and estrogen were released into the environment as a result of the use of the female contraceptive pill. Studies have shown that males have a lower sperm count than expected, and this could be attributed to the presence of estrogen in water systems.

Example 11.4.2.

The diagram shows the structure of a sperm.

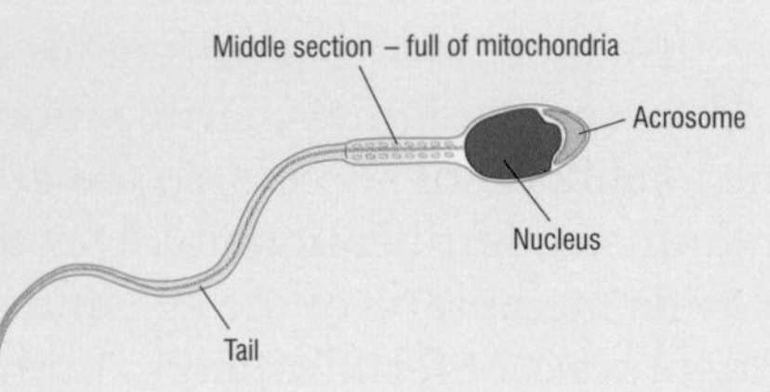

a) Which part is involved in the acrosome reaction?

b) Which part provides energy for movement?

Solution

a) The answer is **D**, as it is showing the acrosome.

b) The answer is **B**, as it is showing the mitochondria.

After fertilization, many cells are produced by mitosis, forming the embryo. The cells in the embryo migrate to form a hollow ball called a blastocyst. The cells that were feeding on nutrients from the ovum now require an external supply of food. Implantation of the blastocyst in the endometrium is essential for the continuation of pregnancy. After the third month of pregnancy the placenta is fully formed. The placenta facilitates the exchange of materials between the mother and fetus.

The pregnancy is maintained by hormonal control. HCG stimulates the ovary to secrete progesterone during early pregnancy, which helps maintain the endometrium. Estrogen and progesterone are secreted by the placenta once it has formed. Progesterone inhibits the secretion of oxytocin by the pituitary gland, thus inhibiting contractions of the uterus during pregnancy. Birth is mediated by positive feedback involving estrogen and oxytocin. Hormones produced by the fetus inhibit the secretion of progesterone. Oxytocin induces the wall of the uterus to contract. The contractions are perceived by sensory neurons that send messages to the pituitary gland to produce more oxytocin through positive feedback. As the cervix dilates and the amniotic sac is broken, birth will take place as the baby is pushed out through the cervix and vagina.

Example 11.4.3.

The graph shows the relationship between offspring body mass at birth and gestational period in some mammals.

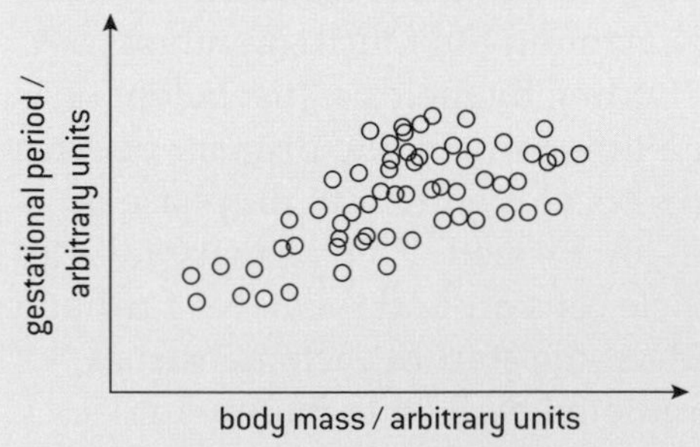

What does this graph show?

A. Body mass and gestational period are not related.

B. The longer the gestational period, the smaller the size of the baby.

C. There is a correlation between body mass and the development of the young at birth in mammals.

D. There is a correlation between gestational period and the number of offspring per gestation.

Solution

The correct answer is **C**. It can be seen in the graph that a larger body mass at birth means a longer gestational period for that organism. A longer gestational period is correlated with a greater development of the young at birth. Answer **A** is not correct, as it can be seen that in smaller organisms there is no correlation, but using all the data, one could draw an increasing trendline. Answer **B** is the opposite of what is shown in the graph. Answer **D** is incorrect because not only is this not shown in the graph, but usually the opposite occurs.

Practice problems for Topic 11

Problem 1

A blood test was performed in a laboratory to test blood groups. The picture shows some results obtained for different blood groups with different antibodies. Where there are open circles the results are not shown.

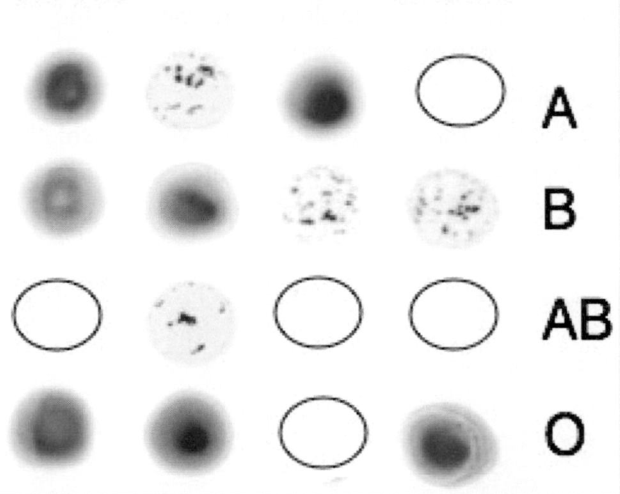

a) Determine the results that would be expected when mixing the given antibodies and blood groups by completing the table.

Antibody				Blood group
anti-O	**anti-A**	**anti-B**	**anti-AB**	
no reaction	reaction	no reaction		A
no reaction	no reaction	reaction	reaction	B
	reaction			AB
no reaction	no reaction		no reaction	O

b) Explain the reason this test needs to be performed before a blood transfusion.

Problem 2

The diagram shows some of the muscles in the leg of an insect.

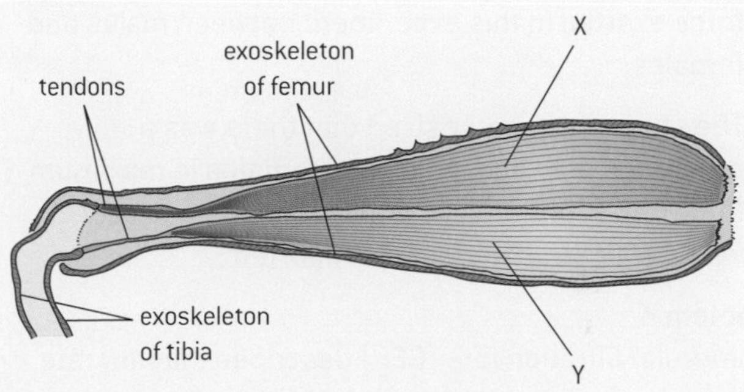

a) State the function of the exoskeleton in movement.

b) **(i)** Label muscles X and Y.

(ii) Muscles X and Y are said to be antagonistic. Explain the term antagonistic using these muscles as an example.

c) Describe how the contraction of a muscle is achieved.

Problem 3

An experiment using a grip strength data logger to assess muscle fatigue was performed. The maximum force exerted and the endurance after many contractions was measured in females and males.

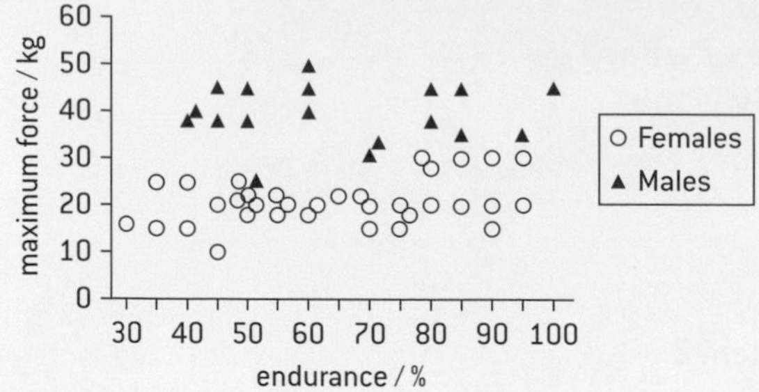

a) Calculate the difference in the highest maximum force exerted in this experiment between males and females.

b) The students hypothesized that there was a difference between males and females in maximum force and in endurance. Analyse the data to see whether the hypothesis is supported.

Problem 4

Glomerular filtration rate (GFR) describes the flow rate of fluid that is filtered through the kidney. One method of determining GFR is to determine the amount of creatinine in blood plasma. The graph represents the relationship between plasma creatinine concentration and GFR.

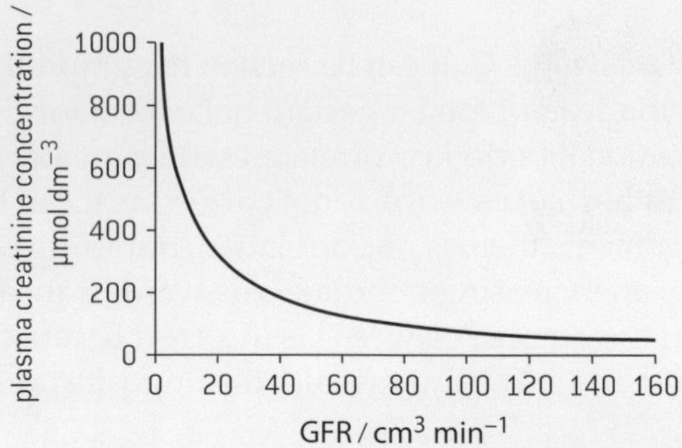

a) State the plasma creatinine concentration at a GFR of 40 $cm^3\ min^{-1}$.

b) Describe the relationship between GFR and plasma creatinine concentration.

c) Explain what could be happening to a patient if the levels of creatinine in blood plasma were high.

Problem 5

Compare and contrast the blood composition in the renal artery and the renal vein.

Problem 6

Describe the exchange of materials between the mother and the fetus in the placenta.

12 DATA-BASED AND PRACTICAL QUESTIONS (SECTION A)

You should know:

- ✔ experimental data can be obtained using a variety of biological techniques.
- ✔ quantitative data are commonly recorded and communicated in the form of a table or graphs.
- ✔ graphical techniques are an effective way of representing qualitative and quantitative data.
- ✔ quantitative data are always associated with random errors caused by the apparatus and human limitations.
- ✔ experimental procedures often lead to systematic errors in measurement.
- ✔ the experimental set-up affects the precision, accuracy and validity of the data obtained.

You should be able to:

- ✔ interpret and analyse experimental data presented in tabular and graphical form.
- ✔ identify dependent and independent variables and their relationships.
- ✔ identify qualitative trends and sketch graphs using correctly labelled axes.
- ✔ estimate unknown quantities and their uncertainties from plots and diagrams.
- ✔ discuss the sources of errors in a laboratory experiment and estimate their impact on the experimental results.
- ✔ suggest how the accuracy and precision of a particular experiment can be improved.
- ✔ suggest and evaluate hypotheses using experimental evidence.
- ✔ analyse historical experiments producing well known biological facts.

In Section A of Paper 3, you will be asked two or three questions on general experimental skills or techniques that all students should know how to carry out in the laboratory, whether in reality or virtually. It is important to stress accurate measurements and controls that must be maintained in any experiments. Questions can also include the theoretical background to these experiments. Reference to actual or past models can also be included.

There are several specific experiments you may be assessed on in Section A of Paper 3, along with other experimental skills from the IB guide. *It is important to remember that these are **not** the only experiments that are going to be tested.*

The specified practicals are:

1. Use of a light microscope to investigate the structure of cells and tissues, with drawing of cells, along with calculation of the magnification of drawings and the actual size of structures shown in drawings or micrographs.
2. Estimation of osmolarity in tissues by bathing samples in hypotonic and hypertonic solutions.
3. Experimental investigation of a factor affecting enzyme activity.
4. Separation of photosynthetic pigments by chromatography.

5. Setting up sealed mesocosms to try to establish sustainability.
6. Monitoring of ventilation in humans at rest and after mild and vigorous exercise.
7. (HL only) Measurement of transpiration rates using potometers.

There are many experiments and models that can be tested in Section A of Paper 3.

Example 12.1.

The electron micrograph shows mesophyll cells in a young leaf of *Zinnia elegans*.

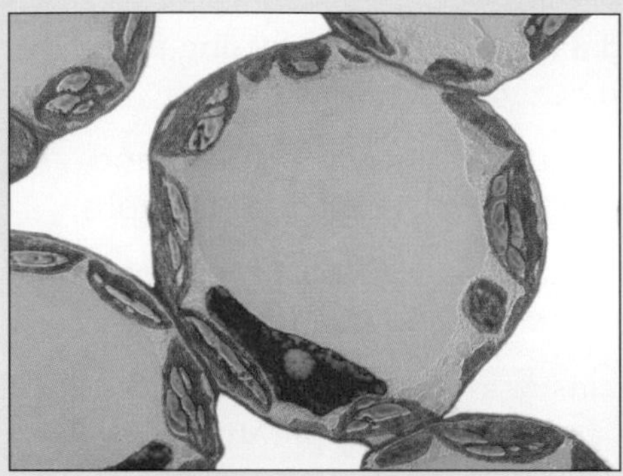

a) The actual diameter of the central mesophyll cell is 20 μm. Calculate the magnification of the electron micrograph. Show your working.

b) Identify **two** structures seen in this cell that cannot be seen in an animal cell.

c) Outline an advantage and a disadvantage of using an electron microscope over a light microscope.

Solution

a) You first measure the diameter of the central mesophyll cell using a ruler.

For example, if the diameter measures 4 cm, you first have to change this to micrometres. One centimetre is equal to 10,000 micrometres, thus 4 cm is 40,000 μm.

You then calculate the magnification using the formula:

$$M = \frac{\text{image size}}{\text{actual size}}$$

$$M = \frac{40{,}000\ \text{m}}{20\ \text{m}} = 2{,}000$$

The magnification of the electron micrograph is × 2,000.

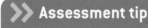

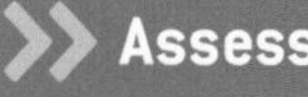 **Assessment tip**

Remember this is a practical question, therefore the answer must include structures that can be seen at this magnification.

a) Chloroplasts, cellulose cell wall, starch granules and large vacuole. (Only two needed.)

 **Assessment tip**

Only one advantage and one disadvantage are required, so do not spend time giving more than one.

b) An advantage of the electron microscope over the light microscope is that it has a greater magnification and a higher resolution, allowing the observation of ultrastructures. A disadvantage is that the cells need to be killed in the preparation of the slides and the artefacts resulting from specimen preparation. Another important disadvantage is the cost, size and training required to use this type of microscope compare to the light microscope.

Example 12.2.

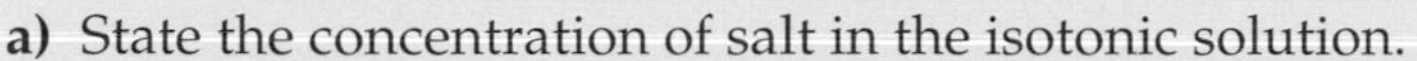

The graph shows the change in length of cytoplasm in cells placed in salt solutions of different concentrations compared with the mean length of the cells in an isotonic solution.

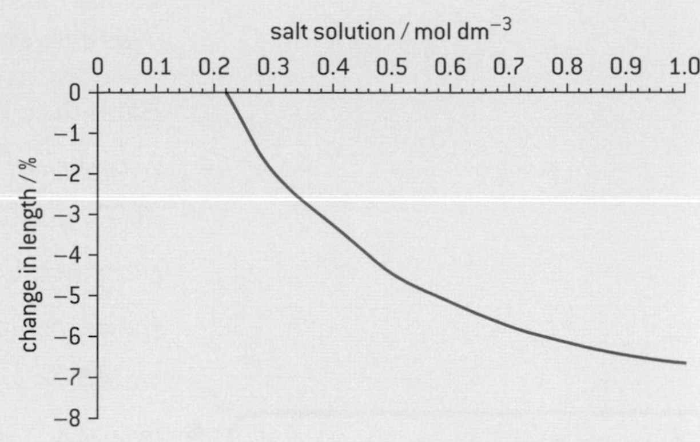

a) State the concentration of salt in the isotonic solution.

b) Determine the percentage change in length for cells placed in a 0.5 mol dm^{-3} salt solution.

c) Identify the dependent and independent variables in this experiment.

Solution

a) 0.22 mol dm^{-3}

b) –5%

c) The independent variable (what is varied) is the salt concentration and the dependent variable (what is measured) is the percentage change in length in the diameter of cells.

SAMPLE STUDENT ANSWER

A common reaction occurring in cells is the decomposition of hydrogen peroxide, which breaks down into oxygen (producing the foam) and water. Catalase (also known as hydrogen peroxide oxidoreductase) protects human cells from the toxic effects of hydrogen peroxide, which is a powerful oxidizer produced as a biochemical by-product.

Catalase is present in pieces of potato. In an experiment to see the effect of enzyme concentration on the rate of reaction, discs of potato were placed in hydrogen peroxide and the height of the foam was measured with a ruler. The first conical flask had 5 discs, the second 10 and the third 20 potato discs.

a) State **two** conditions of the **potato discs** that must be kept constant. [2]

This answer could have achieved 0/2 marks:

Number of potato discs in the replicates and time of the experiment.

▼ The experiment mentioned here does not refer to replicates, so this answer is not correct, as it is not answering the question. The duration of the experiment does not refer to the potato discs directly. To score a mark the answer had to mention that the surface area or volume of each disc must be the same. All the discs must be from the same potato. The discs must not be touching; they must have their surface in contact with the peroxide. (Only two conditions needed.)

b) State **two** conditions not related to the potato discs that need to be kept constant in this experiment. [2]

This answer could have achieved 1/2 marks:

The amount of light and the temperature.

▲ The temperature must be kept constant.

▼ The light will not affect this experiment directly, therefore has not been considered as correct. Other conditions that need to be constant are the volume and concentration of hydrogen peroxide. The volume and shape of the conical flasks need to be the same.

c) Suggest one reason a conical flask is not appropriate for this experiment. [1]

This answer could have achieved 1/1 marks:

The conical flask is not straight, so the volume is not constant all along the height. A test tube would have been more appropriate.

Assessment tip

In Example 12.2 part (b), it is important to either use the negative sign or mention decrease in length.

Assessment tip

Make sure the answer refers directly to the question.

d) Explain the differences in enzyme-catalysed reactions in the three conical flasks. [2]

This answer could have achieved 2/2 marks:

Each disc has an approximate amount of enzymes. If the number of discs is different, the amount of enzyme is different in the three flasks. The greater the amount of enzyme, the faster the rate of reaction, thus forming more foam. More enzymes means they have more chances of collision with the substrate (peroxide) and therefore more oxygen is formed, forming more foam.

▲ The student explains clearly why there are differences in the reactions of the conical flasks. It was important to realize that each disc represents an amount of enzyme and that increasing the number of discs means increasing the amount of enzyme.

e) Predict what would happen if the discs were boiled before the experiment. [1]

This answer could have achieved 0/1 marks:

Boiling the potato discs will kill the potato, so no reaction occurs.

▼ The answer is too vague. Boiling the potato discs would denature the catalase in them, causing the rate of reaction to be nil, so no foam will be formed.

Example 12.3.

Melvin Calvin, James Bassham and Andrew Benson discovered the light-independent reactions of photosynthesis. In an experiment similar to that performed by Calvin, green algae (*Chlorella*) were placed within a thin transparent container (the lollipop container) and radioactive carbon-14 was added to the apparatus. At 10-second intervals the samples of algae were dropped into hot methanol. Samples were analysed using two-dimensional paper chromatography, which separates out the different carbon compounds. The results of these experiments showed that carbon dioxide was converted to carbohydrates during the light-independent reactions of photosynthesis.

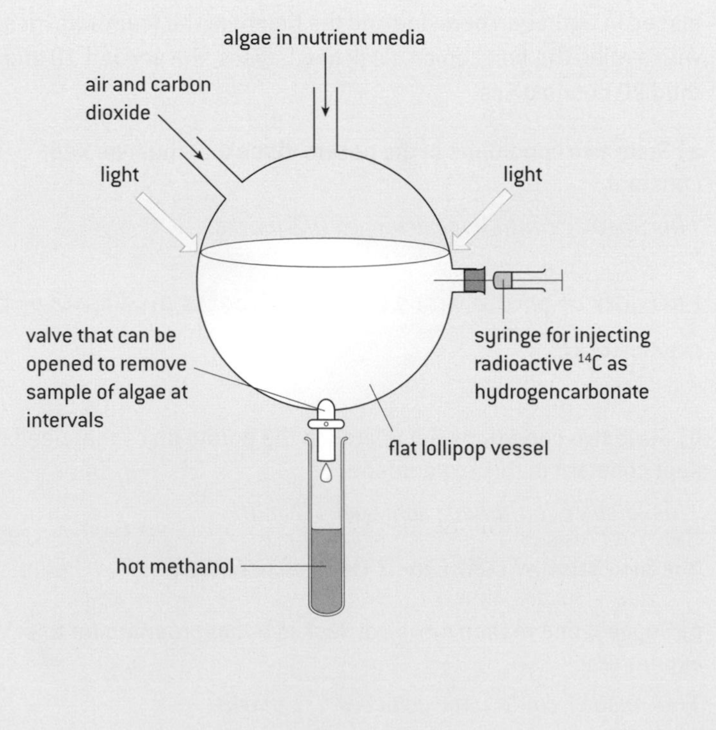

a) Determine a reason for the use of transparent and not dark glass in the lollipop vessel.

b) Suggest **one** reason for the use of hot methanol.

c) The substances found at different times were recorded in a table. Identify the substances detected at 10 seconds and at 30 seconds that are missing in the table.

Time / seconds	Radioactive molecules
0	carbon dioxide
10	
20	triose phosphate glycerate 3-phosphate
30	
40	glycerate 3-phosphate triose phosphate glucose ribulose bisphosphate

A mathematical model was made to provide a description of the effects of variations in CO_2 concentrations on the assimilation of carbon atoms at increasing rubisco enzyme concentrations. The dashed line indicates the average rubisco concentration in algae.

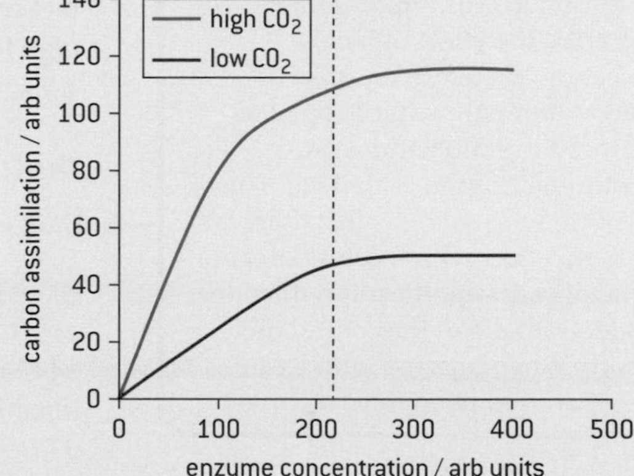

d) Experimental results show that rubisco is the limiting factor at low CO_2 levels, but not when the CO_2 concentration is high. Suggest with reasons whether this mathematical model matches the experimental results.

Solution

a) Transparent glass allows light to pass through. Light is needed for the production of hydrogen ions to produce reduced NAD ($NADH_2$) and ATP required in the light-independent reactions.

b) Hot methanol is used to stop chemical reactions or to kill algae.

c) At 10 seconds glycerate 3-phosphate appears (some carbon dioxide might still remain) and at 30 seconds glycerate 3-phosphate, triose phosphate and glucose might be found.

d) The graph shows that at low CO_2 concentrations, as the enzyme concentration increases, the carbon assimilation increases, matching the experimental results. It also shows that the same occurs for high CO_2 concentrations, which does not match the experimental results. At enzyme concentrations greater than normal for the algae, the increase in enzyme concentration does affect the assimilation (matching the experimental results) but it also has no effect on the assimilation of carbon for low CO_2 concentrations, not matching the experimental results.

SAMPLE STUDENT ANSWER

In an experiment to determine the effect of diet on response to leptin, mammals were fed a control diet or a high fructose diet for 8 months. One group was injected with leptin while the other remained as a control. The food intake of the different groups was monitored over a period of 48 hours.

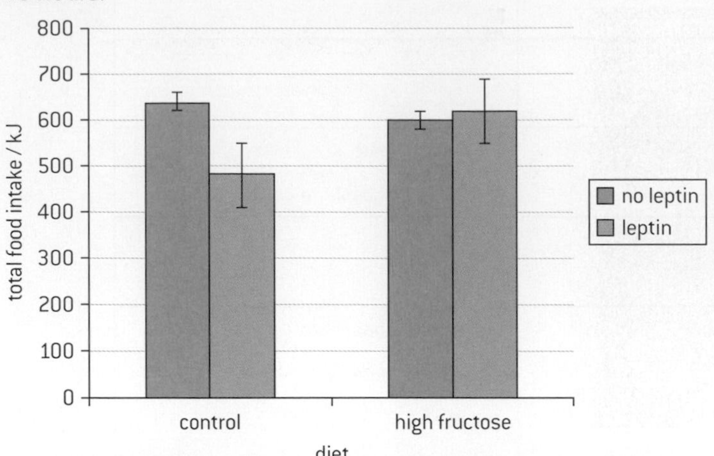

a) Distinguish between the effect of leptin injection on the to tal food intake in the mammals fed the control diet and the mammals fed the high fructose diet. [1]

This answer could have achieved 0/1 marks:

The leptin injection given to the control diet group resulted in lower food intake compared to that of the high fructose diet group.

▼ This does not answer the question. It was important to describe the effect of leptin injection; therefore the answer had to compare what happened without injection and with leptin injection in both diets. The expected answer was that there was no effect with fructose diet but (statistically significant) reduction in food intake in the control diet group when injected with leptin compared to no leptin.

b) Discuss the implications of these results for recommending leptin injections as an appetite suppressant for humans. [2]

This answer could have achieved 2/2 marks:

From these results it implies that leptin would be beneficial in suppressing appetite, as it lowered food intake compared to those not injected with leptin. However, on a high fructose diet, leptin would in fact increase intake, suggesting that its effectiveness on these mammals is debatable. Furthermore, leptin injections have a different effect on humans as they do on these mammals, with previous experiments suggesting some humans have gained weight when injected leptin.

▲ This answer scored full marks because it mentioned that the effectiveness of leptin depends on diet and that results for these mammals may not be the same for humans. The mark scheme also awarded marks for saying that if obese people have a high fructose diet, then it will not suppress appetite and that for obese people with a control diet, it will suppress appetite.

c) Leptin is a hormone. Hormones are chemicals produced in one part of the body that have an effect in another part of the body. State the tissue that produces leptin in humans. [1]

This answer could have achieved 1/1 marks:

Adipose tissue.

d) State the target that leptin normally acts on. [1]

This answer could have achieved 1/1 marks:

Hypothalamus.

Assessment tip

This type of question is about scientific research and its biological background rather than experiments you might have done in the laboratory.

Example 12.4.

Non-pathogenic *E. coli* bacterial colonies were grown on culture media at different pH values. The Petri dishes show two examples of the results obtained. Each dot is one bacterial colony.

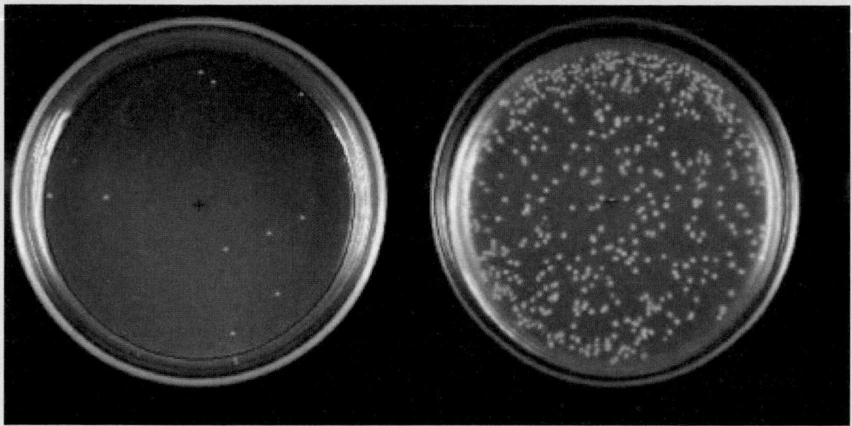

The table shows the mean number of colonies (plus or minus one standard deviation) measured in 10 replicates at each pH value.

pH	Number of colonies
4	11 (±2)
5	138 (±4)
6	361 (±7)
7	558 (±11)
8	310 (±51)
9	191 (±155)
10	121 (±21)

a) Identify the independent variable in this experiment.

b) Identify a variable that needs to be controlled.

c) State the optimum pH for the growth of *E. coli*.

d) Describe a method to count the number of colonies in the dishes.

e) Suggest with a reason which results are not reliable.

Solution

a) The independent variable is the pH in the growth media.

b) Volume of bacterial culture placed on the dish; spreading of bacteria should be even; temperature at which bacteria were grown; amount of growth medium used; time left in incubator. (Only one needed.)

c) pH = 7 (or close to this value).

d) The dishes can be divided into quadrats (for example squares of 1×1 cm). The colonies are counted in squares chosen randomly and the mean value that is obtained is multiplied by the total number of squares. The divisions can also be shown as sections in a pie chart.

e) Results at pH = 9 have a large standard deviation, therefore are not reliable. This means that there is a great variability between replicates.

Assessment tip

In (b) just mentioning temperature would not score a mark, as it is the temperature of the incubator or at which they were grown that matters and not the temperature when the colonies are counted.

Assessment tip

Although pH = 7 might seem the most obvious answer, as the number of colonies is the greatest, the optimum should really be determined by drawing the curve. In the exam students are not expected to draw graphs.

Example 12.5.

In an experiment to demonstrate the movement of molecules through osmosis, Visking tubing containing a strong sugar solution coloured with food dye was attached to a clear capillary tube and submerged in a beaker of water. The apparatus was left for several minutes and the height of coloured liquid in the tube was recorded. The experiment was repeated with different sugar concentration solutions.

a) State the independent variable in this experiment.

b) Determine **two** conditions that need to be kept constant in this experiment.

c) Explain the reason coloured water goes up the tube.

Solution

a) The independent variable is the sugar concentration.

b) Conditions that need to be kept constant are size of Visking tube (and type), volume of water in beaker, diameter of glass tube, time the Visking tube is left, environmental conditions such as temperature and pressure. (Only two conditions needed.)

c) Osmosis is the net movement of solvent molecules across a semi-permeable membrane in a direction that equalizes the solute concentrations on either side. As the sugar solution is more concentrated, water passes into the tubing. This causes an increase in the volume of liquid in the tubing, increasing the pressure and forcing the coloured solution up the capillary tube.

Example 12.6.

A model showing the relaxed and contracted state of the sarcomere was made by students.

a) Identify the structure modelled in pale blue.

b) Suggest **one** structure that could be represented by the silver balls.

c) Describe the contraction of muscle sarcomeres as shown in the model.

d) Identify with a reason **one** error in this model.

Solution

a) Z-line

b) Myosin heads or calcium or ATP

c) Myosin forms thick fibres in the sarcomere (pink). The myosin heads (silver) are attached to the thin actin filaments (dark blue). They are connected creating cross-bridges using ATP. During muscle contraction, these heads pull the actin filaments towards the centre of the sarcomere, shortening it. As the filaments stretch all along the muscle, the length of the total muscle is shortened.

d) There are several errors in this model. The myosin heads should be in contact with the actin filaments. They should also be evenly spread out. In the contracted model, the myosin heads should have moved towards their tails. The myosin fibre is too short with respect to the actin fibre. Troponin is not represented. (Only one error is needed.)

Example 12.7.

A model of lung ventilation was made using a plastic bottle, straws and balloons. A hole was drawn into the lid of the bottle to pass the straw.

a) State the muscle represented by the rubber.

b) State the structure represented by the straw with its top part outside the bottle.

c) Explain what happens when the rubber is pulled out in terms of lung ventilation.

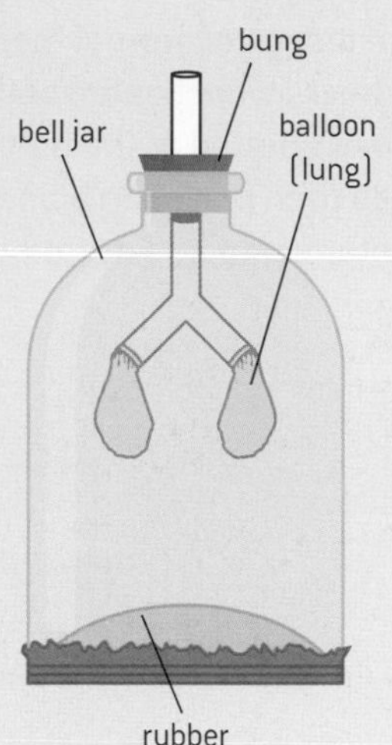

Solution

a) Diaphragm

b) Trachea

c) The rubber being pulled out represents the contraction of the diaphragm. When the diaphragm contracts, the volume of the thoracic cavity increases, lowering the pressure in the lungs compared to the atmospheric air. This causes air to enter through the airway passages to the alveoli in the lungs. In this model, pulling the rubber will cause the balloons to inflate.

Practice problems for Chapter 12

Problem 1

The electron micrograph shows human DNA from a cancer cell, showing a stage of DNA replication. The strand of DNA has a replication bubble, where the DNA has unwound into two single strands forming a Y-shaped molecule termed a replication fork. It is here that daughter strands form as the parent DNA acts as a template for the construction of a new matching strand in DNA replication.

a) Estimate the length of one strand of DNA in the replication bubble.

b) John Forster Cairns demonstrated by autoradiography that the DNA of bacteria was a single molecule that replicated at two replicating forks (unlike humans that have only one) at which both new DNA strands are being synthesized. Describe Cairns' technique for measuring the length of DNA molecules by autoradiography.

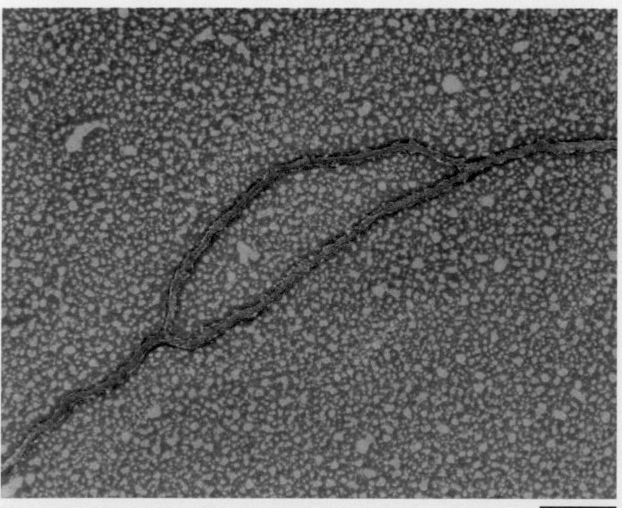

Problem 2

The picture shows a thin layer chromatogram of an extract of leaves of annual meadow grass (*Poa annua*). Plastic sheets were coated with silica gel. A drop of extract was spotted at the bottom of the sheet. The sheet was then placed in a beaker of solvent (75% acetone and 25% petroleum ether).

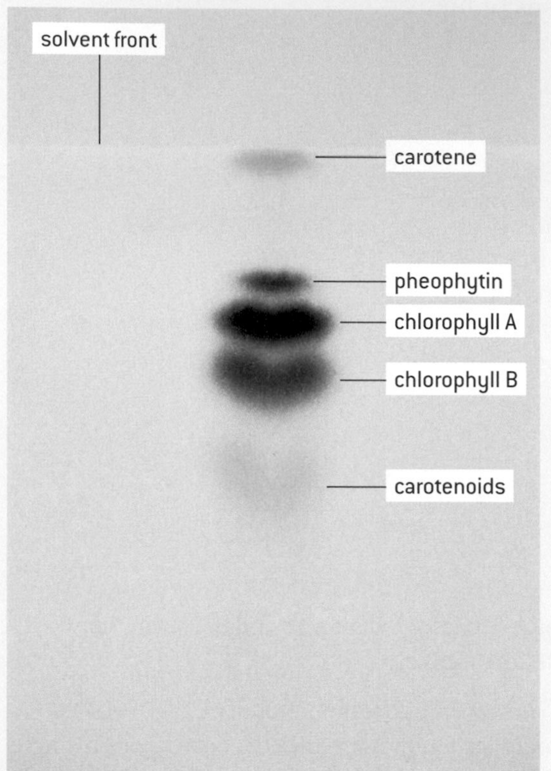

a) Identify the most soluble pigment.

b) Calculate the R_f for pheophytin. Show your working.

c) Suggest what would have happened to the pigments if the solvent had been different.

Problem 3

Glass jars containing cultures of *Chlorella vulgaris* microalgae were used as mesocosms. The carbon dioxide released from the combustion of the waste was pumped into the jars to promote photosynthesis and therefore growth of the algae. The jars were kept at 25°C.

a) State **one** other condition required in this mesocosm.

b) Identify, with a reason, whether this is an open or closed mesocosm.

c) One of the jars was kept at a temperature of 30°C. Suggest a difference that could be seen in this jar with respect to the rest of the jars.

Problem 4

Describe how the ventilation rate of someone breathing can be monitored.

A NEUROBIOLOGY AND BEHAVIOUR

A.1 NEURAL DEVELOPMENT

You should know:

- the neural tube of embryonic chordates is formed by infolding of ectoderm followed by elongation of the tube.
- neurons are initially produced by differentiation in the neural tube.
- immature neurons migrate to a final location.
- an axon grows from each immature neuron in response to chemical stimuli.
- some axons extend beyond the neural tube to reach other parts of the body.
- a developing neuron forms multiple synapses.
- synapses that are not used do not persist.
- neural pruning involves the loss of unused neurons.
- the plasticity of the nervous system allows it to change with experience.

You should be able to:

- describe neurulation.
- annotate diagrams of embryonic tissues in *Xenopus*, used as an animal model, during neurulation
- explain neural pruning.
- state that cultural experiences, including the acquisition of a language, results in neural pruning.
- describe the plasticity of the nervous system.
- outline how events such as strokes may promote reorganization of brain function.
- describe how incomplete closure of the embryonic neural tube can cause spina bifida.

The modification of neurons starts in the earliest stages of embryogenesis and continues to the final years of life. The neural tube of embryonic chordates is formed by infolding of ectoderm followed by elongation of the tube.

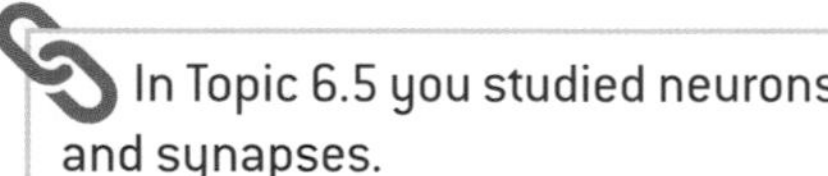

In Topic 6.5 you studied neurons and synapses.

Xenopus is a genus of African frogs that are commonly known as the African clawed frogs. Their worldwide use in research is due to the fact that it is easy to obtain large amounts of their eggs inexpensively, which are easily manipulated. The embryonic tissues of *Xenopus* are a good model for human disease because most essential cellular and molecular mechanisms are the same as in humans. Nematode worms (*Caenorhabditis elegans*), fruit flies (*Drosophila melanogaster*), zebrafish (*Danio rerio*), chickens (*Gallus gallus*) and mice (*Mus musculus*) have also been used to discover the molecular mechanisms fundamental to life, thereby providing a shortcut to understanding human biology.

- **Neurulation** is the development of the dorsal nerve cord by the infolding of the neural plate.
- **Neural pruning** is the elimination of unused neurons or synapses.
- **Neural plasticity** is the ability of the brain to change in time by gain or loss of neurons, parts of neurons or synapses.

Example A.1.1.

The diagram shows the stages of neurulation during the first two weeks in the African clawed frog (*Xenopus laevis*).

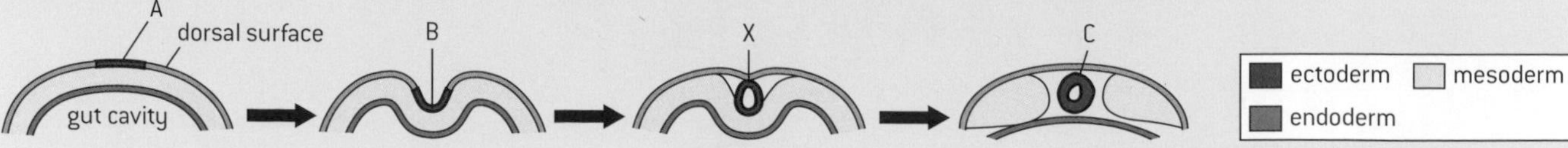

a) Label A to C.

b) Annotate the diagram to explain what is occurring at X.

c) Suggest what happens in humans if neurulation does not occur properly.

Solution

a) and **b)**

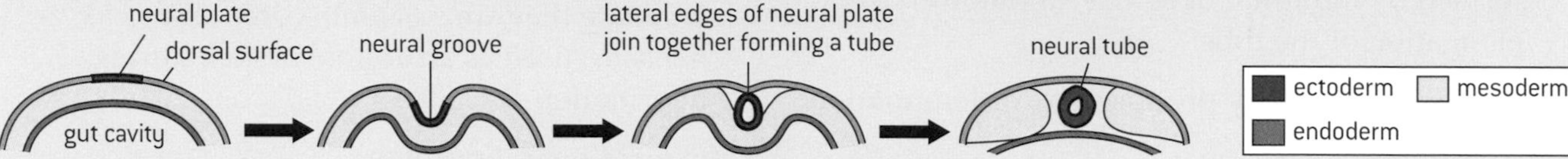

c) Incomplete closure of the embryonic neural tube can cause spina bifida.

Modification of neurons starts in the earliest stages of embryogenesis and continues to the final years of life. Neurons are initially produced by differentiation in the neural tube. Immature neurons migrate to a final location. An axon grows from each immature neuron in response to chemical stimuli. Some axons extend beyond the neural tube to reach other parts of the body. A developing neuron forms multiple synapses.

Developing neurons form multiple synapses, so there are more connections between neurons. Synapses that are not used do not persist, so there is less interference between stimuli.

Neural pruning involves the loss of unused neurons, again reducing interference between stimuli. Cultural experiences, including the acquisition of a language, result in neural pruning.

Neurons and synapses that are used a lot are reinforced so information is accessed faster via these neurons. This process gives the nervous system neural plasticity. Events such as strokes may promote reorganization of brain function.

SAMPLE STUDENT ANSWER

The synaptic density is the number of synapses per unit volume. The graph shows the synaptic density for a human from birth to 4 years old.

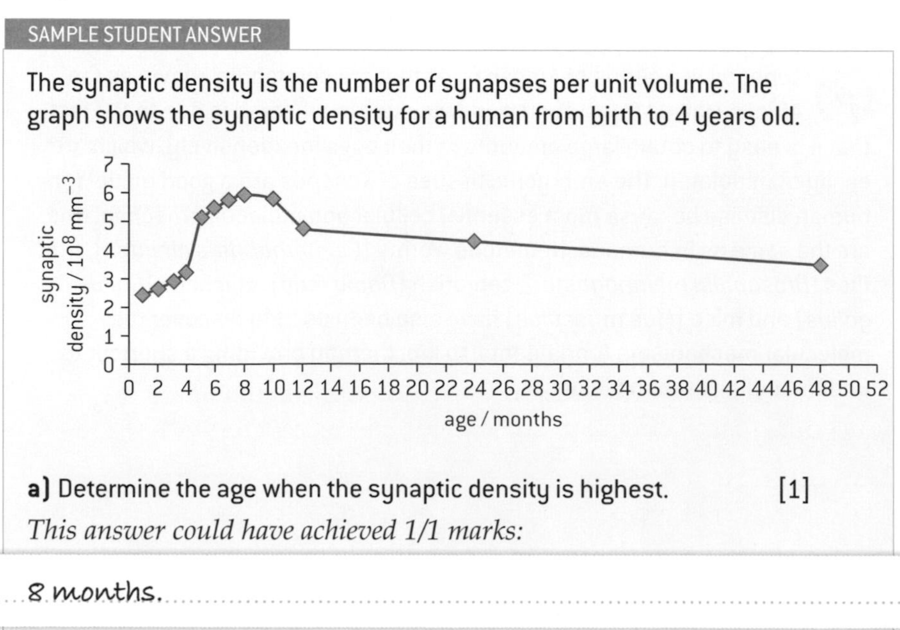

a) Determine the age when the synaptic density is highest. [1]

This answer could have achieved 1/1 marks:

8 months.

Assessment tip

Remember to write the units, otherwise no mark is given.

b) Explain how the synaptic density decreases after the age determined in (a). [3]

This answer could have achieved 3/3 marks:

During neural pruning, if they are not used, there is loss of neurons through apoptosis, which is programmed cell death. There is also loss of dendrites and axon branches. Synapses are also eliminated. This can also happen due to old age, damage to the brain by strokes or by chemical abuse.

A.2 THE HUMAN BRAIN

You should know:

- the anterior part of the neural tube expands to form the brain.
- different parts of the brain have specific roles.
- the autonomic nervous system controls involuntary processes in the body using centres located mainly in the brain stem.
- the cerebral cortex forms a larger proportion of the brain and is more highly developed in humans than in other animals.
- the human cerebral cortex has become enlarged principally by an increase in total area, with extensive folding to accommodate it within the cranium.
- the cerebral hemispheres are responsible for higher order functions.
- the left cerebral hemisphere receives sensory input from sensory receptors in the right side of the body and the right side of the visual field in both eyes, and vice versa for the right hemisphere.
- the left cerebral hemisphere controls muscle contraction in the right side of the body and vice versa for the right hemisphere.
- brain metabolism requires large energy inputs.

You should be able to:

- identify parts of the brain in a photograph, diagram or scan of the brain including the medulla oblongata, cerebellum, hypothalamus, pituitary gland and cerebral hemispheres.
- outline the functions of the visual cortex, Broca's area and nucleus accumbens as areas of the brain with specific functions.
- describe swallowing, breathing and heart rate as examples of activities coordinated by the medulla.
- analyse correlations between body size and brain size in different animals.
- describe Angelman syndrome as a genetically inherited condition that is diagnosed from characteristically abnormal patterns on an electroencephalogram.
- explain the use of the pupil reflex to evaluate brain damage.
- explain the use of animal experiments, autopsy, lesions and fMRI to identify the role of different brain parts.
- discuss how the definition of living varies depending on local and national laws and culture.

The brain and spinal cord are mainly formed of neurons. These neurons communicate with other neurons through synapses. Synaptic communication between neurons leads to the establishment of functional neural circuits that mediate sensory and motor processing.

In Topic A.1 you studied the formation of the neural tube.

The cerebral hemispheres form the largest part of the human brain. The outer part of the cerebral hemispheres is called the cerebral cortex. In the posterior part of the cerebral hemispheres we can find the brainstem and the cerebellum. The brain is connected to the rest of the body through the spinal cord.

Assessment tip

The part labelled as A could be cerebral cortex or cerebral hemisphere. If in doubt, it is better to be as precise as possible. You could have answered: cerebral cortex of the cerebral hemispheres.

- The **medulla oblongata** controls breathing and reflexes such as swallowing, coughing, sneezing and vomiting.
- The **cerebellum** controls of equilibrium and posture.
- The **hypothalamus** regulates metabolic processes and hormones.
- The **pituitary gland** secretes hormones.
- The **cerebral hemispheres** are involved in learning and memory.
- The **visual cortex** processes the light images perceived in the eyes.
- **Broca's area** controls speech.
- The **nucleus accumbens** is the pleasure reward centre.

The neuroscientist Wilder Penfield experimented on human brains, trying to identify the areas responsible for epilepsy. He stimulated the surface of the brain cortex using a small electric current and asked his patients what they felt. With all the information gathered, he mapped the areas of the cortex that control different parts of the body. Penfield developed a homunculus, a cartoon drawing of a human body, sized proportionally to the amount of brain space devoted to processing motor functions, or sensory functions, for different parts of the body.

Example A.2.1.

The diagram shows a human brain.

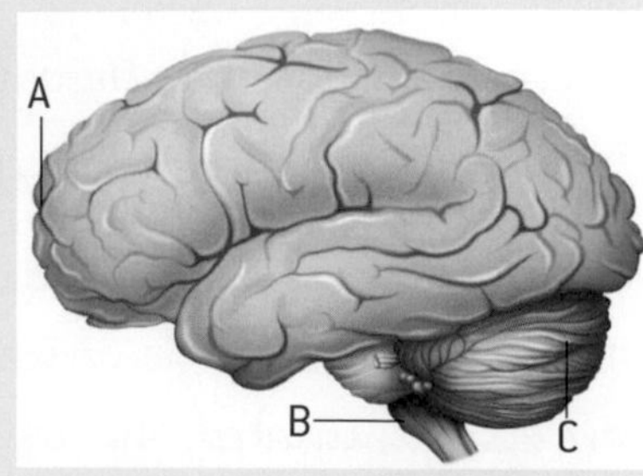

a) Label A to C shown in the diagram.

b) Outline the function of the Broca's area and nucleus accumbens.

c) On the diagram label the visual cortex.

Solution

a) A: cerebral cortex, B: medulla oblongata and C: cerebellum.

b) The Broca's area is a region in the frontal lobe of the human brain with functions linked to speech production and language processing. The nucleus accumbens is a group of cells in the human brain associated with pleasure and reward. It is also involved in laughter, fear, aggression and addiction.

c) Area of the brain above the cerebellum shown and labelled.

The autonomic nervous system controls involuntary processes in the body using centres located mainly in the brain stem. The autonomic system is divided into parasympathetic and sympathetic, which have many opposing functions. The sympathetic system is considered the "fright and flight" system, as it prepares the body for dangerous situations, while the parasympathetic system prepares the body to "rest and digest". For example, the sympathetic system accelerates the heart rate while the parasympathetic system slows it down.

The cerebral cortex is the outer layer of the cerebral hemispheres; it forms a larger proportion of the brain and is more highly developed in humans than in other animals. The human cerebral cortex has become enlarged principally by an increase in total area, with extensive folding to accommodate it within the cranium.

Example A.2.2.

The graph shows the relationship between body mass and brain mass in some mammals.

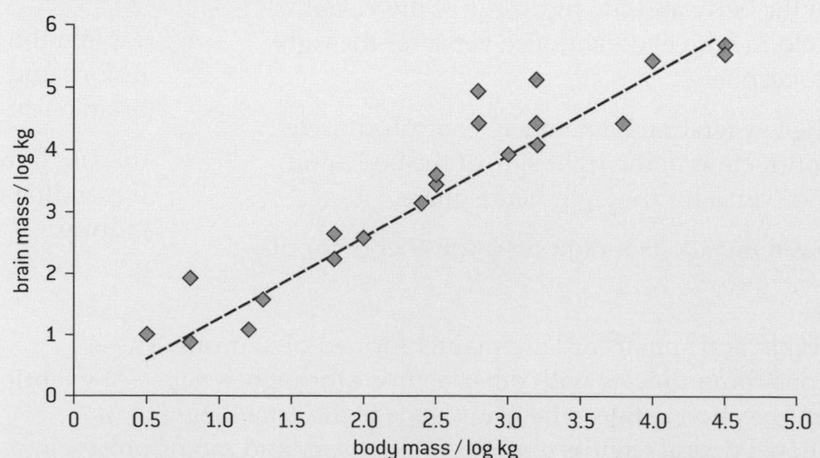

a) Describe the relationship between body mass and brain mass.

b) Discuss whether the data provides evidence that larger animals are more intellectually developed.

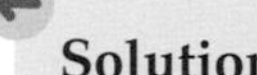

Solution

a) There is a positive correlation between body mass and brain mass: the larger the animal, the larger the brain.

b) This graph does not provide any information about development. The positive correlation between body mass and brain mass does not mean that the organism is more developed, as many large animals that have a large brain are not as developed as smaller organisms that have a small brain. Development is determined by the infolding of the cerebral cortex, which increases the brain surface area:volume ratio.

Example A.2.3.

Explain contralateral processing of images.

Solution

The left cerebral hemisphere receives sensory input from sensory receptors in the right side of the body and the right side of the visual field in both eyes, and vice versa for the right hemisphere. Contralateral processing is due to the optic chiasma, where the right brain processes information from the left visual field and vice versa.

Nowadays, animal experiments, lesions and functional magnetic resonance imaging (fMRI) scanning can be used in the identification of the brain part involved in specific functions. Although specific functions can be attributed to certain areas, brain imagery shows that some activities are spread over many areas and that the brain can even reorganize itself following a disturbance such as a stroke.

Brain metabolism requires large energy inputs. Although the brain is served by blood vessels that supply the neurons with glucose for energy, the high metabolic rate of these cells means they require the help of other cells called glial cells or astrocytes to supply more energy. Glucose taken up by glial cells undergoes glycolysis which generates ATP to meet these extra energy requirements. The lactate that this process generates is sent back to neurons, which use it aerobically in the Krebs cycle.

A.3 PERCEPTION OF STIMULI

You should know:

- ✔ receptors detect changes in the environment.
- ✔ rods and cones are photoreceptors located in the retina.
- ✔ rods and cones differ in their sensitivities to light intensities and wavelengths.
- ✔ bipolar cells send the impulses from rods and cones to ganglion cells.
- ✔ ganglion cells send messages to the brain via the optic nerve.
- ✔ the information from the right field of vision from both eyes is sent to the left part of the visual cortex and vice versa.
- ✔ structures in the middle ear transmit and amplify sound.
- ✔ sensory hairs of the cochlea detect sounds of specific wavelengths.
- ✔ impulses caused by sound perception are transmitted to the brain via the auditory nerve.
- ✔ hair cells in the semicircular canals detect movement of the head.

You should be able to:

- ✔ explain the detection of chemicals in the air by the many different olfactory receptors.
- ✔ label a diagram of the structure of the human eye.
- ✔ annotate diagrams of the retina to show the cell types and the direction in which light moves.
- ✔ explain how light is detected by the eye.
- ✔ describe red–green colour-blindness as a variant of normal trichromatic vision.
- ✔ label a diagram of the structure of the human ear.
- ✔ explain how sound waves are detected by the ear.
- ✔ explain the use of cochlear implants in deaf patients.

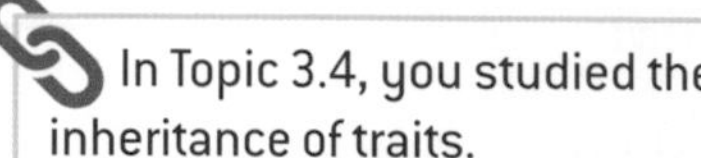
In Topic 3.4, you studied the inheritance of traits.

Living organisms have receptors that are able to detect changes in the environment. Humans' sensory receptors are mechanoreceptors, chemoreceptors, thermoreceptors and photoreceptors. For example, olfactory receptors in the nose are chemoreceptors, as they detect chemicals in the air.

- **Mechanoreceptors** detect mechanical forces and movement.
- **Chemoreceptors** detect chemicals.
- **Thermoreceptors** detect differences in temperature.
- **Photoreceptors** detect light.

SAMPLE STUDENT ANSWER

Odorants are very small molecules with varied chemical formulae. Volatile odorants that enter the nose are detected by millions of olfactory neurons. The discriminatory capacity of the mammalian olfactory system is so specific that a vast number of volatile chemicals are perceived as having distinct odours.

The table shows the response of olfactory receptors to different substances. Acids and alcohols with the same number have the same formula except for the first carbon, where one has an acid and the other an alcohol group.

Chemical stimulus (odorant)	Olfactory receptors									
	1	2	3	4	5	6	7	8	9	10
acid 1	■			■		■				
alcohol 1		■		■			■	■		
acid 2	■		■	■	■					
alcohol 2				■						
acid 3	■			■		■				
alcohol 3							■	■		
acid 4				■		■				■
alcohol 4		■		■				■		

Key

response ■

no response □

a) State the name given to the type of receptor the olfactory receptors belong. [1]

This answer could have achieved 0/1 marks:

a) Smell receptors.

▼ This answer is too vague—the correct answer is chemoreceptors.

b) Identify the receptor that: [2]

(i) detects most odorants

(ii) does not perceive the odorants shown.

This answer could have achieved 2/2 marks:

(i) 4

(ii) 9

Assessment tip

Remember that in the options the examiners are expecting more in-depth knowledge of the topics.

c) Identify the chemical that is the least perceived by these receptors. [1]

This answer could have achieved 1/1 marks:

Alcohol 2

▲ Only two receptors show a response to this chemical.

d) A slight alteration of the odorant will be perceived as a completely different smell. Comment on this statement referring to the results obtained in this experiment. [2]

This answer could have achieved 2/2 marks:

A substance with the same chemical formula as another, apart from, changing the first carbon group from acid to alcohol (for example, acid 1 and alcohol 1), is perceived by completely different receptors, thus smelling completely different. If a change in group can be considered a slight alteration, then this is a true statement.

▲ Recognizing that the acid and alcohol with the same number were the ones that had to be compared scored one mark. The fact that both substances were perceived by completely different receptors gained the second mark. Another mark could have been scored for commenting that perhaps this change was not only a slight alteration.

The eye is the organ that can receive light stimuli and provide the organism with vision. The information from the right field of vision from both eyes is sent to the left part of the visual cortex, and vice versa.

Example A.3.1.

The drawing shows a cross section of the human eye.

a) Label I to IV.

b) Outline the function of I.

c) State the name of the tissue containing photoreceptors.

I
II
III
IV

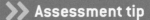

Solution

a) I: lens; II: retina, III: optic nerve; IV: cornea.

b) The function of the lens is to bend the light rays so they are focused on the retina.

c) Retina

Assessment tip

In exams you may be asked to identify some parts of the eye including the sclera, cornea, conjunctiva, eyelid, choroid, aqueous humour, pupil, lens, iris, vitreous humour, retina, fovea, optic nerve and blind spot.

Rods and cones are photoreceptors located in the retina. Rods and cones differ in their sensitivities to light intensities and wavelengths. Impulses travel from rods and cones to bipolar cells. From the bipolar cells they pass to the ganglion cells that carry the information to the brain through the optic nerve.

- **Rods** are photoreceptors that detect monochromatic images.
- **Cones** are photoreceptors that detect colour.
- **Bipolar cells** send the impulses from rods and cones to ganglion cells.
- **Ganglion cells** send messages to the brain via the optic nerve.

Example A.3.2.

The diagram shows part of the retina of the human eye.

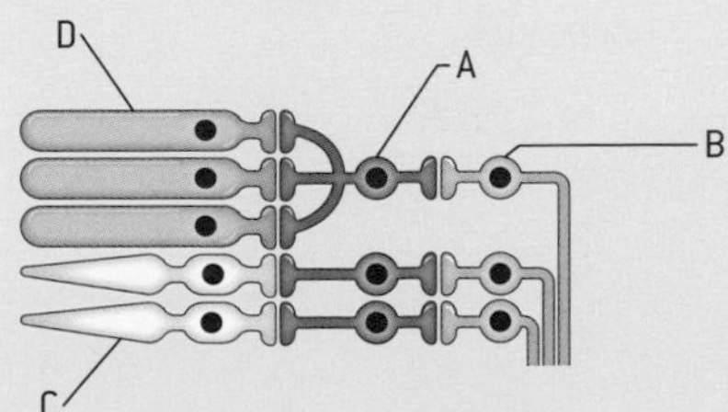

a) Label parts A to D in the diagram.

b) Draw an arrow showing the direction of light.

c) Annotate the diagram to show the direction of the nerve impulse from the eye to the brain.

d) Describe red–green colour-blindness as a variant of normal trichromatic vision.

Solution

a) A: bipolar cell; B: ganglion cell; C: cone; D: rod

b) Arrow from left to right

c) Arrow from ganglion cell from top to bottom

d) Trichromacy means that three types of cones are present in the retina, giving the person normal vision. In red–green colour-blindness the cones that detect red light and green light do not work properly and the person has difficulty in distinguishing these colours. Because the wavelengths of red and green light are very close together in the spectrum, most affected people have problems with both these colours.

- **Ossicles** are the bones of the middle ear (hammer, anvil and stirrup) that transmit and amplify sound.
- **Sensory hairs** of the cochlea detect sounds of specific wavelengths.
- The **auditory nerve** sends impulses caused by sound perception to the brain.
- **Hair cells** in the semicircular canals detect movement of the head.

The ear is the organ of hearing. Sound waves are detected by the ear through the movement of the eardrum and ossicles. The eardrum and ossicles amplify the sound. This causes the vibration of cochlear fluid. All along the basal membrane of the cochlea there are sensory hair cells. These hair cells are in different positions and have cilia of different lengths. These cilia vibrate at different wavelengths, each sending different nerve signals in the auditory nerve. Impulses caused by sound perception are transmitted to the brain via the auditory nerve. Hair cells in the semicircular canals detect movement of the head.

The discovery that electrical stimulation of the auditory system can create a perception of sound has had a huge impact on the treatment of hearing loss and impairment. Cochlear implants can now be used to help those who are deaf to hear. Sounds are detected by an external microphone placed on the external ear. These sounds are transmitted through wires to electrodes that are implanted in the cochlea of the inner ear. The electrodes then transmit the sound as electrical impulses to the auditory nerve, which carries the message to the brain, allowing hearing.

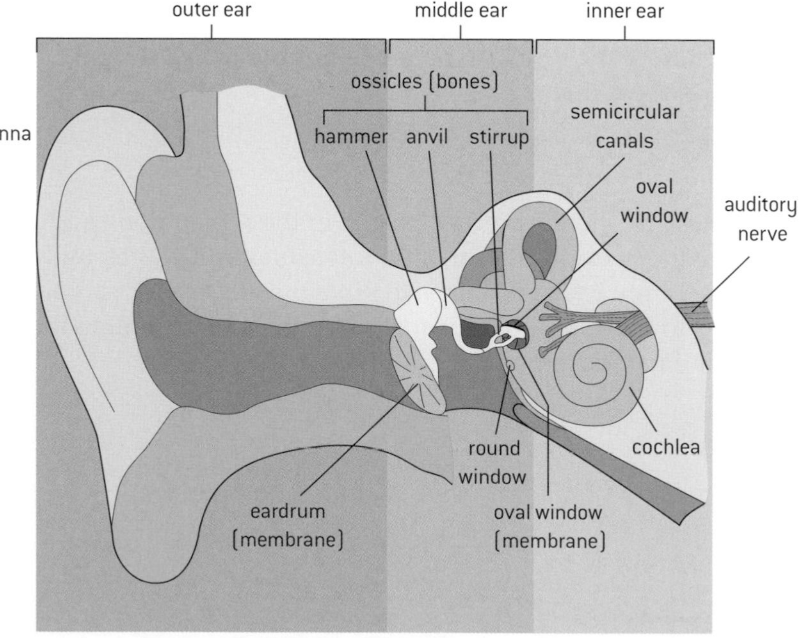

▲ Figure A.3.1. Structure of the ear

A.4 INNATE AND LEARNED BEHAVIOUR (AHL)

You should know:

- ✔ innate behaviour is inherited from parents and so develops independently of the environment.
- ✔ autonomic and involuntary responses are referred to as reflexes.
- ✔ reflex arcs comprise the neurons that mediate reflexes.
- ✔ reflex conditioning involves forming new associations.
- ✔ learned behaviour develops as a result of experience.
- ✔ imprinting is learning occurring at a particular life stage and is independent of the consequences of behaviour.
- ✔ operant conditioning is a form of learning that consists of trial and error experiences.
- ✔ learning is the acquisition of skill or knowledge.
- ✔ memory is the process of encoding, storing and accessing information.

You should be able to:

- ✔ explain the withdrawal reflex of the hand from a painful stimulus.
- ✔ describe Pavlov's experiments into reflex conditioning in dogs.
- ✔ explain the role of inheritance and learning in the development of birdsong.
- ✔ analyse data from invertebrate behaviour experiments in terms of the effect on chances of survival and reproduction.
- ✔ draw a labelled diagram of a reflex arc for a pain withdrawal reflex.

Behavioural patterns can be inherited or learned. Inherited behaviours are said to be innate. Learned behaviour develops as a result of experience.

Example A.4.1.

Compare and contrast learned and innate behaviour.

Solution

Feature	Innate	Learned
type of behaviour in humans	yes	yes
inherited	yes	no
instinctive	yes	no
affected by the environment	no	yes
can be modified by experience	no	yes
produces variability in a population	no	yes

Reflexes are an example of innate behaviour. Autonomic and involuntary responses are referred to as reflexes. Reflex arcs comprise the neurons that mediate reflexes. Reflex conditioning involves forming new associations.

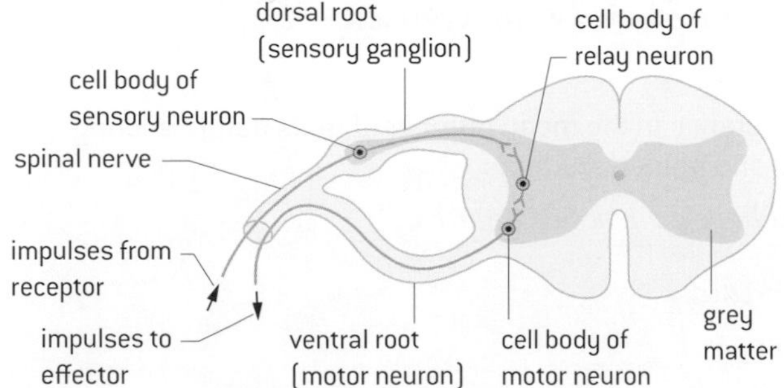

▲ Figure A.4.1. The reflex arc

Learned behaviour develops as a result of experience.

In Topics 3.3 and 3.4 you studied inheritance of alleles through meiosis and in Topic 6.6 you studied reproduction.

Laboratory experiments and field experiments both have their strengths and limitations. Laboratory experiments have the advantages that it is much more possible to precisely control the variables and it is far easier to replicate the experiment. However, the artificial setup of a laboratory study can produce unnatural behaviour or biased results. Field studies have the advantage that the results are more likely to reflect real life and therefore have more validity. Predictions and results obtained in the laboratory do not always correspond with studies performed on organisms in field conditions.

- **Imprinting** is learning occurring at a particular life stage and is independent of the consequences of behaviour.
- **Operant conditioning** is a form of learning that consists of trial and error experiences.
- **Learning** is the acquisition of skill or knowledge.
- **Memory** is the process of encoding, storing and accessing information.

Example A.4.2.

a) With respect to Pavlov's experiments with dogs, distinguish between a conditioned and an unconditioned stimulus.

b) Outline the differences between classic conditioning, as seen in Pavlov's dogs, and operant conditioning.

c) Describe an example of imprinting.

Solution

a) An unconditioned stimulus triggers a response automatically; therefore innately. The sight or smell of food naturally triggers salivation in dogs. The conditioned stimulus is a previously neutral stimulus, for example the sound of a bell, that becomes associated with the unconditioned stimulus and is learned. The conditioned stimulus triggers a conditioned response, for example salivation in response to hearing a sound before food.

b) Both classic and operant conditioning involve behaviours controlled by the environment. Classic conditioning depends on reflexes in reaction to a stimulus while in operant conditioning there is a response to either reward or punishment which will therefore control the learned behaviour. Operant conditioning is a form of learning that consists of trial and error experiences. It is a voluntary learned behaviour while classic conditioning is a learned behaviour depending on reflexes.

c) Imprinting is learning occurring at a particular life stage and is independent of the consequences of behaviour. An example is a duckling that first sees a human after hatching and follows that human as it would follow its mother duck.

SAMPLE STUDENT ANSWER

Students were asked to score a basket using a ball . The number of trials before they scored was recorded. They were then asked to train for two weeks and the same test was performed again. The results are shown in the graph.

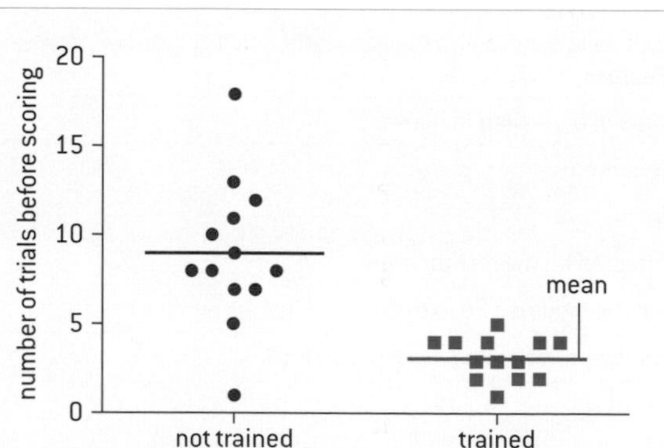

a) State the highest number of trials attempted by an untrained student and by a trained student before scoring a basket. [1]

This answer could have achieved 1/1 marks:

18 trials for untrained and 5 for trained

b) Calculate the difference in the mean number of trials before scoring between untrained and trained students. [1]

This answer could have achieved 1/1 marks:

Untrained = 9 trials

Trained = 4 trials

Difference = 5 trials

Assessment tip

This graph does not show the subdivisions on the Y axis, therefore it is hard to give the exact answer. The markscheme will have a range, but you should not give a range, but an exact number. The reason for this is that if your range is larger and does not match the range in the markscheme, you do not score a mark.

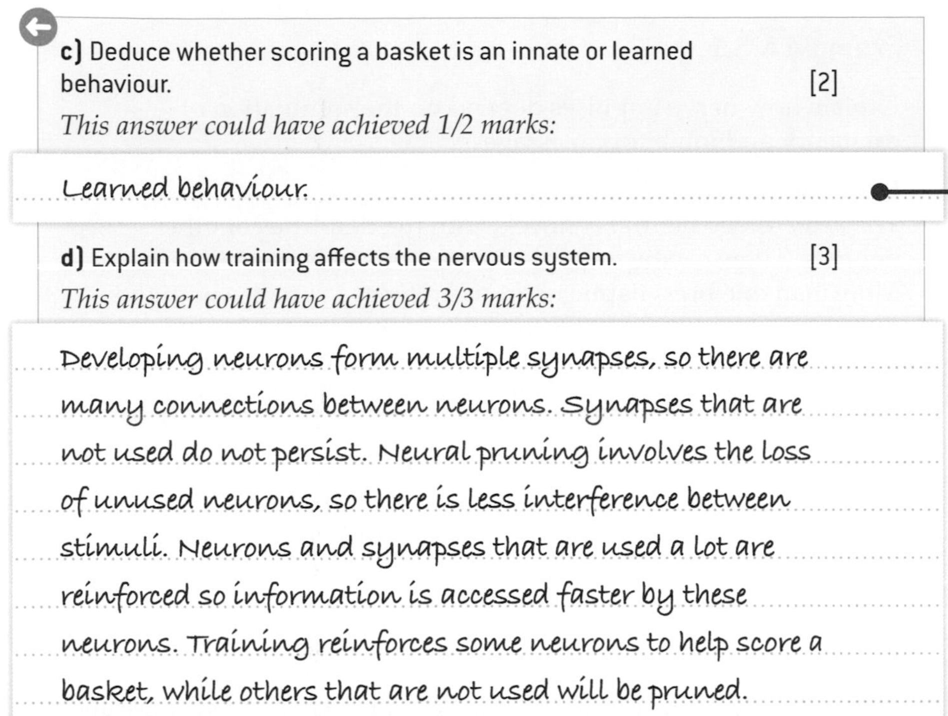

c) Deduce whether scoring a basket is an innate or learned behaviour. [2]

This answer could have achieved 1/2 marks:

Learned behaviour.

d) Explain how training affects the nervous system. [3]

This answer could have achieved 3/3 marks:

Developing neurons form multiple synapses, so there are many connections between neurons. Synapses that are not used do not persist. Neural pruning involves the loss of unused neurons, so there is less interference between stimuli. Neurons and synapses that are used a lot are reinforced so information is accessed faster by these neurons. Training reinforces some neurons to help score a basket, while others that are not used will be pruned.

▼ This answer is incomplete. A reason must be given for the choice of answer to a "deduce" question. In this case, the behaviour is learned because once students were trained, the number of trials required to score a basket was nearly half the number before training.

A.5 NEUROPHARMACOLOGY (AHL)

You should know:

- ✔ communication between neurons can be altered through the manipulation of the release and reception of chemical messengers.
- ✔ some neurotransmitters excite nerve impulses in postsynaptic neurons and others inhibit them.
- ✔ nerve impulses are initiated or inhibited in postsynaptic neurons as a result of the summation of all excitatory and inhibitory neurotransmitters received from presynaptic neurons.
- ✔ many different slow-acting neurotransmitters modulate fast synaptic transmission in the brain.
- ✔ memory and learning involve changes in neurons caused by slow-acting neurotransmitters.
- ✔ psychoactive drugs affect the brain by either increasing or decreasing postsynaptic transmission.
- ✔ anesthetics act by interfering with neural transmission between areas of sensory perception and the CNS.
- ✔ stimulant drugs mimic the stimulation provided by the sympathetic nervous system.
- ✔ addiction can be affected by genetic predisposition, social environment and dopamine secretion.

You should be able to:

- ✔ explain the action of excitatory and inhibitory neurotransmitters.
- ✔ explain the action of slow-acting neurotransmitters in memory and learning.
- ✔ explain the effects on the nervous system of two stimulants and two sedatives.
- ✔ describe the effect of anesthetics on awareness.
- ✔ explain how endorphins can act as painkillers.
- ✔ evaluate data showing the impact of MDMA (ecstasy) on serotonin and dopamine metabolism in the brain.

Neurotransmitters are chemicals released by vesicles in the presynaptic neuron that enable neurotransmission. They are chemical messengers which transmit signals from one neuron to another across the neural synapse. Some neurotransmitters excite nerve impulses in postsynaptic neurons and others inhibit them. Nerve impulses are initiated or inhibited in postsynaptic neurons as a result of the summation of all the excitatory and inhibitory neurotransmitters received from presynaptic neurons.

In Topic 6.5 you studied neurons and synapses.

An innovative method of treating cancer patients is through immunotherapy. Because it is a relatively new treatment, there is little data about the long-term effects of immune cell therapy. There are still concerns about long-term survival as well as pregnancy complications in female patients treated with these cells. Some immune therapy drugs have been approved, although they produce serious side effects. Patient advocates have pressed for the speeding up of drug approval processes, encouraging more tolerance of risk.

- **Stimulants** are excitatory neurotransmitters. They excite nerve impulses in postsynaptic neurons.
- **Sedatives** are inhibitory neurotransmitters. They inhibit nerve impulses in postsynaptic neurons.
- **Summation** is the result of all excitatory and inhibitory neurotransmitters received from presynaptic neurons.
- **Slow-acting neurotransmitters** act through secondary messengers. They modulate fast synaptic transmission in the brain.
- **Memory** and **learning** involve changes in neurons caused by slow-acting neurotransmitters.

Example A.5.1.

Explain how nerve impulses depend on the summation of excitatory and inhibitory messages.

Solution

The axon of one neuron connects with the dendrites of other neurons. There can be multiple connections between neurons. More than one presynaptic neuron can form a synapse with the same postsynaptic neuron. Summation involves combining the effects of excitatory and inhibitory neurotransmitters. The action potentials form depending on the balance of signals of excitatory and inhibitory signals. The membrane potential must rise above a given threshold for the generation of an action potential. The effect of inhibitory neurotransmitters cancels the effect of excitatory neurotransmitters. Inhibitory neurotransmitters can prevent the threshold being reached, when an excitatory neurotransmitter is released, and therefore the action does not occur.

Neurotransmitters can be fast-acting or slow-acting. The fast-acting neurotransmitters take less than a millisecond to cross the synapse and they affect only one neuron. They attach to receptors in the postsynaptic neuron which are protein ion channels. Slow-acting neurotransmitters on the other hand can take hundreds of milliseconds to act. They affect more than one neuron. They do not act through ion channels, but instead cause the release of secondary messengers. Examples of slow-acting neurotransmitters are noradrenalin, dopamine and serotonin.

Psychoactive drugs affect the brain by either increasing or decreasing postsynaptic transmission. Stimulant drugs, such as nicotine, cocaine or amphetamines, mimic the stimulation provided by the sympathetic nervous system. Sedatives such as benzodiazepines, alcohol or tetrahydrocannabinol (THC) mimic inhibition. Anesthetics act by interfering with neural transmission between areas of sensory perception and the central nervous system (CNS). They prevent the transmission of nerve impulses by binding to sodium channels, inhibiting the influx of sodium ions through these channels. Addiction can be affected by genetic predisposition, social environment and dopamine.

Example A.5.2.

a) Outline the effect of benzodiazepines and alcohol on the nervous system.

b) State the effect of MDMA on synapses.

c) Outline the use of anesthetics in surgery.

Solution

a) Both benzodiazepines and alcohol act as sedatives, affecting synaptic transmission. Benzodiazepines increase the effect of receptors for GABA, which is an inhibitory neurotransmitter. Alcohol enhances the effect of GABA and decreases the excitatory neurotransmitter glutamate.

b) MDMA is also called ecstasy. It increases the serotonin and dopamine levels in synapses, enhancing its release and preventing reuptake.

c) Anesthetics act by interfering with neural transmission. Anesthetics such as ketamine prevent glutamate receptors from being active. This interferes with neural transmission between areas of sensory perception and the CNS, preventing pain during surgery.

A.6 ETHOLOGY (AHL)

You should know:

- ✔ behaviour can be innate or learned.
- ✔ ethology is the study of animal behaviour in natural conditions.
- ✔ natural selection can change the frequency of observed animal behaviour.
- ✔ behaviour that increases the chances of survival and reproduction will become more prevalent in a population.
- ✔ learned behaviour can spread through a population or be lost from it more rapidly than innate behaviour.

You should be able to:

- ✔ explain how natural selection favours specific types of behaviour.
- ✔ describe examples of innate behaviours.
- ✔ describe different types of animal learned behaviours such as migration, altruistism, foraging, breeding strategies and courtship.
- ✔ explain migratory behaviour in blackcaps is an example of the genetic basis of behaviour and its change by natural selection.
- ✔ explain how blood sharing in vampire bats as an example of the development of altruistic behaviour by natural selection.
- ✔ explain foraging behaviour in shore crabs as an example of increasing chances of survival by optimal prey choice.
- ✔ explain breeding strategies in Coho salmon populations as an example of behaviour affecting chances of survival and reproduction.
- ✔ explain courtship in birds of paradise as an example of mate selection.
- ✔ explain synchronized oestrus in female lions in a pride as an example of innate behaviour that increases the chances of survival and reproduction of offspring.
- ✔ explain feeding on cream from milk bottles in blue tits as an example of the development and loss of learned behaviour.

Shore crabs (*Carcinus maenas*) choose middle-sized mussels to feed on in preference to large mussels, despite the fact they provide less energy. This is because the time taken to break the harder shells of larger mussels is not worth the trouble. This foraging behaviour is an example of increasing the chances of survival by optimal prey choice.

In Topic 5.2 you studied natural selection.

SAMPLE STUDENT ANSWER

a) The sketch shows the growth rate of two types of the same species of migratory fish (Coho salmon, *Oncorhynchus kisutch*). The mean age at maturity is shown with a circle.

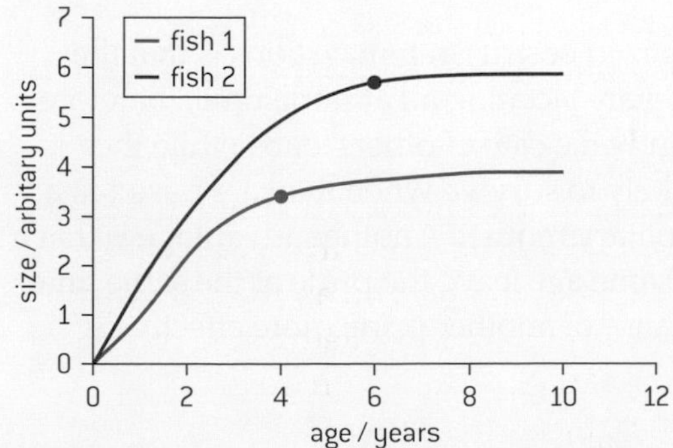

Use the graph to explain breeding strategies in Coho salmon populations as an example of behaviour affecting the chances of survival and reproduction. [6]

▲ This answer is very complete. The student uses the information in the graph to explain the different types of salmon and their breeding strategies.

Experiments to test hypotheses on the migratory behaviour of blackcaps have been carried out. Birds are tagged and followed through GPS recordings. The bird known as the blackcap (*Sylvia atricapilla*) traditionally migrated from its summer breeding grounds in Central Europe to warmer areas in Spain and Portugal for the winter. Lately they have been seen to migrate to the UK. The abundance of garden birds increased with levels of bird feeding. Studies found that blackcaps migrating to the UK from Germany had become adapted to eating food supplied by humans. In contrast blackcaps migrating to Spain had bills adapted to feeding on fruit such as olives. The birds migrating to the UK have an evolutionary advantage, as they need to travel less so expend less energy. Therefore the genetic basis of behaviour has been changed by natural selection.

This answer could have achieved 6/6 marks:

One part of the population stays in one habitat, reaching a small size. These are called jacks. These are represented by the graph that reaches a plateau first (Fish 1). The dot shows when they become adults, which is at a younger age than fish 2. The other fish grow at a slower rate, reaching a larger size before returning to the breeding site. These are called hooknoses. The hooknoses are usually larger in size so they fight to fertilize females, while the smaller jacks do not fight but hide and wait for the right moment to fertilize females. The larger jacks may behave like the hooknoses, as it is difficult for them to hide.

Many birds show sexual dimorphism. This means the males are very different from the females. The males may be very colourful or may have distinctive feathers to attract females. Others show special courtship behaviours such as dances. Birds of paradise show both of these behaviours, in a very exaggerated manner. Females choose their mates according to their plumage and dance. Natural selection has led to the selection of these exaggerated traits.

Example A.6.1.

a) Outline the behaviour of feeding on cream from milk bottles by blue tits.

b) Outline **one** way in which synchronized oestrus in female lions increases the chances of survival and reproduction of offspring.

Solution

a) Blue tits started to feed on cream from milk bottles. This behaviour soon spread throughout Europe. Nowadays there are few blue tits showing this behaviour, probably because milk bottles are no longer delivered door-to-door. This is an example of the development and loss of a learned behaviour.

b) The advantage of synchronized oestrus in female lions is that the females have their cubs and are lactating all at same time, therefore some females can suckle and take care of others' cubs while they hunt. The cubs are more likely to survive when they are raised in a nursery rather than by a solitary mother. Another advantage is that a group of male cubs of the same age leave the pride at the same time, so can compete for dominance of another pride more effectively.

Practice problems for Option A

Problem 1

The growth of the axon and one dendrite were measured in a developing neuron. The graph shows the length of the axon and dendrite from the tip to the cell body over time.

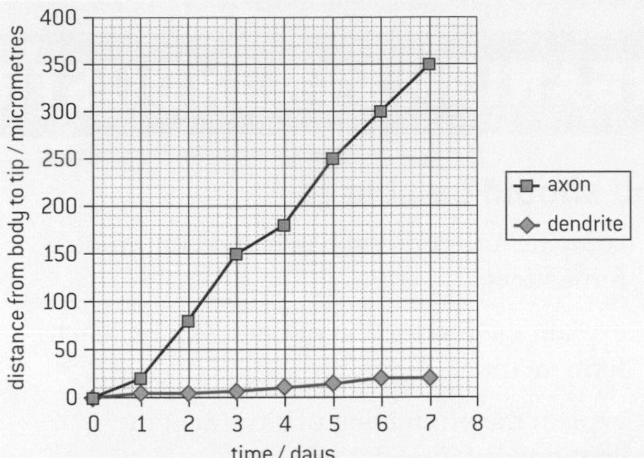

a) State the process occurring in this neuron from day 1 to day 7.

b) Calculate the difference in length of the axon and dendrite after 7 days' growth.

c) Explain the difference in trend of growth of the axon and dendrite.

d) Outline what could happen to the dendrite if it is not used.

Problem 2

The following is a magnetic resonance image (MRI) of a human brain.

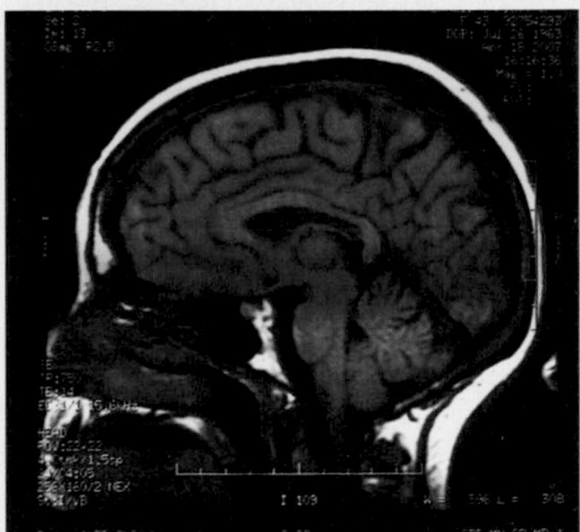

a) On the image, label the medulla oblongata, cerebellum, hypothalamus and pituitary gland.

b) Explain how functional MRI can be used to detect the functions of different parts of the brain.

Problem 3

Explain how, in humans, colour in the environment is detected by the eyes and relayed to the brain.

Problem 4 (AHL)

Outline the processes occurring when a person touches a rough surface with their hand and then moves the hand away.

Problem 5 (AHL)

The volume of the left and right nucleus accumbens was measured in tetrahydrocannabinol (THC) consumers and compared with those of control participants who do not consume cannabinoids. The bar graph shows the results of several recordings.

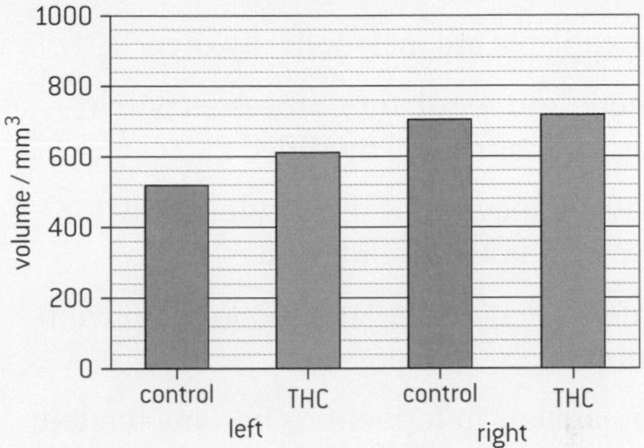

a) State the mean volume of the left nucleus accumbens in THC consumers.

b) Compare the volume of the accumbens in THC consumers and control participants.

c) Explain the effect of THC on the nervous system.

Problem 6 (AHL)

Common vampire bats (*Desmodus rotundus*) feed only on blood and die after 70 h of fasting. Unfed bats often receive food from other bats by regurgitation. The graph shows the changes in weight at the time of feeding compared to the pre-feeding weight (100%) of donor and recipient bats before and after blood donation.

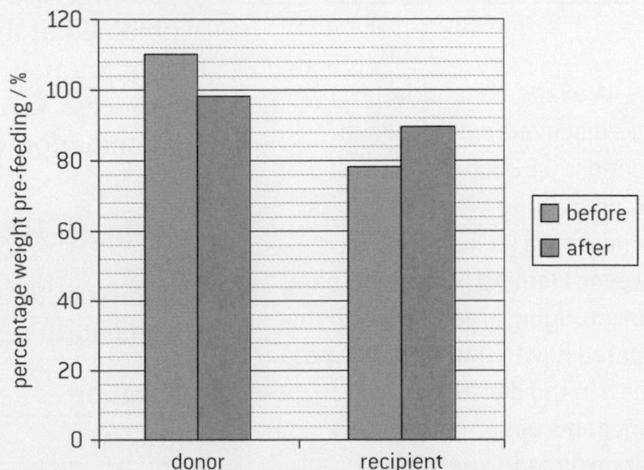

Using the data, explain blood sharing in vampire bats is an example of the development of altruistic behaviour by natural selection.

B BIOTECHNOLOGY AND BIOINFORMATICS

B.1 MICROBIOLOGY: ORGANISMS IN INDUSTRY

You should know:

- ✔ microorganisms are metabolically diverse.
- ✔ the genetic and regulatory processes within microorganisms can be modified.
- ✔ pathway engineering is used industrially to produce metabolites of interest.
- ✔ fermenters are used for large-scale production of metabolites.
- ✔ microorganisms in fermenters become limited by their own waste products.
- ✔ probes are used to monitor conditions within fermenters.
- ✔ conditions are maintained at optimal levels for the growth of the microorganisms being cultured.

You should be able to:

- ✔ compare and contrast batch and continuous fermenters.
- ✔ explain the production of penicillin by batch fermentation.
- ✔ explain the production of citric acid by continuous fermentation.
- ✔ explain how biogas is produced in fermenters.
- ✔ explain zones of inhibition around cultured bacterial growths.
- ✔ describe Gram staining of Gram-positive and Gram-negative bacteria.

In Topic 2.1 you studied molecules involved in metabolism. In Topic 4.3 you have seen carbon cycling. In Topic 6.3 you studied defence against infectious disease.

Alexander Fleming discovered penicillin in England in 1928, on a discarded petri dish. Many perceive this as serendipity, a lucky accident. However, Fleming had spent many years studying organisms and this prepared him to understand what he saw in the dish. He could connect his observations with what he already knew about microorganisms to recognize what was happening.

Microorganisms are small cells that reproduce very quickly. This allows scientists to modify them and use them in different industrial processes. Genes can be introduced into microorganisms by genetic engineering so that they will make a desired product. The modified microorganism is placed in a fermenter. In batch fermentation the substrate is introduced at the beginning of the process and the product is obtained at the end, while in continuous fermentation the substrate is continuously added and the product is harvested throughout the process. Continuous fermentation is more efficient, but any contamination will affect the whole process.

Example B.1.1.

Complete the table to compare and contrast batch and continuous fermentation according to the time when each process occurs.

Solution

Process	Batch	Continuous
substrate added	at the beginning	**all the time**
product harvested	**at the end**	at periodic intervals
oxygen monitored	**all the time**	all the time
production of foam check	all the time	**all the time**

Example B.1.2.

Citric acid can be produced by continuous fermentation of cassava bagasse (CB) using the filamentous fungus *Aspergillus niger*. The starch consumed by the fungal fermentation (left axis) and the changes in pH due to citric acid production (right axis) during the fermentation are shown in the graph.

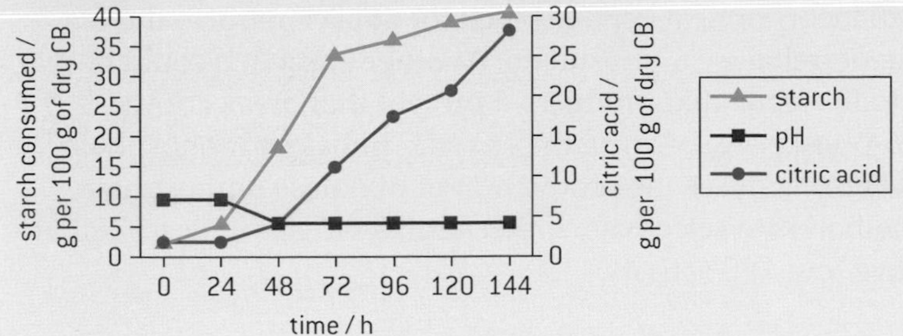

Source of data: F. C. Prado *et al.* (2005), *Brazilian Journal of Chemical Engineering*, 22(4), pp. 547–555.

a) State one use of citric acid in the food industry.

b) (i) State a mode of detection of the pH in the fermenter.

(ii) Describe the changes of pH over time in this fermenter.

(iii) Suggest one reason for the pH value after 144 hours.

c) Explain the production of citric acid in relation to the consumption of starch.

Solution

a) Citric acid can be used as a preservative or flavouring.

b) (i) A probe can be used to detect the pH in the fermenter.

(ii) The pH is alkaline at the beginning of the experiment and decreases to 4 (acid) at 48 hours. It remains constant after 48 h.

(iii) The pH is prevented from becoming more acidic by the addition of alkaline substances such as hydroxide or bicarbonate.

c) Starch is the substrate of this fermentation. It is found in cassava bagasse (CB) as a storage molecule. It is hydrolysed or broken down to maltose and eventually to glucose (both smaller sugar molecules). Citric acid is the product or primary metabolite of the fermentative reaction as it is produced during the Krebs cycle.

Example B.1.3.

Aspergillus niger is used to produce citric acid by continuous fermentation. Glucose is converted to pyruvate by glycolysis. Trehalose 6-phosphate normally inhibits hexokinase, an important enzyme in the glycolysis pathway.

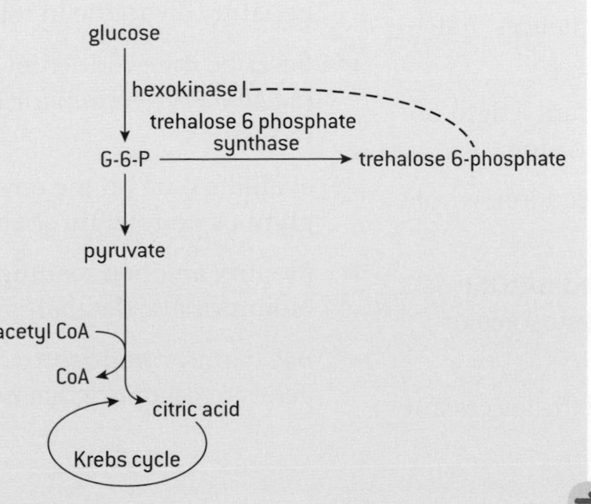

Suggest how pathway engineering could be used to address this factor which reduces yields of citric acid.

Solution

The molecule trehalose 6-phosphate binds to the enzyme hexokinase inhibiting the production of citric acid. This is a type of negative feedback. In order to avoid this, *Aspergllus niger* can be genetically modified to incorporate a gene to block (or at least reduce) the production of trehalose 6-phosphate. Another approach could be to incorporate a gene that produces a protein that breaks down trehalose 6-phosphate. Because the enzyme trehalose 6-phosphate synthase (TPS) catalyses the production of trehalose 6-phosphate, another method is to selectively breed strains of *Aspergllus niger* that already have low TPS activity.

Microorganisms produce metabolites such as acids that can change the pH of the medium. The carbon dioxide produced in respiration dissolves in water, producing carbonic acid which also acidifies the medium. Conditions need to be kept constant in the fermentation tanks, otherwise they can affect microorganism metabolism. These conditions include temperature, pH, cell density, foam and levels of oxygen, nutrients, waste products and carbon dioxide. All these are monitored using probes. These alert the system to, for example, add alkaline substances such as bicarbonate or alkalis to raise the pH when it gets too low.

B.2 BIOTECHNOLOGY IN AGRICULTURE

You should know:

- ✔ organisms can be modified to include DNA from other organisms.
- ✔ these transgenic organisms produce proteins that were not previously part of their species' proteome.
- ✔ genetic modification can be used to produce novel products.
- ✔ plants can be genetically modified to overcome environmental resistance to increase crop yields.
- ✔ bioinformatics can be used in identifying target genes.
- ✔ a target gene is linked to other sequences that control its expression.
- ✔ an open reading frame is a significant length of DNA from a start codon to a stop codon.
- ✔ recombinant DNA can be introduced into whole plants, leaf discs or protoplasts.
- ✔ recombinant DNA must be inserted into the plant cell and taken up by its chromosome or chloroplast DNA.
- ✔ marker genes are used to indicate the successful uptake of a target gene.
- ✔ recombinant DNA can be introduced by direct physical and chemical methods or indirectly by vectors.

You should be able to:

- ✔ explain how transgenic organisms can be produced.
- ✔ describe the use of the tumour-inducing (Ti) plasmid of *Agrobacterium tumefaciens* to introduce glyphosate resistance into soybean crops.
- ✔ describe the genetic modification of tobacco mosaic virus to allow bulk production of hepatitis B vaccine in tobacco plants.
- ✔ describe the production of Amflora potato (*Solanum tuberosum*) for the paper and adhesive industries.
- ✔ evaluate data on the environmental impact of glyphosate-tolerant soybeans.
- ✔ identify an open reading frame (ORF) and use bioinformatic databases to find an ORF.
- ✔ use bioinformatic software to identify similar genes in other organisms (BLASTn).

In order to genetically modify an organism, an open reading frame (ORF) is found using bioinformatics. This target gene sequence is linked to other sequences that control its expression. Recombinant DNA is formed when the target gene forms part of either the nuclear DNA or the mitochondrial or chloroplast DNA. Marker genes such as antibiotic resistance are used to indicate successful uptake of the target gene. When the transgenic organism's cells divide by mitosis, the target DNA is incorporated into the new cells.

- An **open reading frame (ORF)** is a significant length of DNA from a start codon to a stop codon. It contains sufficient nucleotides to code for a polypeptide chain.
- **Transgenic** describes an organism that has foreign DNA included in its own DNA.

Example B.2.1.

Find an open reading frame (ORF) in the following DNA sequence.

GATCCATGCTGCTGCTGACTCGGAGCCCCACAGCTT
GGCACAGGCTCTCTCAGCTCAAGCCTCCGGTCCT
CCCTGGGACCCTGGGAGGCCAGGCCCTGCATCTGAGGT
CCTGGCTTTTGTCAAGGCAGGGCCCTGCAGAG ACAGGT
GGGCAGGGCCAGCCCCAGGGCCCTGGGCTTCGAACCCG
GCTGCTGATCACAGGCCTGTTTAGGC TGGACTCGGT

Solution:
The start codon is ATG, which you can see at 6 nucleotides from the beginning of the sequence. The stop codons are TAG, TAA and TGA. Starting from the ATG you must divide the sequence into groups of three (codons) and see when one of these sequences appears. If there are approximately 20 codons between the start and the stop codons, you have found an ORF.

In Topic 1.5 you studied the origin of cells and in Topic 3.5 you studied genetic modification and biotechnology.

Plants resistant to stressful environmental conditions can be obtained through natural selection, but this takes many generations so requires a large investment of time. Through the use of transgenic tools, scientists can produce plants with desired traits, and can increase yields to produce more crops that keep longer and can withstand pests and disease. Improving resistance to diseases and insects reduces the need for herbicides and pesticides. Some transgenic plants are used to produce useful molecules such as vitamins or high-amylopectin starch. Genetic modification may allow increased tolerance to cold, frost or drought, making a crop easier to grow in a constantly changing environment. The production of edible vaccines has been achieved by the incorporation of hepatitis B antigens in the tobacco mosaic virus, which is then used to infect tobacco plants. The plants eaten by animals induce the formation of antibodies against hepatitis B.

Scientists must assess risks and benefits associated with scientific research, for example, the need to evaluate the potential of herbicide resistance genes escaping into the wild population. Wild relatives of genetically engineered crops can acquire transgenic traits such as herbicide resistance. This can happen by spontaneous transgenic crop–wild hybridization. In agricultural crops, resistance to herbicides is often a beneficial trait, but little is known about possible problems that could result if this trait persists when herbicides are not used. Another problem is the appearance of herbicide-resistant weeds.

Example B.2.2.

Starch from different sources contains different proportions of amylose and amylopectin. Potatoes (*Solanum tuberosum*) have been genetically modified to produce high-amylopectin starch (Amflora potatoes).

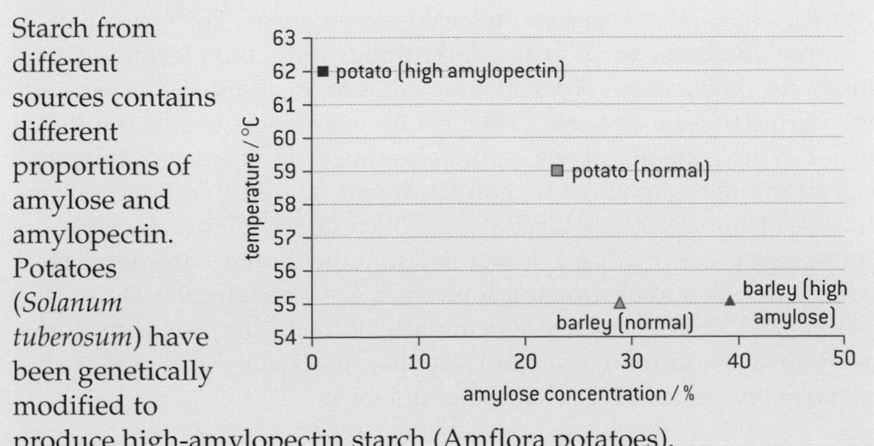

Source of graph data: H. Fredriksson *et al.* (1998), *Carbohydrate Polymers*, **35**, pp. 119–134.

Heat induces starch to form a gel in excess water. The graph shows the gel formation temperature at different amylose concentrations.

a) Discuss the hypothesis that the temperature at which starches form a gel depends on the degree of cross-linking of amylopectin.

b) State one advantage of potatoes with a high amylopectin content.

c) The Amflora potato was approved for industrial applications in the European Union (EU) in 2010 and was withdrawn in January 2012 due to opposition. Discuss reasons for people supporting or opposing the introduction of the Amflora potato in the EU.

Solution

a) High-amylopectin potatoes will have more amylopectin molecules so there is a great chance of cross-linking. High-amylopectin (or low-amylose) potatoes form a gel at a higher temperature than normal potatoes, therefore we can say the hypothesis is supported. However, normal barley and high-amylose barley show very similar gel formation temperature, which does not support the hypothesis.

b) High-amylopectin potato starch is used in adhesives production as it forms a sticky paste. It can be stored for long periods of time.

c) There is some support for the production of Amflora potatoes. They are a cheap and natural source of adhesives, reducing deforestation. This benefits local farmers and reduces pollution. Opposing arguments include that Amflora potatoes are transgenic—that they have foreign DNA in their genome. This could represent a health risk to the population as they could be allergenic or toxic. There is also a risk of horizontal gene transfer, and the future effects of this are unknown.

Assessment tip

When looking at the data, check the labels on the x and y axes and make sure you use the same wording in your answer. The x (independent) variable affects the y (dependent) variable; in this case, the amylose concentration affects the temperature at which the gel is formed. Use all the data to answer the question. In a discussion you should try to include a statement that supports the hypothesis and one that does not support it.

A transgenic crop is a genetically modified organism. A transfer of genes has occurred incorporating DNA from a different species. The gene of interest is amplified or added to a plasmid, and then introduced into the cells of whole plants, leaf discs or protoplasts. The new DNA is incorporated into DNA in the nucleus of the plant cell or the chloroplast DNA. Chemical methods of introducing genes into plants can include calcium chloride or liposomes while physical methods include electroporation, microinjection and biolistics. Recombinant DNA can also be introduced using vectors such as *Agrobacterium tumefaciens* and tobacco mosaic virus. The two most common methods are biolistics and tumour-inducing plasmid (Ti) methods. Biolistics involves sticking DNA to small micro-bullets and then firing these into a cell. This technique is clean and safe but it can result in the gene of interest being rearranged upon entry or the target cell sustaining damage upon bombardment. The Ti method involves the use of *Agrobacterium tumefaciens,* which infects plant cells. The transgenic piece of DNA is integrated into the plant's chromosomes using a Ti plasmid as a vector. A plasmid is a large circular DNA particle, which replicates independently of the bacterial chromosome and can take control of the plant's cellular machinery and use it to make many copies of its own bacterial DNA.

- **Calcium chloride** promotes the binding of plasmid DNA to the cell membrane, alllowing the plasmid to enter the cell using heat shock.
- A **liposome** is a spherical lipid vesicle used as a vehicle to carry transgenic DNA.
- **Electroporation** is the application of an electric field to cells in order to increase their permeability to trangenic DNA.
- **Microinjection** is the use of a glass micropipette to inject the trangenic DNA into cells.
- **Biolistics** is the sticking of DNA to "micro bullets" that are fired into a cell.
- ***Agrobacterium tumefaciens*** is a vector used to infect plant tissue and the tumor-inducing plasmid (Ti) is used to carry transgenic DNA.
- **Tobacco mosaic virus** is a vector used to incorporate recombinant DNA into tobacco plants by infection.

Assessment tip

Limit the chemical methods of introducing genes into plant cells to calcium chloride and liposomes. Limit the physical methods of introducing genes into plant cells to electroporation, microinjection and biolistics. Limit vectors to *Agrobacterium tumefaciens* and tobacco mosaic virus.

SAMPLE STUDENT ANSWER

Succinate is industrially produced by continuous fermentation. It is used as a raw material in the production of flavour enhancers, drugs and industrial chemicals. One method of increasing the production of succinate is to genetically modify *E. coli* to express high levels of formate dehydrogenase (FDH1). This results in the production of higher concentrations of NADH. The engineered pathway is shown as a bold dotted line in the image.

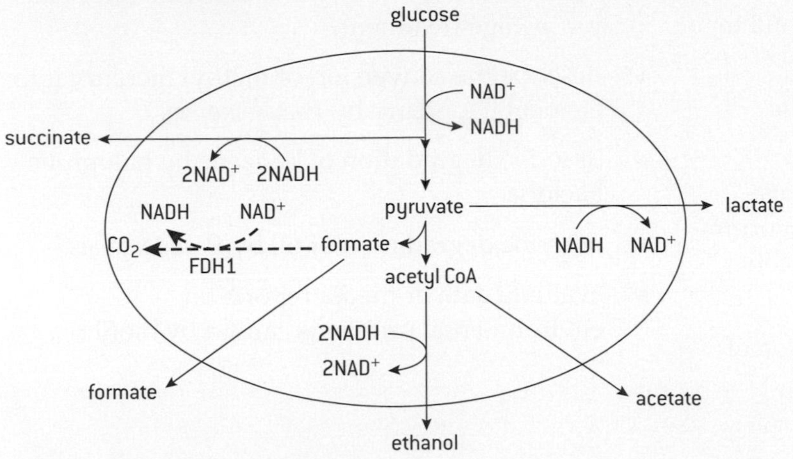

a) Using the diagram, suggest a reason for high concentrations of NADH favouring the production of succinate. [1]

This answer could have achieved 1/1 marks:

High concentrations of NADH favour the production of succinate because succinate is produced using the conversion of NADH to NAD^+ (the oxidation of NADH), so the greater the concentration of NADH the more NADH there is available to be oxidized creating more H^+, e^- and energy to produce more succinate.

b) Predict **one** metabolite other than succinate that will be produced in a greater amount if the amount of NADH available is increased. [1]

This answer could have achieved 1/1 marks:

CO_2, because the reduction of NAD^+ to NADH leads to the formation of CO_2 from formate.

Assessment tip

Remember to include a reason for a "predict" or "suggest" question.

c) Outline one reason this process represents pathway engineering. [1]

This answer could have achieved 1/1 marks:

This represents pathway engineering because E. coli is being altered to express higher levels of FDH1 to form greater concentrations of succinate. Pathway engineering involves altering the original metabolic pathway to achieve better results in terms of yields.

B.3 ENVIRONMENTAL PROTECTION

You should know:

- ✔ biotechnology can be used in the prevention and mitigation of contamination from industrial, agricultural and municipal wastes.
- ✔ bioremediation uses microorganisms to aid in clearing up pollution incidents.
- ✔ some microorganisms metabolize specific pollutants.
- ✔ biofilms are aggregates of microorganisms that have emergent properties such as quorum sensing, formation of a matrix and antibiotic resistance.
- ✔ bacteriophages are used in the disinfection of water systems.

You should be able to:

- ✔ explain the advantages of the emergent properties of microorganisms in biofilms.
- ✔ describe the use of biofilms in trickle filter beds for sewage treatment.
- ✔ describe the conversion of methyl mercury into elemental mercury by *Pseudomonas*.
- ✔ describe degradation of benzene by halophilic bacteria.
- ✔ describe degradation of oil by *Pseudomonas*.
- ✔ evaluate data or media reports on environmental problems caused by biofilms.

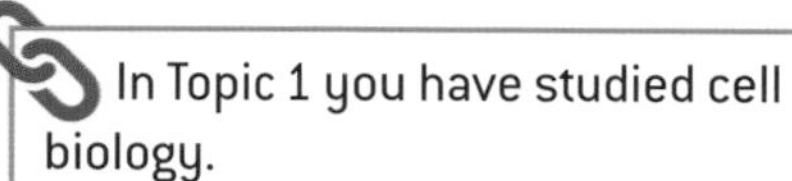
In Topic 1 you have studied cell biology.

Industrial contamination or pollution can affect all of the organisms inhabiting that area. Microorganisms are capable of metabolizing some pollutants, rendering them less harmful to the ecosystem. This process is called bioremediation. Physical and chemical procedures may also be necessary to remove the pollutants from water or soil.

- A **biofilm** is a layer of microorganisms attached to a surface, forming an organized structure with complex structural and functional characteristics.
- **Emergent properties** arise from the interaction of the members of a biofilm that are not present in the single-celled microorganism.

Example B.3.1.

Compounds containing the cyanide group (CN) are used to help extract gold from gold-containing rocks called ore. The process results in waste rock that is contaminated with cyanide, a toxin that can inhibit cellular respiration. The bacterium *Pseudomonas fluorescens* degrades cyanide to ammonia (NH_3), which is less toxic.

cyanide + oxygen + organic carbon source →
carbon dioxide + ammonia + nitrates

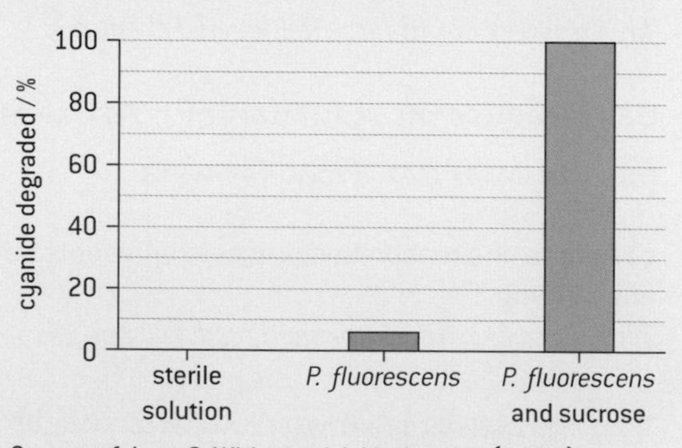

Source of data: C. White and J. Markweise (1994), *Journal of Soil Contamination*, **3**, pp. 271–283.

To research the conditions that result in maximum degradation of cyanide, researchers sprayed different samples of cyanide-processed ore with one of three solutions:

- a sterile solution
- a solution containing a culture of *P. fluorescens*
- a solution containing a culture of *P. fluorescens* and sucrose.

a) Outline the evidence that *P. fluorescens* can degrade cyanide.

b) Sugest how the addition of sucrose promotes the degradation of cyanide.

Solution

a) With the sterile solution, which is the control, there is no degradation of cyanide but there is degradation with *P. fluorescens*.

b) Sucrose provides the energy or carbon source in this reaction. In the solution containing *P. fluorescens* and sucrose, the amount of degradation of cyanide is higher than without sucrose. There was no control with sucrose only—this would help establish causality.

In bioremediation of contaminated soils, microorganisms are used to degrade pollutants such as hydrocarbons, oil, heavy metals, pesticides, herbicides and dyes. This technique takes advantage of the ability of certain microorganisms to metabolize these pollutants. It is an environmentally friendly, low-cost and efficient environmental clean-up technique. A few examples of bioremediation are the conversion of methyl mercury into elemental mercury by *Pseudomonas*, the degradation of benzene by halophilic bacteria and the degradation of oil by *Pseudomonas*.

- **Bioremediation** is the use of microorganisms to degrade contamination.
- **Pseudomonas** is a genus of bacteria used in the bioremediation of oil and mercury.

Industrial and urban activities, such as those involving the metal and plastic industries, fluorescent lights and electrodes, may result in mercury (Hg) contamination, which adversely affects the health of human beings and wildlife. Environmental contamination by mercury is a public health concern because of the neurotoxic substance methyl mercury. This substance can bioaccumulate and biomagnify in food chains. The transformation of methyl mercury to the less toxic element mercury is performed by *Pseudomonas putida*.

Biofilms are populations of microorganisms, generally bacteria, that grow together on a surface which can be solid (living or non-living) or liquid. These biofilms present emergent properties; these are properties that are not present in the individual organism. Biofilms are better able to survive changes in the environment, for example changes in pH. The biofilm organisms have their metabolisms altered so they can produce substances including the matrix they live on, which in many cases is exopolysaccharide (EPS). This matrix acts as a protective layer, preventing antibiotics or bacteriostatic substances from affecting the organisms. They form channels through which food can travel faster. They synchronize their gene expression through a mechanism called quorum sensing; this means one organism senses there are more organisms and only then does it express certain proteins. All of these advantages allow microorganisms in biofilms to survive in different environments, for example, in water pipes and on teeth, catheters and shower caps.

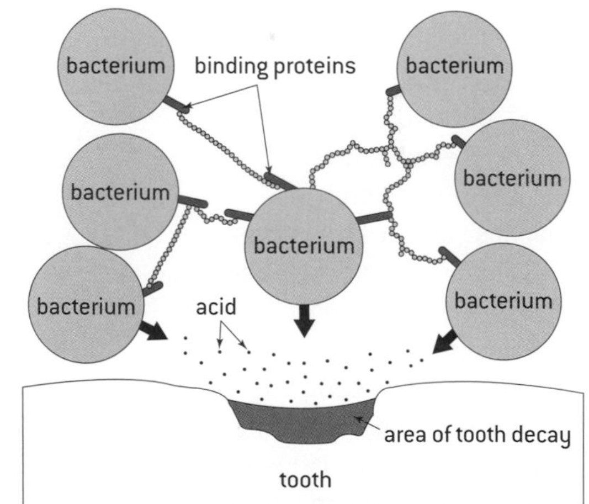

SAMPLE STUDENT ANSWER

Using the diagram on the right, explain the emergent properties of biofilms. [3]

This answer could have achieved 0/3 marks:

You can see that the tooth is attacked because it presents cavities. Each bacterium has proteins that join them together attacking the tooth.

▼ This student described what could be seen, but did not answer the question, therefore scored no marks.

Bacteriophages are viruses that infect bacteria. These viruses are unaffected by antibiotics and are able to kill bacteria within biofilms. Bacteriophages replicate within their host cells, amplifying the bacteriophage numbers which are released within the biofilm. By spreading through the biofilm, eliminating the bacteria that produce the EPS matrix, bacteriophages destroy the biofilm. Bacteriophages can also themselves degrade the EPS. Bacteriophages are used in the disinfection of water systems, as they destroy biofilms that form in pipes and tanks.

B.4 MEDICINE (AHL)

You should know:

- ✔ infection by a pathogen can be detected by the presence of its genetic material or by its antigens.
- ✔ predisposition to a genetic disease can be detected through the presence of markers or metabolites in blood or urine or through DNA microarrays.
- ✔ biopharming techniques are used to produce protein products for therapeutic use.
- ✔ tracking experiments are used to localize and study the interactions of a desired protein.

You should be able to:

- ✔ describe how pathogens can be detected in organisms.
- ✔ describe how genetic predisposition can be detected.
- ✔ explain the use of genetic markers.
- ✔ analyse simple DNA microarrays.
- ✔ explain the use of viral vectors in the treatment of SCID.
- ✔ describe the use of PCR to detect different strains of influenza virus.
- ✔ describe the use of transferin linked to luminescent probes to detect tumours.
- ✔ explain how biopharming can be used in the production of antithrombin.
- ✔ interpret results of ELISA diagnostic tests.

Developments in scientific research and technology have allowed scientists to diagnose and treat diseases. Preventative medicine is becoming a desirable therapeutic approach and patients are beginning to demand fast and early qualitative diagnosis. Rapid developments in technology have facilitated the production of tests and devices that can detect diseases. Using artificial intelligence and web-based knowledge systems we can now monitor disease epidemiology. One example is a blood test that can detect prostate gland cancer and can distinguish between benign cancers and metastatic disease. This test detects molecular changes in the prostate-specific antigen in blood, reducing the need for biopsies.

Infection of an organism by a pathogen can be detected by polymerase chain reaction (PCR) techniques that show the presence of the genetic material of the pathogen. This test is usually performed on a blood sample. Another method of detecting pathogens is by observing the reaction of antibodies specific to the pathogen antigens.

Example B.4.1.

Tobacco plants were transformed to include the hepatitis B virus gene coding for surface antigen. Surface antigen gene expression levels were determined by ELISA.Transformed tobacco plant callus cells were analysed for the integration of the gene by PCR. A band of 681 base pairs confirms the transformation (gel on the right).

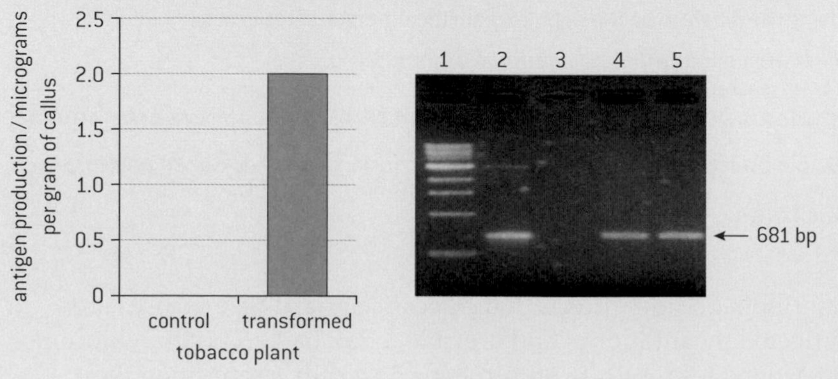

a) Suggest a reason why this is an example of biopharming.

b) Explain how the ELISA test is used to detect the presence of the hepatitis B antigen.

c) Describe the PCR method used in this experiment to detect the expression of the antigen gene.

Solution

a) Biopharming is the production and use of transgenic animals and plants genetically engineered to produce pharmaceutical substances for use in humans or in other animals. In this case the gene for the hepatitis B antigen is inserted in tobacco plants in order to produce large amounts of this protein that will serve as a vaccine in humans.

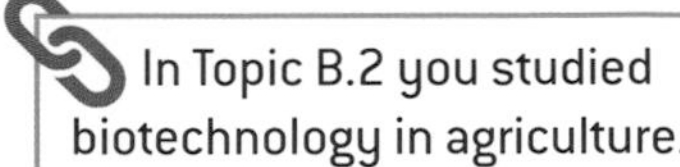
In Topic B.2 you studied biotechnology in agriculture.

b) In this ELISA, hepatitis B surface antigen antibodies are fixed to a surface (usually a special dish with many wells). These antibodies bind to and capture the hepatitis B antigen when a sample of transformed tobacco plants is placed on them. This sample is obtained by crushing calluses of tobacco plants containing transformed cells and by purifying the proteins present in them. Once the antigen has bound to the fixed antibody, the wells are thoroughly cleaned to avoid any free antigen. Next, an antibody against this antigen linked to an enzyme is added. This antibody will attach to any antigen that bound to the fixed antibody. The wells are washed again to remove any free antigen. Lastly, a substrate for the enzyme is added, which changes colour when the enzyme is present. This change in colour is an indication of the presence of the antigen. Different dilutions of the second antibody can be used to detect the concentration of the antigen present in the sample.

c) PCR is a method used to amplify small amounts of DNA. An enzyme, DNA polymerase, is added to a tube containing primers, nucleotides and buffer. The tubes are placed in a thermal cycler. The temperature is increased to separate the DNA strands (melting stage). The complementary primers stick to the DNA because they are complementary. The DNA polymerase will then add nucleotides in a 5′ to 3′ direction, extending the DNA. There is a forward primer which will copy one strand and a reverse primer which will copy in the opposite direction. This process occurs in many cycles, increasing the amount of DNA exponentially.

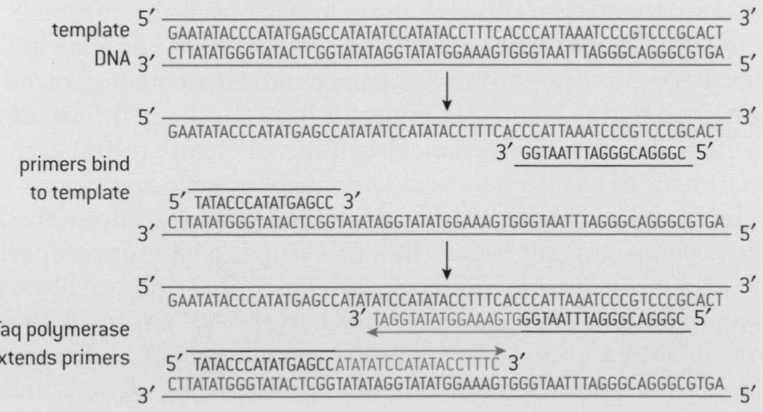

Assessment tip

You can give your answer as an annotated diagram, as shown.

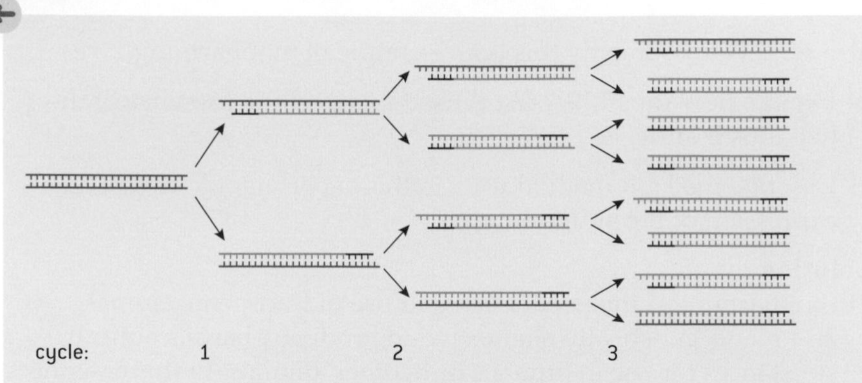

B.5 BIOINFORMATICS (AHL)

You should know:

- ✔ databases store information on DNA, RNA and protein sequences.
- ✔ databases allow scientists easy access to information and the body of data stored in databases is increasing exponentially.
- ✔ BLAST searches can identify similar sequences in different organisms and sequence alignment software allows comparison of sequences from different organisms.
- ✔ BLASTn allows nucleotide sequence alignment while BLASTp allows protein alignment.
- ✔ model organisms can be used to predict gene functions.
- ✔ databases can be searched to compare newly identified sequences with sequences of known function in other organisms.
- ✔ EST is an expressed sequence tag used to identify genes.

You should be able to:

- ✔ deduce information from databases.
- ✔ predict gene functions using bioinformatic tools.
- ✔ compare sequences using alignments.
- ✔ explain how mutating a gene can be used to determine its function.
- ✔ identify genes using EST mining.
- ✔ analyse chromosome 21 genes using bioinformatic tools.
- ✔ use software such as BLASTp to align proteins.
- ✔ design phylogenetic cladograms from sequence alignments.

A bioinformatic database is a collection of data organized in such a way that other scientists can access useful information. There are many different databases, allowing users to perform a chromosome gene search (Embl), view the structure of proteins (PDB) or analyse sequences. Particularly significant sequence databases are organized and maintained by the National Center for Biotechnology Information (NCBI) which is part of the National Institutes of Health (NIH) of the U.S. Department of Health. The NCBI advances science and health research by providing access to biomedical and genomic information. The Basic Local Alignment Search Tool (BLAST) is a tool for comparing an amino acid or nucleotide sequence with an entire sequence library, identifying regions of high sequence similarity. BLASTn is used for sequences of DNA and BLASTp for protein sequences. BLAST software can be used for sequence alignment and to produce phylograms and cladograms.

Example B.5.1.

The image obtained from chromosome 21 in Ensemble is shown. The combination of numbers and letters provides a gene's "address" on a chromosome.

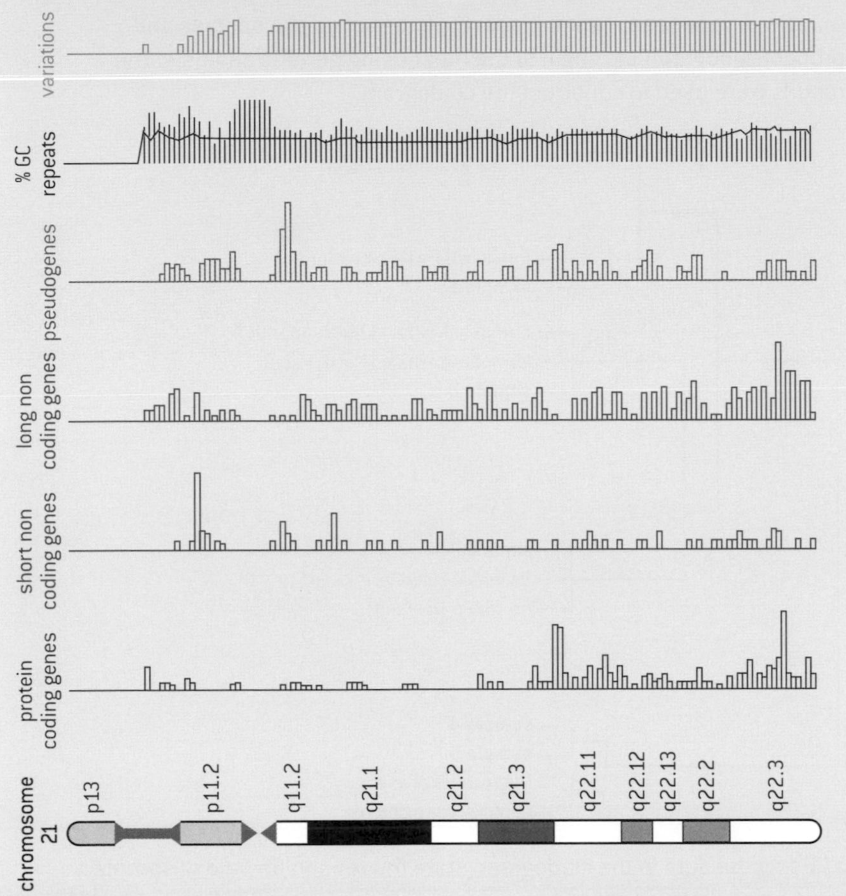

a) State the meaning of the letters p and q on the chromosome.

b) Identify the address of one gene close to the centromere.

c) Explain how expressed sequence tag (EST) mining could be used to find genes.

Solution

a) p signifies the short arm and q the long arm of the chromosome.

b) The genes p11.2 and q11.2 are the closest to the centromere.

c) An EST is a short sequence of complementary DNA (cDNA) of approximately 500 to 800 nucleotides. This cDNA sequence is obtained by reverse transcription of mRNA. Because the DNA is complementary to the mRNA, the ESTs represent portions of expressed genes. They may be represented in databases either as a cDNA/mRNA sequence or as the reverse complement of the mRNA, the template strand. The ESTs can be shown on specific chromosome locations. The EST sequence can be aligned with the genome using bioinformatic software.

SAMPLE STUDENT ANSWER

International agreement limits the hunting of whales. Only the meat of the minke, fin and humpback whales from Southern Hemisphere populations is allowed to be sold on the domestic market in Japan. Scientists obtained five samples of food that were being sold as "whale meat" in a Japanese market place. They identified the species and probable geographical origin of the meat using genetic analysis. The results were used to construct the cladogram.

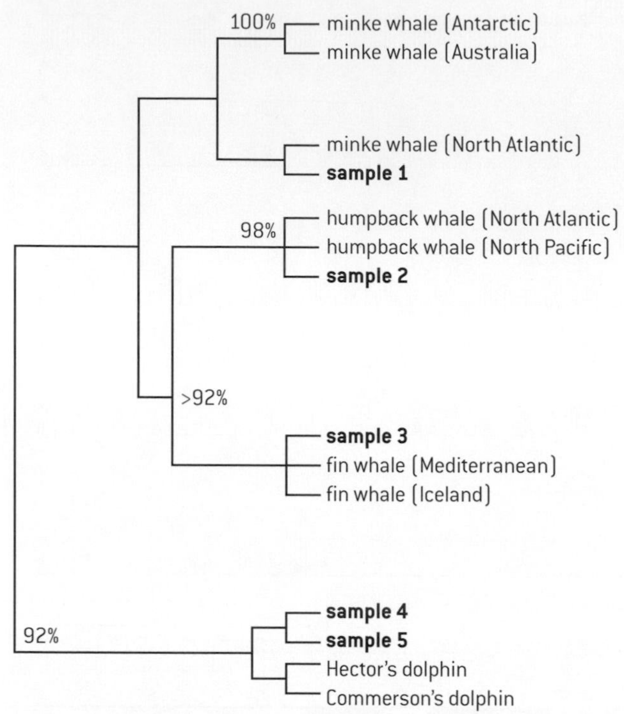

a) Using the data in the cladogram, state the reason for sale of sample 1 meat being illegal in Japan. [1]

This answer could have achieved 1/1 marks:

Sample 1 meat is illegal in Japan since it is closely related to the minke whale from the North Atlantic, suggesting that the Sample 1 organism is likely to be from a nearby area to the North Atlantic. Therefore it is illegal since only minke whale meat from the Southern Hemisphere is allowed to be sold in Japan.

b) Using the data in the cladogram, state the reason for the sale of sample 4 meat being illegal in Japan. [1]

This answer could have achieved 1/1 marks:

Sample 4 is illegal in Japan since it shares a common ancestor with dolphins and since it is located in a different clade than the whales, it is likely not to be a whale. Therefore it is illegal because only whales are allowed to be sold in Japan.

c) Outline how the polymerase chain reaction (PCR) might have been used in this study. [3]

This answer could have achieved 0/3 marks:

Polymerase chain reaction is used to amplify DNA. It could be used alongside gel electrophoresis to identify markers for certain genes. Scientists may have compared a specific gene in the organism's DNA. They could have compared the specific gene by amplifying the DNA, cutting it with restriction enzymes and running the fragments on an electrophoretic gel to compare the fragments. The gene fragments obtained could be compared for similarities and differences with known samples.

▼ The student has correctly described genetic fingerprinting, but the question has not been answered. In this case PCR was used to amplify small amounts of DNA (small had to be included). The primers used were from the "legal" samples of conserved genes such as 18S ribosomes or cytochrome C oxidase. If no amplification occurs then the test was negative. If the gene obtained was very different from the legal gene, then the sample was "illegal".

d) Explain how sequence alignment software might have been used in this study. [2]

This answer could have achieved 2/3 marks:

Sequence alignment software is used to align sequences of DNA, proteins or mRNA to identify similarities and differences between sequences. The DNA of the samples might have been aligned alongside the known sequences of other organisms to determine similarities between the two sequences. The percentage of similarities and certain differences identified through sequence alignment software have been used to construct the cladogram.

▲ The student has correctly identified that sequences of DNA or protein are used to find similarities and differences with a known sequence (in this case the "legal" whale sequence) in order to construct the cladogram.

▼ The student failed to mention the software that could be used, such as BLASTn for DNA sequences or BLASTp for protein sequences.

Gene function can be studied using model organisms with similar sequences by using knockout technology. By causing a specific gene to be inactive in an organism and observing any differences in behaviour or physiology, researchers can infer the probable function of the gene.

Practice problems for Option B

Problem 1

In India around 5% of the biogas consumed is produced in small-scale family-run biogas plants. Describe and explain the production of biogas from organic matter.

Problem 2

Describe an inhibition zone experiment to show the resistance of bacteria to three different antibiotics.

Problem 3

Explain how plants can be genetically modified to overcome environmental resistance to increase crop yields.

Problem 4

Microorganisms such as bacteria can be used in response to pollution incidents. Explain how bacteria can be used in response to an oil spill.

Problem 5 (AHL)

Explain how simple DNA microarrays can be used to detect disease.

Problem 6 (AHL)

Explain the use of viral vectors in the treatment of SCID.

Problem 7 (AHL)

Explain how mutating a gene can be used to determine its function.

C ECOLOGY AND CONSERVATION

C.1 SPECIES AND COMMUNITIES

You should know:

- ✔ the distribution of species is affected by limiting factors.
- ✔ community structure can be strongly affected by keystone species.
- ✔ each species plays a unique role within a community because of the unique combination of its spatial habitat and interactions with other species.
- ✔ interactions between species in a community can be classified according to their effect.
- ✔ two species cannot survive indefinitely in the same habitat if their niches are identical.

You should be able to:

- ✔ describe the distribution of one animal and one plant species to illustrate limits of tolerance and zones of stress.
- ✔ describe local examples to illustrate the range of ways in which species can interact within a community.
- ✔ explain the symbiotic relationship between zooxanthellae and reef-building coral reef species.
- ✔ analyse data sets that illustrate the distinction between fundamental and realized niche.
- ✔ explain the use of transects to correlate the distribution of plant or animal species with an abiotic variable.

In Topic 4.1 you studied species, communities and ecosystems.

- A **keystone species** is one in which its presence has a disproportionate impact on ecosystem and its removal often leads to significant changes.
- A **niche** is the role or function of an organism or species in an ecosystem.
- A **fundamental niche** of a species is the potential mode of existence.
- A **realized niche** is the actual mode of existence, depending on adaptations and competition with other species.
- **Competitive exclusion** is a principle which states that two species cannot occupy the same niche.

The distribution of species is affected by limiting factors. Plant distributions are affected by abiotic factors such as light, temperature, water, pH, salinity and minerals. Animal distributions are affected by food, temperature, water, breeding sites and geographical factors. Aquatic organisms are also affected by light, pH and salinity. There are biotic factors affecting distribution such as predation, herbivory and competition.

The community structure is an emergent property of an ecosystem, which means the whole community has certain properties that only exist because of the interactions of the individual organisms within that community.

Zones of stress and limits of tolerance graphs are models of the real world that have predictive power and explain community structure. Each species has a range of values for biotic and abiotic factors that can be tolerated. Nevertheless, within a population there is variability, where some are more tolerant than others. An example of such limits of tolerance of abiotic factors is shown in populations of the marine fish *Cheilodactylus spectabilis* in the Tasman Sea. Here, increasing temperatures coincide with increased growth but only up to a certain temperature. Increasing water temperatures have pushed the species past the point where warming is beneficial, as these high temperatures result in higher metabolic costs and less availability of energy for growth and reproduction. As a result, *C. spectabilis* will be absent in those areas where water temperatures surpass their limits of tolerance.

The keystone species has a disproportionate effect because even a few organisms affect the environment more than expected. Examples of keystone species are the sea star *Pisaster*, sea otters, elephants, prairie dogs and mountain lions.

SAMPLE STUDENT ANSWER

Outline what is meant by a keystone species. [2]

This answer could have achieved 1/2 marks:

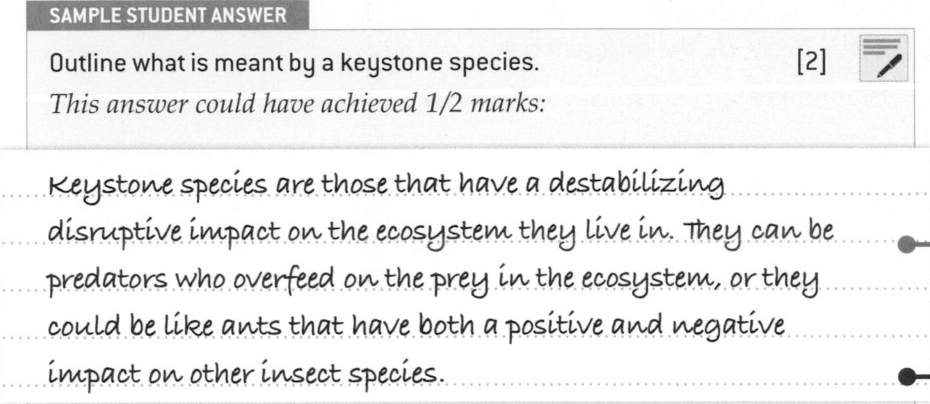

Keystone species are those that have a destabilizing disruptive impact on the ecosystem they live in. They can be predators who overfeed on the prey in the ecosystem, or they could be like ants that have both a positive and negative impact on other insect species.

▲ This answer scored one mark for saying that it could be a species that is a predator that overfeeds on the prey, therefore implying that the impact of the organism is significant.

▼ The answer fails to mention that the removal of the keystone organism has a disproportionate impact on the ecosystem.

Each species plays a unique role within a community because of the unique combination of its spatial habitat and interactions with other species. Interactions between species in a community can be classified according to their effect. These interactions can be competition, predation, grazing, mutualism, symbiosis, parasitism and commensalism.

Competition can be for resources such as food, water, sunlight, breeding sites or territory. Intraspecific competition is between members of a species while interspecific competition is between different species. Two species cannot survive indefinitely in the same habitat if their niches are identical due to competitive exclusion.

Example C.1.1.

Acanthocephalus tahlequahensis is a parasitic worm often found in fish intestines. The percentage of fish infected by *Acanthocephalus* parasites was measured in freshwater fish.

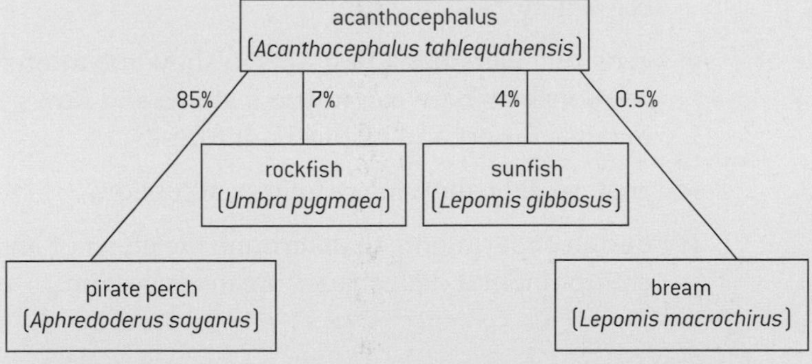

a) Define parasitism using the data provided as examples.

b) State the species where most fish presented parasites.

c) The bream (*Lepomis macrochirus*) are recent invaders of this environment and are relatively resistant to the parasite. Comment on this statement based on the results shown.

d) Suggest with a reason whether the bream (*Lepomis macrochirus*) and the sunfish (*Lepomis gibbosus*) show intraspecific competition.

Assessment tip

Many students just mention that keystone species are very important to the environment. Although this is true, it is not really explaining what it is.

- **Competition** is a situation in which one organism or species contends for the same resources as another.
- **Predation** is when an organism (predator) kills a prey to feed on it.
- **Grazing** is when herbivores feed on plants or algae.
- **Symbiosis** is a long-term interaction between different biological species. It can be mutualism, parasitism or commensalism.
- **Mutualism** is when both organisms benefit. For example birds feeding on ectoparasites of cattle.
- **Parasitism** is when an organism lives on/in a host and feeds from it, causing the host harm. For example, worms in animal intestines.
- **Commensalism** is when one species gains benefits while the other species is neither benefited nor harmed. For example remora fish feed on the parasites found on the skin of sharks.

Solution

a) Parasitism is when one organism feeds off another (host), causing it harm but not normally killing it; therefore the host is harmed and the parasite benefits. The parasitic organism in this case is the acanthocephalus (*Acanthocephalus tahlequahensis*) and the hosts are the different fish.

b) Pirate perch (*Aphredoderus sayanus*).

c) The bream shows the lowest percentage of fish infected by parasites. This could be due to two reasons: one may be that they are resistant to the parasite, therefore supporting the statement. Another reason could be that as they have only recently been introduced in the environment, they have not had time to become infected.

d) They cannot have intraspecific competition because they are not from the same species, they are from the same genus. If they do compete, it would be interspecific competition.

C.2 COMMUNITIES AND ECOSYSTEMS

You should know:

- ✔ most species occupy different trophic levels in multiple food chains.
- ✔ a food web shows all the possible food chains in a community.
- ✔ the percentage of ingested energy converted to biomass is dependent on the respiration rate.
- ✔ the type of stable ecosystem that will emerge in an area is predictable based on climate.
- ✔ in closed ecosystems energy but not matter is exchanged with the surroundings.
- ✔ disturbance influences the structure and rate of change within ecosystems.

You should be able to:

- ✔ analyse food webs.
- ✔ explain conversion ratio in sustainable food production practices.
- ✔ explain how humans interfere with nutrient cycling.
- ✔ compare pyramids of energy from different ecosystems.
- ✔ analyse climographs showing the relationship between temperature, rainfall and the type of ecosystem.
- ✔ construct Gersmehl diagrams to show the inter-relationships between nutrient stores and flows in taiga, desert and tropical rainforest.
- ✔ analyse data showing primary succession.
- ✔ design experiments to determine the effect of an environmental disturbance on an ecosystem.

You studied energy flow in Topic 4.2.

Assessment tip

Examples of aspects investigated in the ecosystem that you could be tested on could be species diversity, nutrient cycling, water movement, erosion and leaf area index, among others.

An ecosystem is a community made up of living organisms and its interaction with the abiotic environment such as air, water and mineral soil.

Changes in community structure affect and are affected by organisms. Most species occupy different trophic levels in multiple food chains; therefore it is not very practical to show these relationships in a chain. A food web is a more realistic model as it shows all the possible food chains in a community.

Example C.2.1.

The diagram is of an ocean food web.

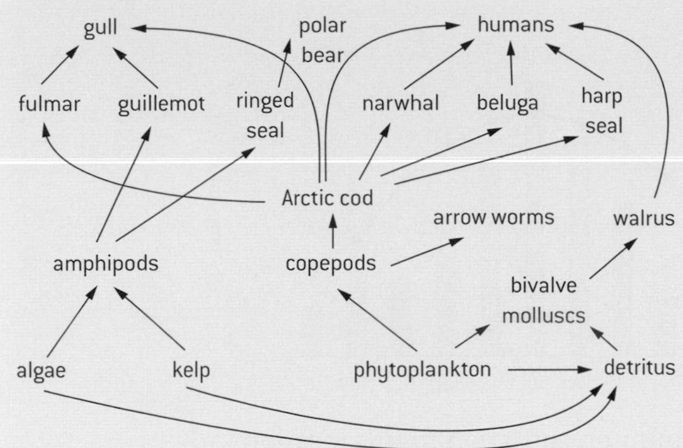

a) State a producer in this web.

b) State the trophic level(s) of the gull.

c) Predict what would happen to this web if a disease killed most guillemots.

Solution

a) Algae, kelp or phytoplankton

b) Third and fourth consumer

c) Amphipods will increase in number as they stop being eaten by guillemots. This means that algae and kelp will decrease, as they will be eaten more. At the same time, as gulls can no longer feed on guillemots, they will feed more on fulmar and Arctic cod, so these will decline.

Assessment tip

In a), only one of these answers is expected and in b), both answers are needed to score the mark.

Pyramids of energy model the energy flow through ecosystems. Energy is lost from one trophic level to the other. Most of the energy is lost as heat. The percentage of ingested energy converted to biomass is dependent on the respiration rate. Energy can also be lost as indigestible material (eg, cellulose and fibre for humans) and as uneaten parts such as hair and teeth. The loss of energy from one trophic level to the other can be shown with a pyramid of energy.

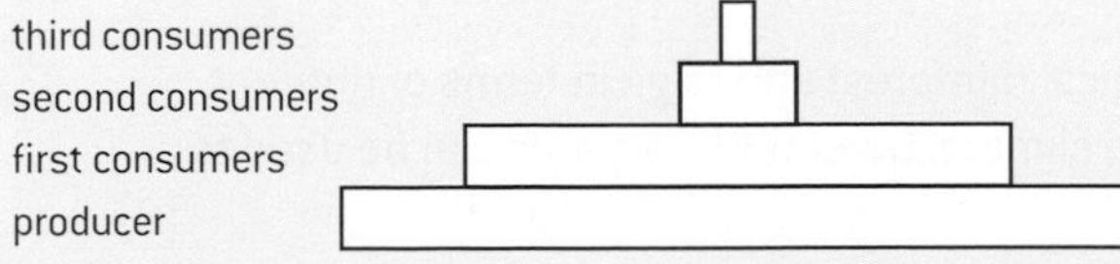

• A **biome** is a community of living organisms that share the same climate.

The type of stable ecosystem that will emerge in an area is predictable based on climate. Climographs represent the monthly average temperature and precipitation of a location. A biome is a community of living organisms living in a common climate. It can comprise many habitats. Examples of biomes are deserts, tropical rainforests and taiga (snow forests).

Example C.2.3.

The diagram shows a climograph.

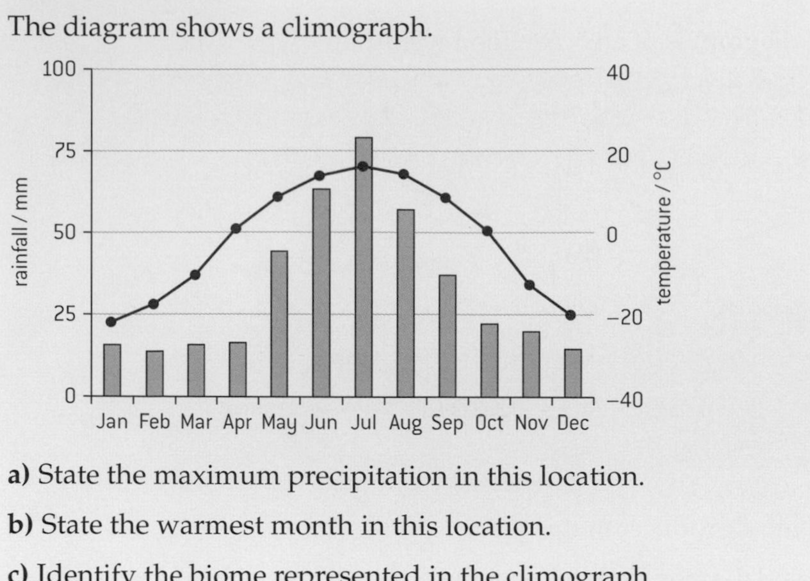

a) State the maximum precipitation in this location.

b) State the warmest month in this location.

c) Identify the biome represented in the climograph.

Solution

a) 78 mm **b)** July **c)** taiga

Gersmehl diagrams are models of the inter-relationships between nutrient stores and flows between biomass, litter and soil. The diameter of the circles is an indication of the amounts stored, the width of the arrows gives an indication of amounts flowing and the arrow shows the direction of the flow.

- A **taiga** is a temperate forest with cold winters and warm summers. There is little precipitation. Its main store is litter (pine needles) with little flow between stores.
- A **desert** is very dry with very hot days and cold nights. Its main store is in soil and there is major flow from biomass to litter.
- A **tropical rainforest** is rainy and hot. Biomass is its main store with a high rate of flow.

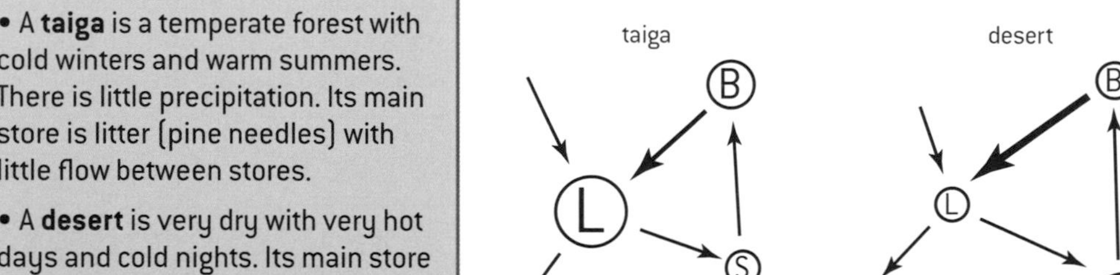

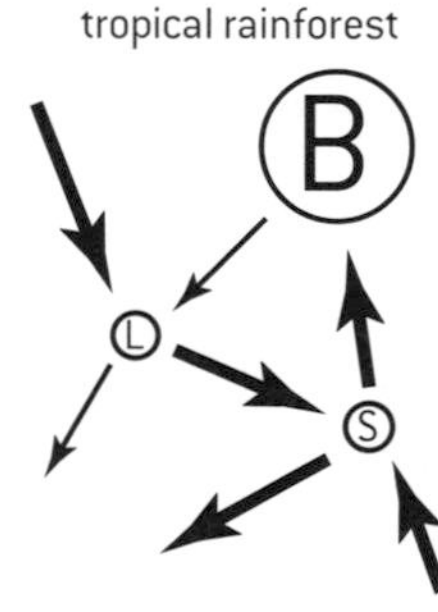

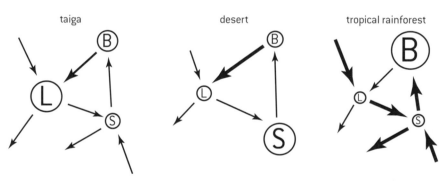

▲ **Figure C.2.1.** Gersmehl diagrams for taiga, desert and tropical rainforest

SAMPLE STUDENT ANSWER

Distinguish between tropical rainforest and taiga in terms of nutrient stores, nutrient flows and climate. Gersmehl diagrams can be used to support your answer. [6]

This answer could have achieved 2/6 marks:

Tropical rainforest has the largest plant species biodiversity of all the land biomes. It has an abundant amount of life, and therefore the amount of nutrient for each energy level is very abundant. Taiga is the largest terrestrial biome, but the diversity of its plant and animal life is much lower, consisting of large mammals and coniferous trees. Because the rainforest is so diverse, the amount of predators and

▲ This answer scored two marks for distinguishing the levels of precipitation and temperature in both biomes.

▼ This answer needed to address the differences in nutrient store, nutrient flow and climate.

prey ensure that there are many trophic levels. Consumers like monkeys can also be secondary consumers and in turn be consumed by tertiary consumers. Taiga does not have as many levels or such complex food webs due to the lesser amount of species. Tropical rainforests have heavy precipitation and the air is usually humid and hotter than many biomes all year round. Taiga is much drier, the temperature is cooler as most taiga are near or in the mountains, and there is heavy snowfall but not as much precipitation.

Assessment tip

The Gersmehl diagrams are based on biomass, nutrient storage in litter and nutrient storage in soil and the flow between these. The answer had to be focused on these, as the question clearly stated you could use these diagrams.

Living organisms in an ecosystem can interconvert energy. Light energy can be converted into chemical energy, which can be converted into kinetic energy. Chemical energy can also be converted into heat energy. Heat energy cannot be converted into any other form of energy; therefore it is lost from the ecosystem. In closed ecosystems energy but not matter is exchanged with the surroundings. The environments can have different energy conversion rates which depend on the organisms involved. These are measured as the average net primary productivity of the environment, which is measured in mass per unit area per unit time (eg, $g/m^2/yr$).

Feed conversion ratio is how efficient the bodies of organisms are at converting their food into the desired output (eg, milk or meat). It is the mass of the input (food given) divided by the output (mass of food produced). Poikilotherms (animals that have a variable body temperature) are more effective producers of protein than homeotherms (animals that maintain a regulated body temperature) as they have a higher rate of conversion of food to biomass.

Example C.2.4.

The efficiencies of production of some food meat products and of crickets are shown in the table.

	Cricket	Beef	Chicken
Feed conversion ratio (FCR)	1.7	10	2.5
Edible portion / %	80	40	55

a) State the least efficient food source.

b) (i) Distinguish between beef and chicken as food sources.

(ii) Explain reasons for the differences.

c) Using the data, discuss whether crickets are a good source of food.

Solution

a) Beef

b) (i) Chicken has a larger percentage of edible parts and a lower FCR than beef.

- **Biomass** is the total dry organic matter.
- **Gross production** is the total amount of organic matter produced per unit area per unit time by a trophic level in an ecosystem.
- **Net production** is the amount of gross production remaining after the amount used for respiration by the trophic level has been subtracted.
- **Feed conversion ratio (FCR)** is the efficiency of production of an animal food source.
- **Poikilotherms** are animals whose body temperature varies with the environment.
- **Homeotherms** are animals that maintain their body temperature.

(ii) Chicken is more efficient as a food source because it has a lower FCR, this mean that it needs to be fed less to obtain the same amount of meat to eat. This is probably because it is a smaller animal therefore requires less energy to maintain its body temperature. Another reason is that chickens may be kept in small cages. If they are not allowed to move around, this saves energy.

c) Crickets are a very good source of food, as they have the lowest FCR. This is mainly due to the fact that most parts of their body are edible, little is lost. One problem is that not everybody likes eating insects, sometimes just because of taste, but other times due to religious issues.

- **Primary succession** occurs in areas that were not inhabited by living organisms.
- **Secondary succession** occurs in areas that were already inhabited but suffered disturbance.

Ecological successions are processes where the species structure of an ecological community changes over time. Primary succession occurs in environments where there were no soil or organisms, such as lava flows or glacier retreats. Secondary succession occurs in areas that previously supported organisms before an ecological disturbance such as a fire, flood or agricultural practices. Disturbance influences the structure and rate of change within ecosystems.

C.3 IMPACTS OF HUMANS ON ECOSYSTEMS

You should know:

- ✔ introduced alien species can escape into local ecosystems and become invasive.
- ✔ competitive exclusion and the absence of predators can lead to reduction in the numbers of endemic species when alien species become invasive.
- ✔ pollutants become concentrated in the tissues of organisms at higher trophic levels by biomagnification.
- ✔ macroplastic and microplastic debris has accumulated in marine environments.

You should be able to:

- ✔ explain the effect of the introduction of an alien species, especially the cane toads in Australia.
- ✔ evaluate eradication programmes and biological control as measures to reduce the impact of alien species.
- ✔ analyse data illustrating the causes and consequences of biomagnification.
- ✔ discuss the trade-off between control of the malarial parasite and DDT pollution.
- ✔ explain the impact of marine plastic debris on Laysan albatrosses and one other named species.

In Topic 4.1 you studied species, communities and ecosystems.

Assessment tip

In Topic 5.4 you studied reclassification of organisms using evidence from cladistics. You will find many books refer to the cane toad as *Bufo marinus*, but it has now been reclassified as *Rhinella marina*.

The use of biological control has associated risk and requires verification by tightly controlled experiments before it is approved. One of the most well-known invasive species was introduced as a biological control. The cane toad (*Rhinella marina*) was introduced to Australia to control the cane beetle (*Dermolepida albohirtum*). The cane toad and their tadpoles are highly toxic to most animals if ingested. As they do not have natural predators, they have become a competitor for food resources and have spread throughout the country. Decreases in the populations of certain species of frogs (*Limnodynastes peronii*), large lizards (*Bellatorias major*, *Intellagama lesueurii*, and *Varanus varius*) and snakes (*Pseudechis porphyriacus*) occurred following the introduction of *R. marina*.

Introduced alien species can escape into local ecosystems and become invasive. Competitive exclusion and the absence of predators can lead to reduction in the numbers of endemic species when alien species become invasive. Biological control has been used to try and control alien species.

Pollutants become concentrated in the tissues of organisms at higher trophic levels by biomagnification. Examples of pollutants are pesticides, industrial wastes, heavy metals and plastics. Macroplastic and microplastic debris have accumulated in marine environments.

• **Biomagnification** is the increasing concentration of toxins in the tissues of organisms at successive higher levels in a food chain.

SAMPLE STUDENT ANSWER

a) Outline the concept of biomagnification of plastic pollution in oceans. [3]

This answer could have achieved 2/3 marks:

Biomagnification is when toxins released from primarily human general waste or antibiotics are passed along to each trophic level. The toxins increase in destructive capacity as they pass from prey to predator, and plastic rafts contain toxins that could have such an effect on the ecosystem.

▲ This answer scored one mark for saying that toxins are concentrated in each successive level up the food chain (although the term destructive is not the ideal). The second mark is for saying that toxins are absorbed by lower trophic level organisms (when they mention from prey to predator).

▼ The question was about plastics, so it should have mentioned that plastics in the ocean can release toxins (not just human wastes). The answer does not mention that microplastics are directly ingested or consumed by animals such as birds. As the toxins are not metabolized by organisms they accumulate in tissues.

b) Other than biomagnification, outline **two** concerns associated with mobility of plastic rafts and the communities they host. [2]

This answer could have achieved 2/2 marks:

The plastic on these rafts is hazardous to other animals that do not know what it is. Birds might consume them as food; ingesting macroplastic can choke the bird or worsen its health. Microplastics can be ingested by animals and the toxins can result in biomagnification. The fact that these islands are motile means this can affect a wide selection of animals.

▲ This answer scored one mark for saying that birds can choke on macroplastics and another for saying that plastics can worsen their health.

▼ The answer failed to mention that these islands can introduce pathogens into areas where the pathogen is not found. Another issue not mentioned is that the introduced species may become invasive.

DDT is a pesticide used to reduce pests that are disease vectors. It was widely used in the reduction of mosquitoes carrying malaria. This reduction of mosquitoes led to a reduction in malaria. Unfortunately, the rates of biomagnification in food chains caused a negative impact on health of the top predators. For example, the thinning of eggshells and reduced reproductive success in birds of prey.

SAMPLE STUDENT ANSWER

State one advantage and one disadvantage regarding the use of DDT. [2]

This answer could have achieved 2/2 marks:

DDT prevents the spread of illnesses such as malaria through insects and then prevents humans getting that illness from them. However, DDT releases toxins as soon as those infected with it are consumed, and biomagnification ensures that it detrimentally impacts other species.

▲ This answer scored one mark for mentioning the reduction in the levels of the malarial parasite and the other mark for biomagnification of DDT. The student achieved maximum marks.

C.4 CONSERVATION OF BIODIVERSITY

You should know:

- ✔ entire communities need to be conserved in order to preserve biodiversity.
- ✔ an indicator species is an organism used to assess a specific environmental condition.
- ✔ relative numbers of indicator species can be used to calculate the value of a biotic index.
- ✔ *in situ* conservation may require active management of nature reserves or national parks.
- ✔ *ex situ* conservation is the preservation of species outside their natural habitats.
- ✔ biogeographical factors affect species diversity.
- ✔ richness and evenness are components of biodiversity.

You should be able to:

- ✔ describe cases of captive breeding and reintroduction of an endangered animal species.
- ✔ analyse the impact of biogeographical factors on diversity limited to island size and edge effects.
- ✔ define biotic index.
- ✔ define biodiversity in terms of richness and evenness.
- ✔ analyse the biodiversity of two local communities using Simpson's reciprocal index of diversity.

In Topic 4.1 you studied species, communities and ecosystems.

- ***In situ* conservation** refers to conservation of species in their own natural habitat, for example in natural reserves or national parks.
- ***Ex situ* conservation** is the preservation of species outside their natural habitats, for example, in zoos, botanical gardens or seed banks.

The preservation of species involves international cooperation through intergovernmental and non-governmental organizations. The Andean condor (*Vultur gryphus*) was in danger of extinction. In 1991, the Andean Condor Conservation Program (PCCA) of Argentina developed artificial incubation programmes and techniques for hand rearing birds using puppets (without human contact) and worked to rescue and rehabilitate wild condors. The venture included public and private funding from different countries. In 20 years more than 160 condors have been released in all South America. Andean condors routinely cross borders between countries and, therefore, it was essential that international stakeholders unite to coordinate in its conservation.

An indicator species is an organism used to assess a specific environmental condition. Relative numbers of indicator species can be used to calculate the value of a biotic index. Indicator species can be pollution tolerant or pollution intolerant. If a pollution tolerant species is the only one found in the environment, there is a high probability that this environment is polluted. If on the other hand, there are pollution intolerant species present, then the environment must be unpolluted. Examples of pollution tolerant organisms are leeches and worms while pollution intolerant ones include stonefly larvae and caddisflies.

The formula for biotic index is the following:

$$\text{Biotic index} = \frac{\sum_{n=1}^{1}(n_i \times a_i)}{N}$$

Where n_i is the number of individuals of a species and a_i is the level of tolerance of that given species. The sign Σ means "the sum of"; this means that for each species its number is multiplied by the tolerance and this value is added to the next species and so on, until all i number of species have been added. N is the total number of individuals collected.

Example C.4.1.

The graph shows the evolution of the percentage of different ecological groups (from very tolerant to very sensitive) and the biotic index (BI) from 1990 to 2005 recorded in an estuary.

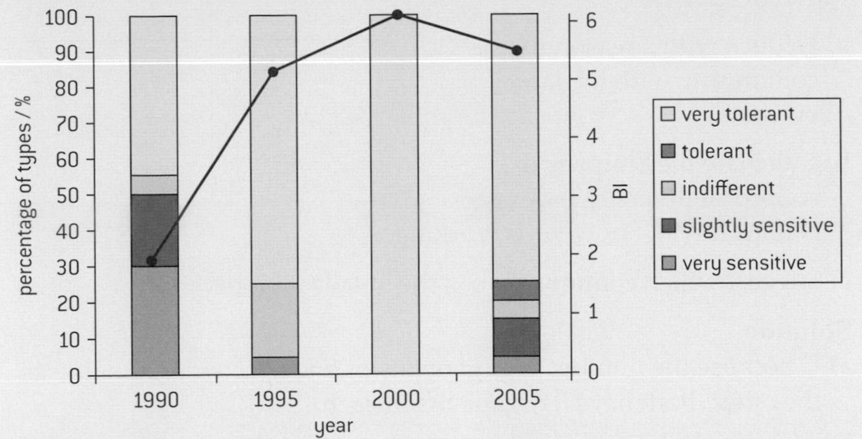

a) State the year with greatest amount of very sensitive species in the estuary.

b) Using the data, explain how the biotic index gives an idea of the quality of the water.

c) Suggest a reason for the lowering of the BI from the year 2000 to 2005.

Solution

a) 1990 (with 30%)

b) In this data, in the year 2000, one can see only very tolerant species and no sensitive species, therefore in this year the pollution was the greatest. When there are many very sensitive species, this means there is little pollution, therefore they can thrive, for example in 1990. The higher the biotic index, the higher the pollution.

c) In 2005, the very sensitive and slightly sensitive species reappeared (there were none in 2000). This is probably due to measures taken to stop pollution, for example banning a factory from sending their chemicals into the estuary.

Assessment tip

Any other reasonable suggestion will be accepted.

Biogeographical factors affect species diversity. For example, the larger a nature reserve is, the more efficient it is in preserving biodiversity and small reserves that are connected by corridors are more effective than those that are not connected. The shape of a reserve also affects biodiversity. Biodiversity differs close to the edge of a reserve and the shape determines the amount of edge there is, so, ideally, a reserve should have a large area and a small perimeter (like a circle). Richness and evenness must be considered when calculating biodiversity. Simpson's reciprocal index of biodiversity considers both of these numbers.

$$D = \frac{N(N-1)}{\sum n(n-1)}$$

Where D is the diversity index, N is the total number of species found and n is the number of organisms of each individual species. A large value for D means there is a lot of biodiversity.

- **Indicator species** are organisms that indicate the health of an ecosystem or level of pollution.
- **Biodiversity** measures the variation of species in a community.
- **Richness** is the number of different species present in a community.
- **Evenness** is how close in numbers each species in an environment is – how equal the community is numerically.

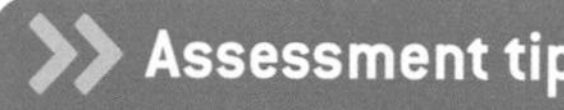

Assessment tip

Always show your working.

Example C.4.2.

The diagram represents four different communities. Each geometric figure represents a different species.

a) Deduce with a reason the community with the largest evenness.

b) Calculate the Simpson's reciprocal diversity index for community B. Show your working.

c) Suggest which community has the smallest biodiversity.

Solution

a) D because the number of organisms of one species is the same as the other (both have 3 organisms in each).

b) $D = \frac{4 \times 3}{(2 \times 1 + 1 \times 0 + 1 \times 0)} = 12/2 = 6$

c) C (with a Simpson's diversity index of 1). This is because it has a low richness.

C.5 POPULATION ECOLOGY (AHL)

You should know:

- ✔ dynamic biological processes impact population density and population growth.
- ✔ sampling techniques are used to estimate population size.
- ✔ the exponential growth pattern occurs in an ideal, unlimited environment.
- ✔ population growth slows as a population reaches the carrying capacity of the environment.
- ✔ the phases shown in the sigmoid curve can be explained by relative rates of natality, mortality, immigration and emigration.
- ✔ limiting factors can be top-down or bottom-up.

You should be able to:

- ✔ evaluate the methods used to estimate the size of commercial stock of marine resources.
- ✔ describe the use of the capture-mark-release-recapture method to estimate the population size of an animal species.
- ✔ discuss the effect of natality, mortality, immigration and emigration on population size.
- ✔ analyse the effects of population size, age and reproductive status on sustainable fishing practices.
- ✔ explain bottom-up control of algal blooms by shortage of nutrients and top-down control by herbivory.
- ✔ design models of the growth curve using a simple organism such as yeast or species of *Lemna*.

In Topic 4.1 you studied species, communities and ecosystems.

Sampling techniques are used to estimate population size. The capture-mark-release-recapture method is frequently used to estimate the population size of an animal species.

SAMPLE STUDENT ANSWER

Explain how the Lincoln index is used to determine population size. [4]

This answer could have achieved 4/4 marks:

The Lincoln index is calculated from the capture-mark-release-recapture method. It is calculated according to the number of individuals caught and marked initially and those recaptured in the second sample divided by the number of the recaptured which are marked.

$$\text{Population size} = \frac{N_1 \times N_2}{N_3}$$

Where N_1 is the number caught and marked initially, N_2 is the total number caught in 2nd sample and N_3 is the number of marked individuals recaptured.

▲ This answer covers all the points of the Lincoln diversity index.

The population sizes of commercial stock of marine resources can be estimated by echo sonography or by inferring it according to the catch age distribution. In the fishing industry, yields depend on natural processes such as growth, mortality and reproduction. Therefore it is important to allow fish to grow to a reasonable size before being caught. It is important to predict the effect of changes in size of net and of mesh and the amount of fishing on the rates of reproduction and survival. This can help prevent the thinning of fishing stocks.

Exponential population growth occurs in an ideal, unlimited environment. The phases shown in the sigmoid curve can be explained by relative rates of natality, mortality, immigration and emigration. In the sigmoid curve, population growth is exponential and then it slows until the population reaches the carrying capacity of the environment.

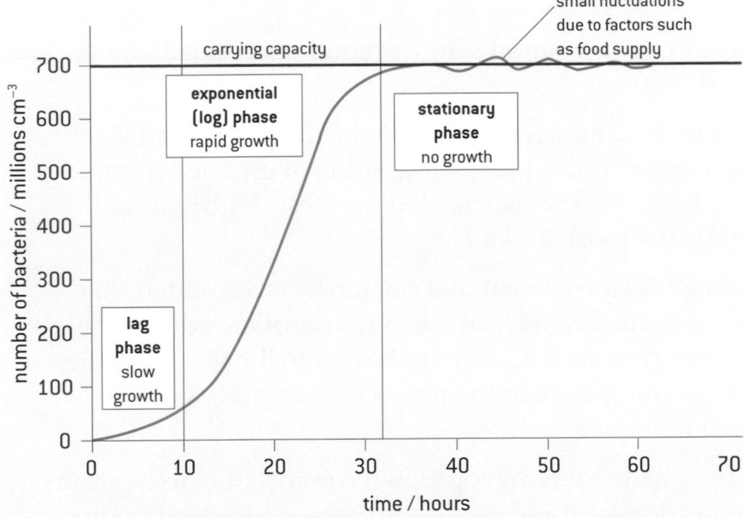

▲ **Figure C.5.1.** Sigmoid growth curve

- The **Lincoln index** uses data obtained from the capture-mark-release-recapture method to estimate the size of a population.
- A **sigmoid growth curve** shows changes in population over time.
- **Natality** is number of births.
- **Mortality** is number of deaths.
- **Immigration** is when organisms arrive at an environment.
- **Emigration** is when organisms leave an environment.
- **Carrying capacity** is the maximum population an environment can hold due to food, territory or breeding sites.

Assessment tip

You can be asked to draw this curve for different organisms.

- **Top-down factors** affect the population from a trophic level above it.
- **Bottom-up factors** affect the population from a lower trophic level or from a nutrient store.

Example C.5.1.

An experiment was performed to examine top-down and bottom-up control on salt marsh grass (*Spartina densiflora*). The grass was grown with (+) and without (–) a herbivore planthopper (*Prokelisia salina*) in mesocosms with or without salt. The experiment was repeated with addition of nutrient.

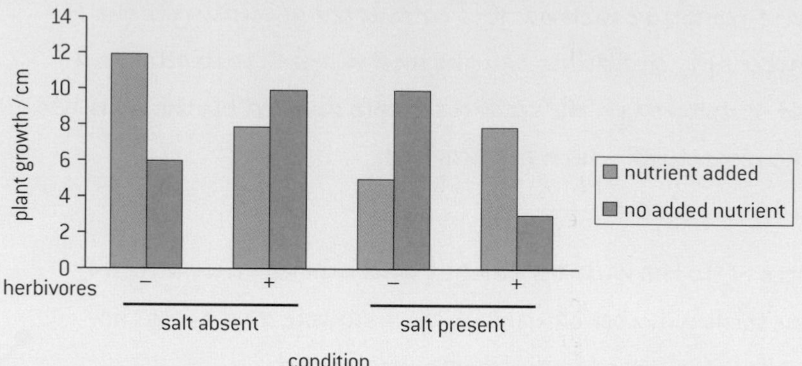

a) State the conditions where the marsh grass grows the most.

b) Outline the mode by which nutrients are bottom-up controls of plant growth.

c) Outline the effect of herbivores on the growth of marsh grasses in these mesocosms without salt.

d) Use the information to explain the top-down effect of planthopper herbivores on marsh grass grown in mesocosms with salt.

Solution

a) Marsh grass grown without salt, with nutrients added and without herbivores.

b) Nutrients include mineral ions needed for plant growth. For example nitrogen is used in the formation of nucleic acids and proteins. Phosphorus is also needed for these. Nutrients are absorbed by the roots of plants.

c) When herbivores are present and no nutrients are added, the marsh grasses grow more. But when nutrients are added, the marsh plants grow less. This could be due to the fact that other plants might compete with the marsh grasses and, with the nutrients, grow more.

d) Planthopper herbivores are top-down controls of marsh grasses when no nutrients are added. This is because they feed on the marsh grasses, destroying plant leaves. When nutrients are present, the presence of herbivores increases marsh grass growth, probably because nutrients allow other plants to grow in the environment, reducing the damage done by the planthoppers to the marsh grass.

C.6 NITROGEN AND PHOSPHORUS CYCLES (AHL)

You should know:

- ✔ nitrogen-fixing bacteria convert atmospheric nitrogen to ammonia.
- ✔ *Rhizobium* associates with roots in a mutualistic relationship.
- ✔ in the absence of oxygen denitrifying bacteria reduce nitrate in the soil.
- ✔ phosphorus can be added to the phosphorus cycle by application of fertilizer or removed by the harvesting of agricultural crops.
- ✔ the rate of turnover in the phosphorus cycle is much lower than the nitrogen cycle.
- ✔ availability of phosphate may become limiting to agriculture in the future.
- ✔ leaching of mineral nutrients from agricultural land into rivers causes eutrophication and leads to increased biochemical oxygen demand.

You should be able to:

- ✔ draw a labelled diagram of the nitrogen cycle.
- ✔ explain the impact of waterlogging on the nitrogen cycle.
- ✔ describe insectivorous plants as an adaptation to low nitrogen availability in waterlogged soils.
- ✔ design experiments to assess the nutrient content of a soil sample.
- ✔ outline how crop rotations allow the renewal of soil nutrients by allowing an area to remain "fallow".
- ✔ describe the phosphorus cycle.
- ✔ explain the reason availability of phosphate may become limiting to agriculture in the future.
- ✔ explain eutrophication.

The nitrogen cycle is very important because nitrogen availability can affect the rate of key ecosystem processes.

In Topic 4.1 you studied species, communities and ecosystems.

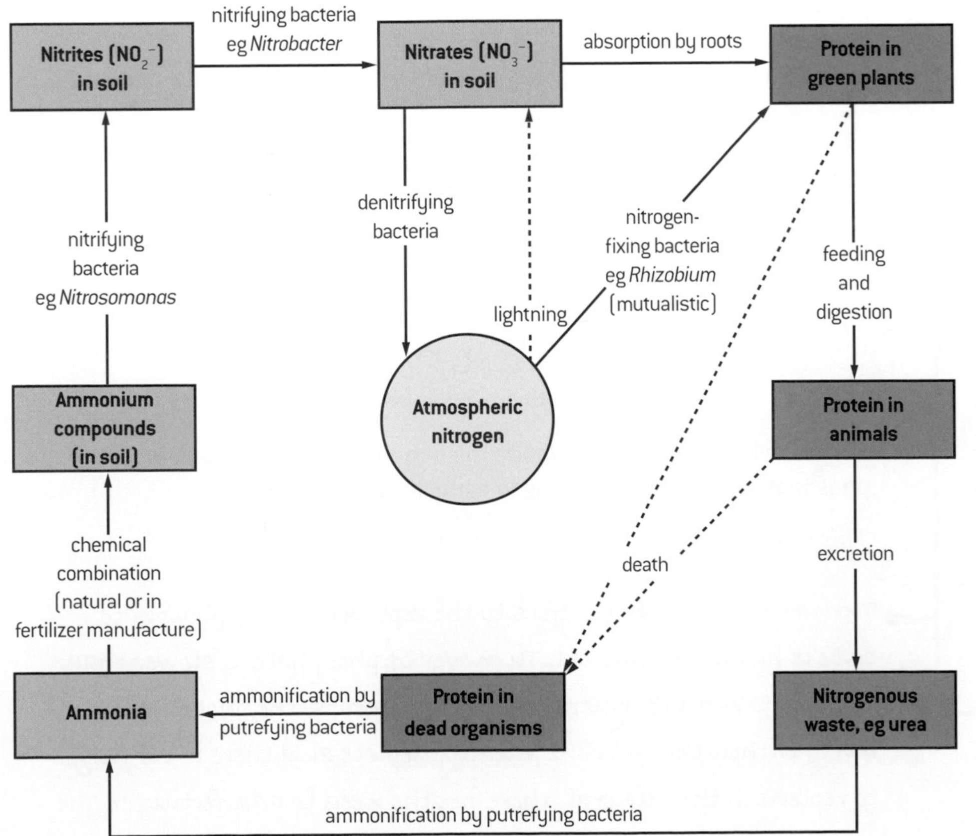

▲ **Figure C.6.1.** The nitrogen cycle

- **Denitrification** is the turning of nitrates to atmospheric nitrogen.
- **Nitrification** is the transformation of ammonium compounds to nitrites and nitrates.
- **Ammonification** is the transformation of proteins to ammonia.
- **Nitrogen fixation** is the transformation from atmospheric nitrogen to protein.

Our huge population growth has led to an increase in the demand for food, which, in turn, has led to intensive methods of agriculture. Intensive agriculture involves labour-intensive and costly practices in order to obtain high yields per unit of land. Intensive animal farming methods are often criticized for involving low standards of animal welfare, which rely on the overuse of antibiotics to be viable. Issues associated with intensive crop farming methods are the use of toxic pesticides which are often carcinogenic and bioaccumulate, harming non-target species, and the use of chemical fertilizers which can cause eutrophication of waterways.

Example C.6.1.

a) Outline the impact of waterlogging on the nitrogen cycle.

b) Describe insectivorous plants as an adaptation for low nitrogen availability in waterlogged soils.

Solution

a) Waterlogging fills the air gaps in the soil with water. In the absence of oxygen denitrifying bacteria reduce nitrate in the soil to atmospheric nitrogen. An example of denitrifying bacteria is *Pseudomonas denitrificans* which uses nitrate as an electron acceptor to produce ATP.

b) Insectivorous plants obtain their food through photosynthesis and make organic products using light. Nevertheless these carnivorous plants live in nutrient-poor habitats where there are almost no nitrates available, therefore need to trap insects in order to obtain their nitrogen.

Phosphorus is present in all living organisms. It is present in the phospholipids of cell membranes, is required for making ATP, DNA and RNA, is found in proteins and carbohydrates after phosphorylation, and is needed to form bones and teeth. Phosphorus can be added to the phosphorus cycle by applying fertilizer or removed by harvesting agricultural crops. The rate of turnover in the phosphorus cycle is much lower than in the nitrogen cycle therefore its availability may become limiting to agriculture in the future.

- **Eutrophication** is the enrichment with nutrients of a body of water leading to the excessive growth of plants and algae, causing oxygen depletion.
- **Biochemical oxygen demand** is the amount of dissolved oxygen needed by aerobic organisms to break down organic material.

SAMPLE STUDENT ANSWER

The graph shows the global phosphorus fertilizer use from 1960 to 2011.

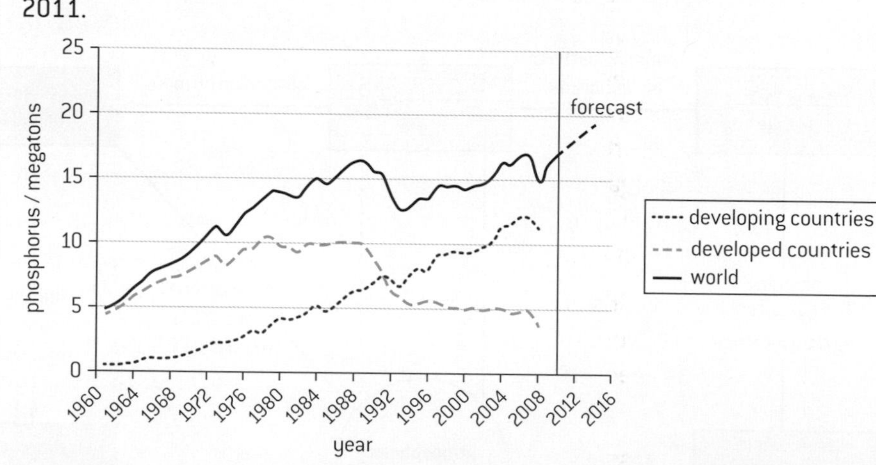

Suggest whether the data supports the hypothesis that availability of phosphate may become limiting to agriculture in the future. [6]

This answer could have achieved 5/6 marks:

Turnover of phosphate refers to the replacement of phosphate once it has been removed. Turnover of phosphate is slower than nitrogen. When farmers remove their crops, the phosphate levels within the soil continually deplete and there is no way to replenish the mineral, therefore the need to add fertilizer. The

▲ The answer addresses most of the issues associated with the availability of phosphates in the future for agriculture. The definition of turnover and fact that the turnover is slow; that fertilizers are needed for crops and the use of fertilizers increased in the last years; and that phosphorus is removed from the soil in order to make fertilizers, are enough for five marks.

phosphate removed by crops is replaced by fertilizers, which at the same time are obtained from sedimentary rocks. Human activity is removing the phosphate from the land to produce this fertilizer at far greater rate than it can possibly be turned over. The graph shows that the use of phosphorus fertilizers has increased, especially after 1940, but has levelled off in the last years. If the turnover is slower than the depletion, the availability of phosphate may become a limiting factor to agriculture in the future, thus supporting the hypothesis.

▼ The answer did not make reference to the fact that in developed countries the amount of P fertilizers used is decreasing. If this trend is transferred to the developing countries, the predicted use of P fertilizers could eventually be reduced. The answer also failed to mention that phosphates are removed from the soil to produce detergents and that these detergents flow into the rivers and lakes and produce eutrophication.

Leaching of mineral nutrients from agricultural land into rivers causes eutrophication and leads to increased biochemical oxygen demand.

Practice problems for Option C

Problem 1

Most reef-building corals contain photosynthetic algae, called zooxanthellae, that live in their cells. Coral bleaching can occur as a result of human-induced changes, leading to the zooxanthellae being ejected from the coral.

a) State the type of interaction that occurs between zooxanthellae and reef-building corals.

b) State the trophic level of zooxanthellae.

c) When coral is bleached, certain organisms become more common in the ecosystem such as the cnidarian *Gorgonia*, the echinoderm *Diadema*, other algae and certain sponges. State the term that is used for organisms whose presence provides evidence of the existence of a particular environmental condition.

d) A coat of algae builds up on coral reefs as a consequence of eutrophication. Explain the relationship between eutrophication and algae growth.

e) Explain how excessive growth of algae on coral reefs can be controlled by top-down factors.

Problem 2

The pie charts show the nutrient levels in two biomes.

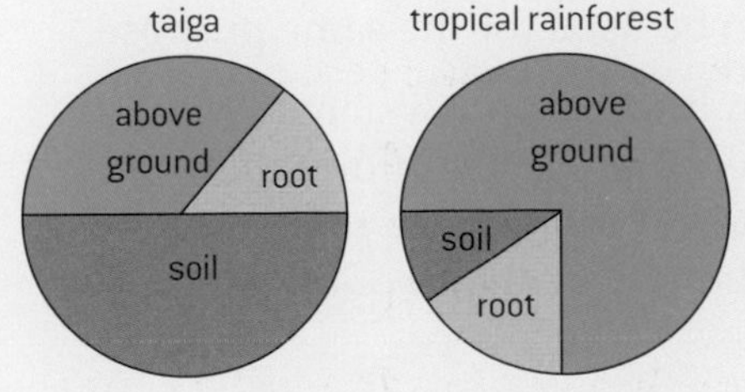

a) Compare and contrast the nutrient stores in taiga and tropical rainforest.

b) Describe the **flow** of nutrients in taiga.

c) Draw a Gersmehl diagram for the tropical rainforest.

d) Suggest **one** reason the nutrient store above the ground is so high in both biomes.

Problem 3

Explain the effect alien species can have on an ecosystem.

Problem 4

Explain the impact of island size and edge effects on community structure.

Problem 5 (AHL)

Explain how the carrying capacity of the environment affects population growth.

Problem 6 (AHL)

Explain the reason for the presence of insectivorous plants in waterlogged soils.

D HUMAN PHYSIOLOGY

D.1 HUMAN NUTRITION

You should know:

- ✔ essential nutrients cannot be synthesized by the body, therefore they have to be included in the diet.
- ✔ dietary minerals are essential chemical elements.
- ✔ vitamins are chemically diverse carbon compounds that cannot be synthesized by the body.
- ✔ some fatty acids and some amino acids are essential.
- ✔ lack of essential amino acids affects the production of proteins.
- ✔ malnutrition may be caused by a deficiency, imbalance or excess of nutrients in the diet.
- ✔ a lack of vitamin D or calcium can affect bone mineralization and cause rickets or osteomalacia.
- ✔ appetite is controlled by a centre in the hypothalamus.
- ✔ overweight individuals are more likely to suffer hypertension and type II diabetes.
- ✔ starvation can lead to breakdown of body tissue.

You should be able to:

- ✔ explain the need for a balanced diet.
- ✔ explain the production of ascorbic acid by some mammals, but not others that need a dietary supply.
- ✔ describe causes of malnutrition.
- ✔ explain appetite control.
- ✔ explain how obesity can lead to hypertension and type II diabetes.
- ✔ outline how cholesterol in blood is an indicator of the risk of coronary heart disease.
- ✔ explain the cause and treatment of phenylketonuria (PKU).
- ✔ explain breakdown of heart muscle due to anorexia.
- ✔ design experiments to determine the energy content of food by combustion.

Topic 2.3 covered the determination of body mass index by calculation or using a nomogram and in Topic 6.1 digestion and absorption were studied.

A balanced diet is essential to human health. Essential nutrients are those that cannot be synthesized by the body and therefore they have to be included in the diet. Certain fatty acids, amino acids and dietary minerals are essential. Vitamins are chemically diverse carbon compounds that cannot be synthesized by the body (with the exception of vitamin D). Non-essential nutrients can either be made by the body or there is another nutrient that can be used for the same purpose.

The Vitamin and Mineral Nutrition Information System (VMNIS), formerly known as the Micronutrient Deficiency Information System (MDIS), was established in 1991 following a request by the World Health Assembly to strengthen surveillance of micronutrient deficiencies at the global level.

- **Balanced diet** is one that contains the accepted and defined proportions of carbohydrates, fats, proteins, vitamins, minerals and water essential to maintain good health.
- **Malnutrition** is caused by eating in excess or undereating or by a diet lacking in an essential element. It may lead to severe deficiencies, illness or even death.
- **Starvation** is the deficiency of nutrients in a diet.
- **Body mass index (BMI)** is the body mass measured in kilograms divided by the square of the body height measured in metres.
- **Obesity** is a medical condition where the BMI is above 30 kg m^{-2}.

The nine essential amino acids humans cannot synthesize are phenylalanine, valine, threonine, tryptophan, methionine, leucine, isoleucine, lysine and histidine. Threonine can be synthesized by the human body when phenylalanine is present in the diet. Arginine cannot be synthesized by infants so it must be present in the diet of small children. Essential fatty acids are, for example, omega-3 and omega-6. These fatty acids can be found in fish and in seeds.

Scurvy is caused by a deficiency of dietary ascorbic acid (vitamin C) since humans are unable to metabolically make this chemical. The symptoms of scurvy are abnormal bleeding. Ascorbic acid is involved in the synthesis of collagen fibres; therefore the lack of this vitamin will cause defective connective tissues and thus fragile capillaries. Scurvy was thought to be specific to humans, because attempts to induce the symptoms in laboratory rats and mice were unsuccessful.

Example D.1.1.

Vitamin C is also known as l-ascorbic acid. Scurvy is a disease caused by the lack of ascorbic acid. In an experiment, rats (*Rattus rattus*) and guinea pigs (*Cavia porcellus*) were fed a restricted diet to avoid ascorbic acid production. On day 2 animals were injected with a hexose sugar, a precursor in the synthesis of ascorbic acid. The amount of ascorbic acid was measured in urine every day.

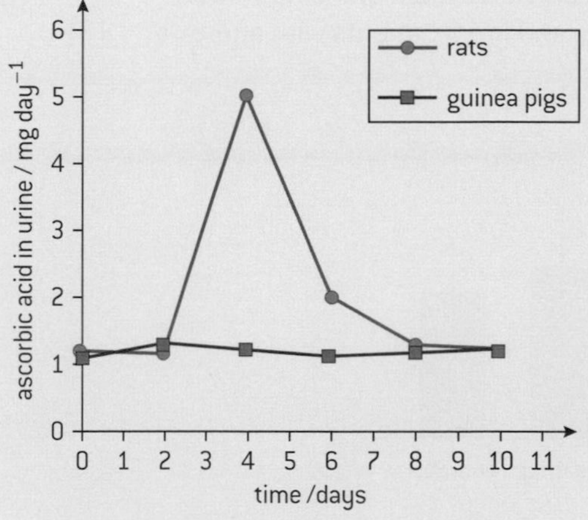

a) List **two** symptoms of scurvy.

b) Suggest a health reason sailors used oranges and lemons in their trips.

c) An attempt to induce the symptoms of scurvy in laboratory rats was entirely unsuccessful but was successful in guinea pigs. Suggest whether the data supports this statement.

d) Explain the reason guinea pigs develop scurvy.

Solution

a) Scurvy is a disease characterized by apathy, weakness, easy bruising with tiny or large skin hemorrhages, friable bleeding gums and swollen legs. Untreated patients may die.

b) Sailors used citrus fruits to cure and prevent scurvy as these fruits are rich in ascorbic acid (vitamin C) which in adequate quantities prevents and cures the disease.

c) The data supports the statement because, using a precursor sugar, the rats were able to synthesize ascorbic acid while guinea

Assessment tip

The ethics of the use of animals in experiments could be discussed. In this case, only 10 days experimentation will cause no harm to the guinea pigs.

Assessment tip

You should give only two symptoms, as only the first two will be marked. Just writing "death" will not score a mark.

- **Scurvy** is caused by lack of vitamin C (ascorbic acid).
- **Rickets** and **osteomalacia** are caused by lack of vitamin D.
- **Phenylketonuria (PKU)** is caused by a deficiency of an enzyme that converts phenylalanine into tyrosine, causing excess phenylalanine in blood.
- **Type II diabetes** is caused by resistance to insulin.
- **Hypertension** is blood pressure above the norm.

pigs were not. The levels of ascorbic acid were reduced a few days after the injection, showing that the rats require these precursors to synthesize ascorbic acid.

d) Guinea pigs cannot synthesize ascorbic acid which is needed for the production of collagen found in skin, connective tissues, tendons and blood vessels. Ascorbic acid cannot be synthesized by guinea pigs because they have a mutation in the gulonolactone oxidase (GLO or GULO) gene which codes for an enzyme that is needed for the last step in the synthesis of l-ascorbic acid. (The same is true in humans.)

Malnutrition may be caused by a deficiency, imbalance or excess of nutrients in the diet.

Lack of essential amino acids affects the production of proteins. This can be seen in cases of kwashiorkor or marasmus. Kwashiorkor is caused by insufficient protein consumption (even if the energy consumption is sufficient) and produces swelling of ankles and feet, a distended abdomen and a large liver. There is thinning of hair, dermatitis and teeth loss. Marasmus is caused by an energy deficient diet including a lack of protein. People suffering from marasmus are extremely thin (emaciated), with muscular and fat loss. In both cases, they have a long-term effect on physical and mental development and in severe cases can lead to death.

Assessment tip

Make sure you read the question properly. In this case, what was expected was a discussion on the data itself, not on the biological effect of anorexia on the heart. When you discuss, you must offer a considered and balanced review that includes a range of arguments, factors or hypotheses. Opinions or conclusions should be presented clearly and supported by appropriate evidence.

SAMPLE STUDENT ANSWER

The diameter of the left ventricle and left atrium of the heart was measured in 20 healthy subjects and in 19 patients with anorexia. [2]

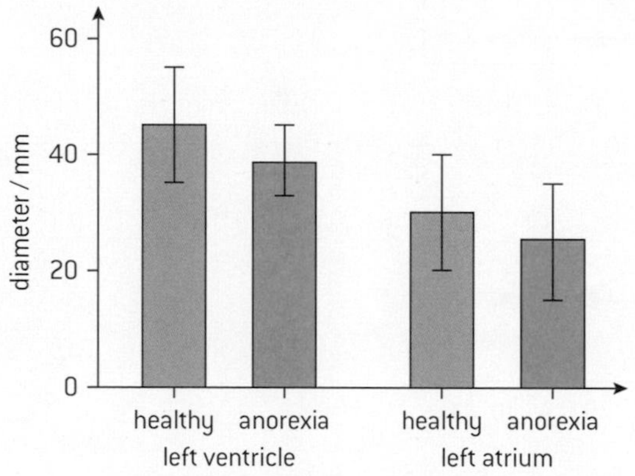

Discuss the support provided by the data for the claim that anorexia leads to the breakdown of heart tissue. [3]

This answer could have achieved 1/3 marks:

In all structures that were measured, each showed a decrease in diameter in anorexia subjects compared to healthy subjects. This is most likely due to malnutrition among the anorexic that is due to lack of carbohydrates, fats and protein, forcing the body of the anorexic to resort to their muscles as a source of energy for respiration. Although the body would focus on degrading skeletal muscle rather than cardiac muscle, the cardiac muscle will still gradually degenerate, as is evidenced in the decreased diameters of the hearts of the anorexic.

▲ This answer scored one mark for mentioning that the data supports this statement as both structures are smaller in anorexic patients.

▼ Although this student does mention some important facts of anorexia, the question is not really answered. The statement mentions that in anorexia there is breakdown of the heart. The answer should have mentioned that although there are differences between healthy and anorexic, there is overlap in error bars and that it may not be reliable because the size of the sample is too small. The fact that there is correlation does not necessarily establish causality.

Lack of vitamin D or calcium can affect bone mineralization and cause rickets or osteomalacia. There are positive effects of exposure to sun such as the production of vitamin D as well as health risks associated with exposure to UV rays which can cause skin cancer such as melanoma.

Appetite is controlled by a centre in the hypothalamus. It receives information from the digestive system and adipose tissue to either stimulate or inhibit appetite.

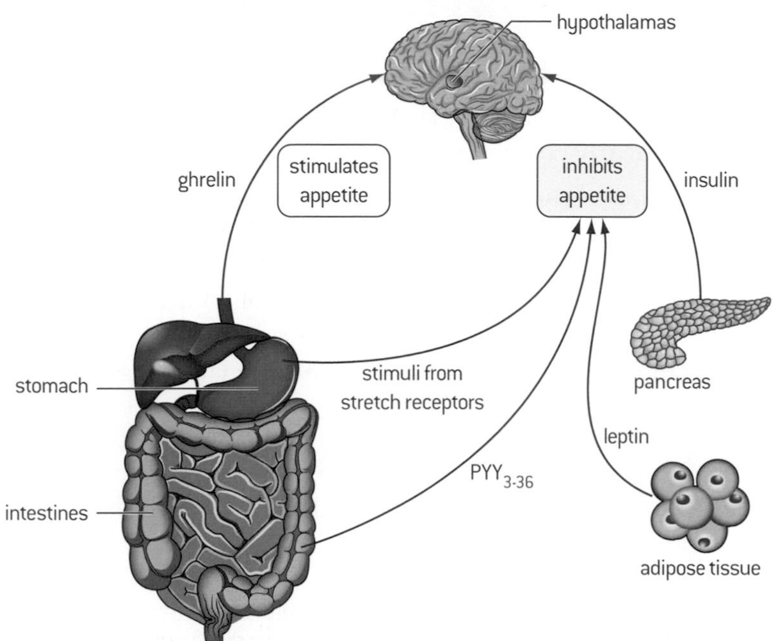

▲ **Figure D.1.1.** Appetite control

BMI values are a good way of assessing if someone is under or overweight. Underweight subjects typically have a BMI under 18.5 kg m^{-2}, overweight subjects a BMI of between 25.1 and 29.9 kg m^{-2} and obese subjects a BMI above 30.0 kg m^{-2}. Obese and overweight individuals are more likely to suffer hypertension and type II diabetes. Hypertension is when a patient's blood pressure is above normal. Normal blood pressure is around 120 mmHg for systolic pressure and 70 mmHg for the diastolic pressure. Hypertension can occur in those with a diet rich in sodium salt or due to genetic factors. Type II diabetes, also known as non-insulin-dependent, is the most common form of diabetes. The β cells in the pancreas still produce insulin but these patients develop insulin resistance, which means that the insulin receptors on the cells are no longer stimulated by insulin.

Example D.1.2.

The incidence of coronary heart disease (CHD) was investigated amongst 20,000 subjects. Baseline parameters of waist to hip ratio and BMI were recorded. The evidence of CHD was identified in follow-up interviews. The bar chart shows the results according to the values of the measured parameters.

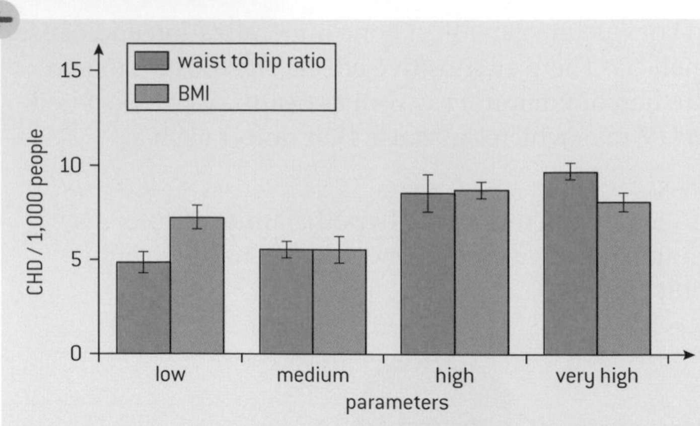

a) (i) Deduce with a reason whether the waist to hip ratio **or** the BMI most clearly correlates to incidence of CHD.

(ii) Suggest whether the parameter chosen in (i) causes CHD.

b) Explain the reason overweight and obese people have a higher risk of CHD.

Solution

a) (i) Waist to hip ratio as increasing ratio shows increasing CHD incidence while increasing BMI does not.

(ii) A correlation does not necessarily mean a causation.

b) Overweight and obese people have a BMI larger than 25 kg m^{-2}. These people usually have high concentrations of cholesterol, high concentrations of LDL and low concentrations of HDL. Overweight people can have hypertension as they have a higher cardiac output which can raise blood pressure. Waistline obesity can increase vascular resistance with arteries becoming stiffer and narrower which can raise blood pressure.

Phenylketonuria (PKU) is the most common congenital disorder of amino acid metabolism. It is caused by a deficiency of phenylalanine hydroxylase, an enzyme that converts phenylalanine into tyrosine. Dietary restriction of phenylalanine is relatively successful in the reduction or prevention of intellectual disability associated with PKU.

SAMPLE STUDENT ANSWER

a) Outline what is meant by the term "essential amino acid". [2]

This answer could have achieved 0/2 marks:

An essential amino acid is one required by the human body in order to remain at a homeostasis state and to carry out functions and processes. The human body requires the essential amino acids along with other substances to conduct metabolic processes.

▼ The answer fails to mention that an essential amino acid must be obtained from the diet as it cannot be synthesized by the body. The body cannot synthesize the essential amino acid from other amino acids.

Phenylalanine is converted into tyrosine by the enzyme phenylalanine hydroxylase.

COOH, NH_2 — phenylalanine —(phenylalanine hydroxylase)→ HO, COOH, NH_2 — tyrosine

b) (i) Deduce the reason that tyrosine is considered to be a "conditionally" essential amino acid. [1]

This answer could have achieved 0/1 marks:

Tyrosine is considered to be a conditionally essential amino acid as it may be required for a process but is not required in a general sense or for the entire body. Tyrosine is also present in food ingested and can aid or provide assistance for the human body.

▼ Tyrosine is a non-essential amino acid. It is only necessary in cases of PKU. The answer fails to mention that tyrosine can only be synthesized when phenylalanine is in the diet.

(ii) When infants with the condition PKU are left untreated, they have a build-up of phenylalanine in the blood and high levels of phenylalanine in the urine. State the cause of this condition. [1]

This answer could have achieved 0/1 marks:

The infant's body doesn't have a control or regulation of phenylalanine.

▼ The answer is too vague. It fails to mention that PKU is a recessive inherited genetic condition where there is lack of the enzyme phenylalanine hydroxylase or the gene for this enzyme is mutated.

In order to determine the energy content of food by combustion an easy experiment can be performed. The initial mass or volume of water and its temperature is measured. The mass of the food is also measured. The food is ignited and placed under a container of water. The final temperature of the water is recorded and the change in temperature of the water calculated. As the heat gained by the water is equal to the heat lost by the food, the energy is calculated as the mass of water times the temperature rise in water times the specific heat capacity of water all divided by the mass of food.

D.2 DIGESTION

You should know:

- ✔ nervous and hormonal mechanisms control the secretion of digestive juices.
- ✔ exocrine glands secrete to the surface of the body or the lumen of the gut.
- ✔ the volume and content of gastric secretions are controlled by nervous and hormonal mechanisms.
- ✔ acid conditions in the stomach favour some hydrolysis reactions and help to control pathogens in ingested food.
- ✔ the structure of cells of the epithelium of the villi is adapted to the absorption of food.
- ✔ the rate of transit of materials through the large intestine is positively correlated with their fibre content.
- ✔ materials not absorbed are egested.

You should be able to:

- ✔ describe the formation of gastric ulcers.
- ✔ identify *Helicobacter pylori* infection as a cause of stomach ulcers.
- ✔ explain the reduction of stomach acid secretion by proton pump inhibitor drugs.
- ✔ explain dehydration due to cholera toxin.
- ✔ identify exocrine gland cells that secrete digestive juices and villus epithelium cells that absorb digested foods from electron micrographs.
- ✔ explain adaptations of villus epithelial cells.

Topic 1.2 covered the ultrastructure of cells and in Topic 6.5 the structure of neurons and synapses is studied.

The role of gastric acid in digestion was established by William Beaumont while observing the process of digestion in an open bullet wound. Beaumont extracted gastric juice by introducing an elastic tube through the wound and observed digestion of different foods.

Nervous and hormonal mechanisms control the secretion of digestive juices. The volume and content of gastric secretions are controlled by nervous and hormonal mechanisms.

Exocrine glands secrete substances onto epithelial surfaces of the body via ducts. For example, pancreatic exocrine cells form glandular clusters (known as acini). These cells secrete digestive enzymes that are exported to the duodenum via a pancreatic duct system. Typical of protein-secreting cells, they have a large nucleus and lots of rough endoplasmic reticulum.

Acid conditions in the stomach favour some hydrolysis reactions and help to control pathogens in ingested food. Excess acid can damage the mucus of the stomach that can lead to an ulcer. *Helicobacter pylori* infection is a cause of stomach ulcers. Reduction of stomach acid secretion can be achieved by taking proton pump inhibitor medication.

- The **parasympathetic** system accelerates digestion.
- The **sympathetic** system slows down digestion.
- **Gastrin** is a hormone produced in response to the presence of food and stimulates the production of gastric juice by the stomach.
- **Secretin** is a hormone produced in the small intestine in response to the presence of acid fluid. It stimulates the production of alkali by the pancreas.
- **Somatostatin** is an inhibitory hormone secreted by the gut that inhibits acid production. It also prevents the release of gastrin and secretin, thus slowing down the digestive process.

Assessment tip

Although not asked, it is neat to start the answer describing in one sentence what an ulcer is, as this leads into the explanation on how they are formed. Nevertheless, make sure you don't take too long on this and always focus on what the question is asking. For example this answer would score 4 marks, as it identifies the excess of acid and its damage on the lining and it identifies *H.pylori* and the mode it damages the stomach through the immune response. No mark would have been given for the last sentence, as it refers to the treatment and not the formation of ulcers.

Example D.2.1.

Describe the formation of gastric ulcers.

Solution

Gastric ulcers are sores in the lining of the stomach. They can be formed by excess acid which damages the mucus layer. The symptoms are stomach pain, heartburn, vomiting and sometimes bleeding which appears in the feces. Another cause of gastric ulcers is due to the presence of the bacterium *Helicobacter pylori*. This bacterium produces toxins in the lumen of the stomach that cause inflammation. The inflammatory response by the immune system damages the stomach lining. Ulcers due to *H. pylori* are treated with antibiotics or by proton pump inhibitors (PPIs) which inhibit acidification of the stomach.

SAMPLE STUDENT ANSWER

The diagram shows a cell of the lining of the stomach.

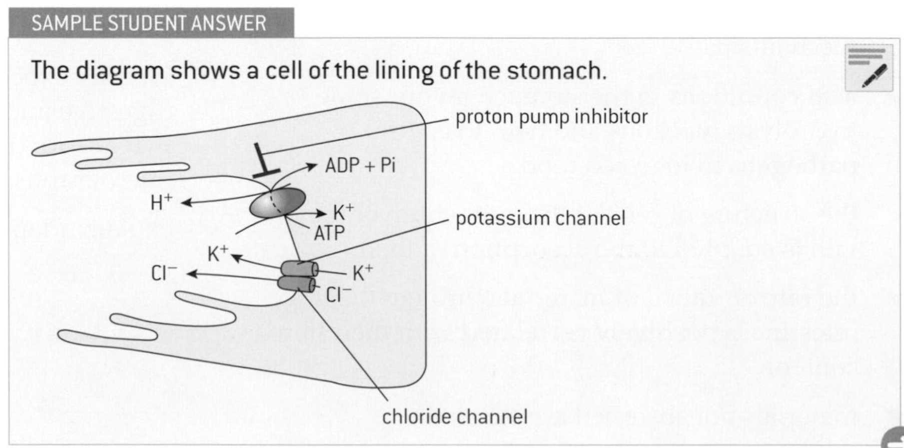

a) Outline the importance of the proton pumps in the digestion of foods. [1]

This answer could have achieved 0/1 marks:

Proton pumps help fasten the process of digestion in the stomach. It pumps nutrients in and out of the cell in the stomach.

▼ The answer should have mentioned that proton pumps pump protons into the stomach, allowing for the production of hydrochloric acid. This acid accelerates digestion as it activates enzymes or gives optimal pH for enzyme (pepsin) digestion.

b) Explain the use of proton pump inhibitors to treat patients complaining of stomach pain. [3]

This answer could have achieved 0/3 marks:

With proton pump, it can help the flow of potassium and chloride into the stomach. The channel protein is between the cells. Avoids the stomach pain.

▼ The answer should have mentioned that the proton pump is a transmembrane protein. When the proton pump inhibitors bind to the proton pump the hydrogen ions are not sent into the stomach lumen. This causes a reduction of gastric acid production, thus increasing the pH of stomach. This decrease will relieve symptoms of acid reflux, gastritis or ulcers.

▼ The student did not score a mark for saying that it reduces stomach pain, as this is already in the question.

The villus epithelium cells absorb digested foods from the lumen of the intestine. Villi are finger-like projections that increase the surface area for the absorption of food. Within the columnar epithelium of the outer surface are goblet cells, which secrete mucus to lubricate food and prevent self-digestion. The central core contains the blood supply that transports the products of digestion and a lacteal that absorbs fats.

The structure of cells of the epithelium of the villi is adapted to the absorption of food. They have microvilli to increase the surface area of absorption. They are also rich in mitochondria, to provide energy in the form of ATP to enable active transport of nutrients into the network of capillaries of the villus or the central lacteal where absorption of fats occurs.

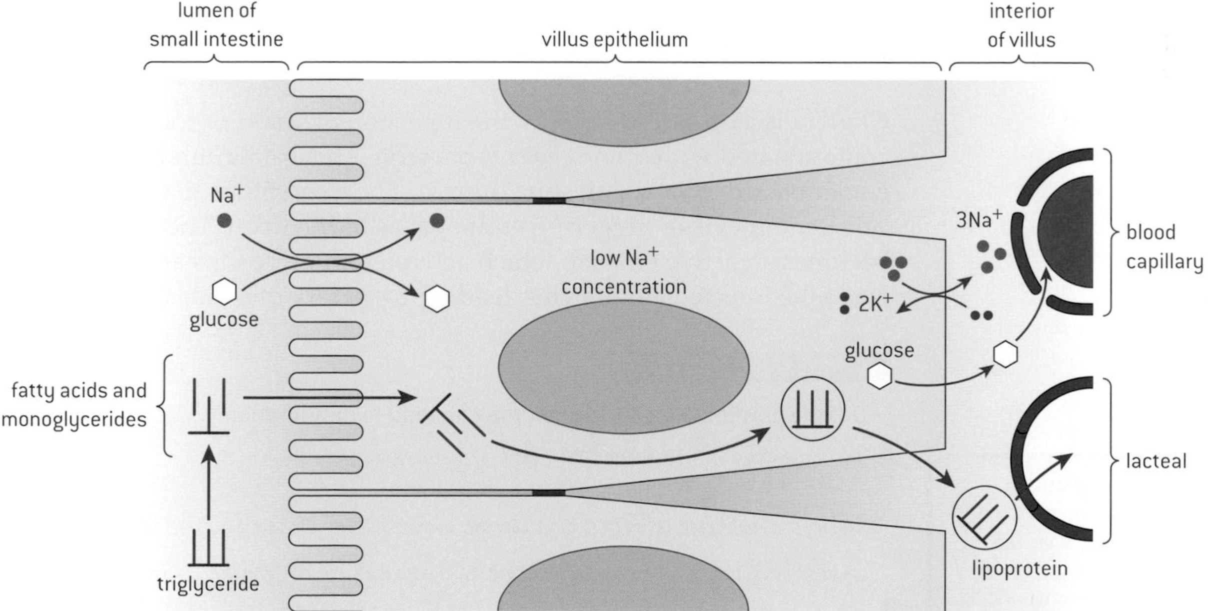

▲ **Figure D.2.1.** Absorption in the villus of the small intestine

The rate of transit of materials through the large intestine is positively correlated with their fibre content. Materials not absorbed are egested.

Assessment tip

The structures in Figure D.2.1 should be seen in intestine electron micrographs.

Example D.2.2.

The graph shows the length of time that the content of a meal takes to pass through the gut as a function of digestible matter content. The more the digestible matter present in the meal, the lower the dietary fibre content.

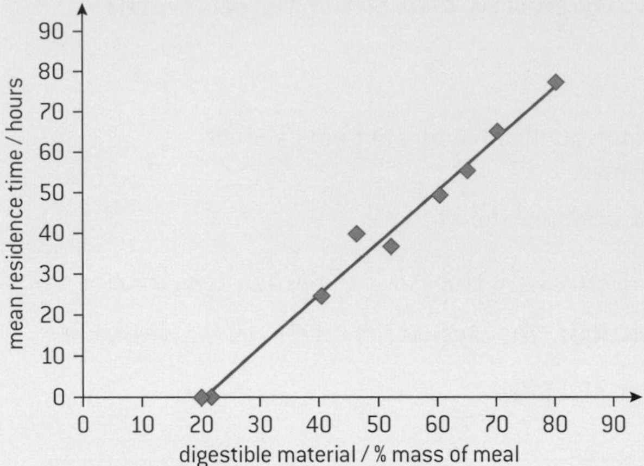

a) Estimate the mean residence time of a meal with 50% digestible matter.

b) Explain the relationship of digestible matter and mean residence time.

Solution

a) 37 hours

b) A higher digestible matter content means a longer residence time. This will derive in a slower bowel movement or peristalsis due to lack of fibre. Fibre is an indigestible material that attracts water to intestinal contents. A higher amount of digestible material increases water absorption into the large intestine. This hardens the stool causing constipation and further increases residence time.

Assessment tip

The mark scheme will probably have a range for this answer, for example from 35 to 39 hours. In your answer, it is better not to write a range, as if your range is outside the accepted range, you will score no mark. It is always better just to write one value and its units.

Cholera is a disease caused by the ingestion of water or food contaminated with *Vibrio cholerae* bacteria. This bacterium releases cholera toxin into the intestinal lumen. The toxin binds to epithelial cells (enterocytes), triggering endocytosis of the toxin. The toxin then becomes an active enzyme which activates chloride ions and water to leave the infected enterocytes, leading to watery diarrhoea.

SAMPLE STUDENT ANSWER

Outline how infection by *V. cholerae* can lead to dehydration. [3]

This answer could have achieved 3/3 marks:

Vibrio cholerae affects the large intestine. Once in the large intestine cells it promotes the pumping of chloride ions into the lumen of the intestine. Via osmosis, water follows the chloride. Via osmosis once more, water leaves the blood from capillaries into the intestine cells to replenish water.

▲ This is a very concise and complete answer that scored full marks. Another mark could have been scored for saying that *V. cholerae* produces a toxin. Another mark could have been obtained for stating that the loss of water to the intestine leads to diarrhoea and vomiting that lead to dehydration.

D.3 FUNCTIONS OF THE LIVER

You should know:

- ✔ the liver removes toxins from the blood and detoxifies them.
- ✔ components of red blood cells are recycled by the liver.
- ✔ the breakdown of erythrocytes starts with phagocytosis of red blood cells by Kupffer cells.
- ✔ iron is carried to the bone marrow to produce hemoglobin in new red blood cells.
- ✔ surplus cholesterol is converted to bile salts.
- ✔ endoplasmic reticulum and Golgi apparatus in hepatocytes produce plasma proteins.
- ✔ the liver intercepts blood from the gut to regulate nutrient levels.
- ✔ some nutrients in excess can be stored in the liver.

You should be able to:

- ✔ explain the dual blood supply to the liver.
- ✔ compare and contrast sinusoids and capillaries.
- ✔ explain how the liver regulates toxin levels in blood.
- ✔ explain recycling of erythrocytes.
- ✔ explain nutrient regulation by the liver.
- ✔ explain causes and consequences of jaundice.
- ✔ label microscopic mounts of hepatocytes.

The liver is a large organ that performs many functions. It is in charge of detoxifying the blood; it maintains blood temperature and pressure; breaks down erythrocytes; regulates blood clotting; converts ammonia to urea; stores glycogen, vitamins and iron; synthesizes plasma proteins and phospholipids; and produces bile. The oxygenated hepatic artery flows into the liver and the deoxygenated hepatic vein leaves the liver. The hepatic portal vein reaches the liver carrying food from the intestines.

Topic 6.1 covered digestion and absorption and in Topic 6.2 the blood system is studied.

- **Hepatocytes** are liver cells that perform functions such as: detoxification; metabolism of carbohydrates; secretion of bile; and synthesis of proteins, cholesterol and phospholipids.
- **Kupffer cells** are macrophages in charge of erythrocyte breakdown.

SAMPLE STUDENT ANSWER

Outline the ways in which the liver regulates the chemical and cellular composition of the blood. [6]

This answer could have achieved 2/6 marks:

The liver regulates chemical and cellular composition of the blood by creating a filtration. Blood is first taken into the liver by the vein and taken out through the artery and carried back through the body. The liver intercepts blood from the gut to regulate nutrient levels.

Some nutrients in excess can be stored in the liver. The liver filters all matter not required in the blood. It also filters chemicals not required or that can be harmful to the human body. One example of that is alcohol. The liver filters out cellular composition not required in the body. Blood is filtered flowing through the liver. Levels of minerals and proteins are filtered through the liver.

▲ One mark is scored for saying that the liver filters chemicals not required, and one mark for saying the liver detoxifies blood by removing toxins from the blood. The rest of the answer is too vague; therefore no more marks are scored.

▼ The answer expected was that the liver stores glucose as glycogen or releases it under the influence of insulin and glucagon respectively depending on blood glucose levels. Some nutrients in excess can be stored in the liver (such as vitamin A or D). Kupffer cells engulf bacteria, therefore removing pathogens. Kupffer cells break down erythrocytes by phagocytosis. Hemoglobin is split into heme and globin. Iron from the heme group of hemoglobin is carried to the bone marrow to produce new hemoglobin in new erythrocytes. Excess iron is stored in the liver. Surplus cholesterol is converted to bile salts. Hepatocytes produce plasma proteins such as albumin.

Assessment tip

This type of question can be answered without the table, but you must make sure you compare the same feature in both vessels. For example, if you say capillaries are small in size and sinusoids have fenestra, you do not score any marks.

Scientific studies have shown that high-density lipoprotein (HDL) could be considered "good" cholesterol. HDL is the smallest lipoprotein and it is the densest because it contains the highest proportion of protein to lipids. HDL has been shown to help prevent and even improve atherosclerosis by removing harmful low-density lipoprotein (LDL) from blood.

Example D.3.1.

a) Compare and contrast sinusoids and capillaries.

b) Outline adaptations of hepatocytes to their function.

Solution

a)

	Vessel	
	Capillary	**Sinusoid**
Content	blood	
Size	small	a bit larger
Pores	very small	large (called fenestra)
Basement membrane	continuous	discontinuous
Intracellular space	small	large
Lumen	round	no definite shape
Kupffer cells	absent	present

b) Hepatocytes are involved in many metabolic processes. They are rich in mitochondria for energy. There are many Golgi complexes, rough endoplasmic reticulum and ribosomes for protein synthesis. Glycogen granules and lipid droplets are for nutrient storage.

D.4 THE HEART

You should know:

- ✔ internal and external factors influence heart function.
- ✔ structure of cardiac muscle cells allows propagation of stimuli through the heart wall.
- ✔ signals from the sinoatrial node that cause contraction cannot pass directly from atria to ventricles.
- ✔ there is a delay between the arrival and passing on of a stimulus at the atrioventricular node.
- ✔ this delay allows time for atrial systole before the atrioventricular valves close.
- ✔ conducting fibres ensure coordinated contraction of the entire ventricle wall.
- ✔ normal heart sounds are caused by the atrioventricular valves and semilunar valves closing causing changes in blood flow.

You should be able to:

- ✔ explain the use of artificial pacemakers to regulate the heart rate.
- ✔ explain the cardiac cycle to a normal ECG trace.
- ✔ analyse systolic and diastolic blood pressure measurements.
- ✔ explain causes and consequences of hypertension and thrombosis.
- ✔ describe the use of defibrillation to treat life-threatening cardiac conditions.
- ✔ design experiments to measure and interpret heart rate under different conditions.
- ✔ analyse epidemiological data relating to the incidence of coronary heart disease.

Topic 6.2 covered the blood system and in Topic 11.2 movement is studied.

The structure of cardiac muscle cells (myocytes) allows propagation of stimuli through the wall of the heart. Cardiac muscle is a striated muscle that has thick and thin muscle fibres with myofibrils containing myofilaments similar to those found in skeletal muscle. The striations are given by the structure of the sarcomeres. The nucleus of cardiac muscle cells is found in the centre of the cell and there are many mitochondria and glycogen granules found adjacent to the myofibrils. Unlike skeletal muscle, which has multinucleate cells, the cardiac muscle has numerous short, cylindrical cells arranged end-to-end, resulting in long, branched fibres.

Example D.4.1.

Explain the reason electrical impulses can pass rapidly from one cardiac muscle cell to the other.

Solution

The cardiac muscle has numerous short, cylindrical cells arranged end-to-end, resulting in long, joined, branched fibres. The cardiac cells are joined by intercalated discs containing adhering junctions. These gap junctions are arrays of densely packed protein channels that permit intercellular passage of ions and small molecules. Electrical activation of the heart requires cell-to-cell transfer of current via gap junctions. This characteristic structure of cardiac muscle allows electrical impulses to pass rapidly from cell to cell, so the linked cells contract almost simultaneously.

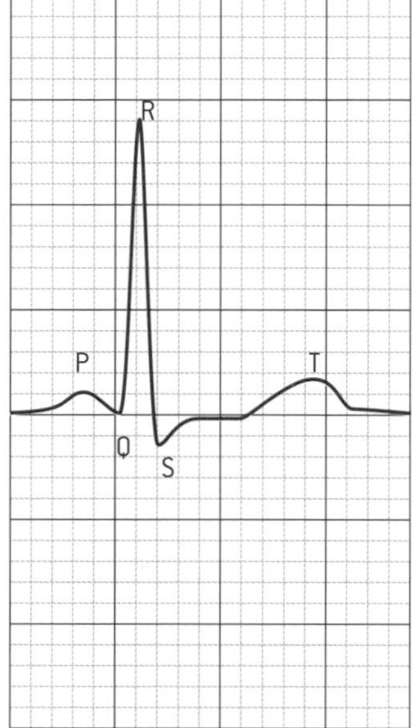

▲ **Figure D.4.1.** A normal ECG trace showing the cardiac cycle

The sinoatrial (SA) node consists of involuntary myogenic muscle cells that contract rhythmically at around 70 beats per minute. Following each contraction, an impulse spreads through the atria to the atrioventricular (AV) node. Contraction of the atria causes blood to move into the ventricles. This contraction is seen as the P wave on the electrocardiograph (ECG). The AV node transmits the impulse to the ventricles, via the bundle of His. The ventricles contract sending blood to the lungs and to the rest of the body. The ventricular contraction is seen as the QRS wave complex on the ECG. The T wave is the relaxation of the ventricles.

SAMPLE STUDENT ANSWER

Outline the events of the cardiac cycle that are occurring during the QRS interval. [2]

This answer could have achieved 1/2 marks:

An electric signal is sent through the atrium to the ventricle. The signal then continues through the connecting fibres to cause contraction of the ventricles.

▲ This answer scored one mark for mentioning that the ventricle contracts.

▼ The answer fails to mention that the septum depolarizes and a signal is sent from the AV node. The conducting fibres carry impulses through the ventricle wall and the bundle of His (Purkinje fibres). The ventricles depolarize and the AV valves close (bicuspid and tricuspid valves). The atria repolarize and the systole is initiated (the ventricles contract).

Signals from the sinoatrial node that cause contraction cannot pass directly from atria to ventricles. There is a delay between the arrival and passing on of a stimulus at the AV node. This delay allows time for atrial systole before the AV valves close. Conducting fibres ensure coordinated contraction of the entire ventricle wall. Artificial pacemakers can be used in cases of SA problems to regulate the heart rate.

>> **Assessment tip**

Do not feel tempted to write about the P and T waves, as no marks will be given for this.

The invention of the stethoscope led to improved knowledge of the workings of the heart. The stethoscope is an acoustic device that allows doctors to listen to heart sounds. Normal heart sounds are caused by the AV valves (lub sound) closing before the ventricles contract and semilunar valves closing (dub) after the ventricles are emptied.

SAMPLE STUDENT ANSWER

The diagram shows the use of a sphygmomanometer in the measurement of blood pressure.

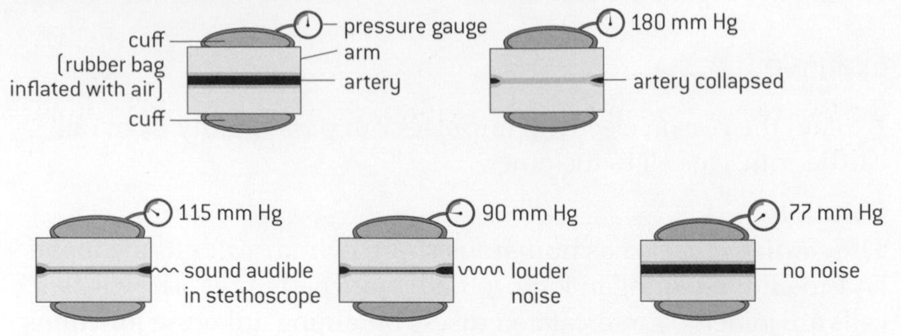

a) Identify the systolic pressure and diastolic pressure for this adult male. [1]

This answer could have achieved 0/1 marks:

Systolic pressure (mmHg): 180 mmHg

Diastolic pressure (mmHg): ~~77~~ mmHg

> ▼ Although the diastolic pressure is correct, both were needed for a mark, and the systolic pressure was 115 mmHg.

> **Exam tip**
>
> In this case the units are already given for the answer so they are not needed. Nevertheless, it is important to remember to add the units in answers if they are not explicit.

b) Explain the meaning of systolic and diastolic pressure. [3]

This answer could have achieved 0/3 marks:

Systolic pressure is the contracting pressure and diastolic pressure is the relaxing pressure. Contracting of the artery makes systole and diastole is the opposite.

> ▼ This answer confuses the contraction of the artery with that of the ventricles. To score the three marks the answer should have said that pressure is the force of blood on arteries. The systolic pressure is measured when the ventricle contracts, when blood is being pumped out of the heart. Diastolic pressure is measured when the ventricles are filled with blood and the heart is relaxed.

> **Assessment tip**
>
> Be careful to write that it is the ventricles that are contracting, not the heart.

c) During cardiac arrest, the ventricles of the heart might begin to contract in an uncoordinated fashion. Outline the treatment used for this condition. [1]

This answer could have achieved 0/1 marks:

A treatment for this condition is that purposefully sending signals to the atrium.

> ▼ This answer is too vague for a mark. It should have mentioned the use of a defibrillator to restore or reset a normal rhythm. This gives an electric shock or discharge to the chest (or indirectly to the heart), restoring the heartbeat or natural rhythm.

> **Assessment tip**
>
> Because it is an "outline" question, just mentioning the defibrillator is not enough to score a mark.

Coronary heart disease (CHD) refers to the reduction of blood flow in the coronary arteries to the heart muscle. The lack of oxygen causes muscle cells to die. Risk factors include high blood pressure, smoking, a high lipid diet, obesity, lack of exercise, high blood cholesterol, diabetes and alcoholism. Atherosclerosis is the main cause of heart attacks. It is the narrowing of an artery due to fatty deposits on its inner walls. The deposits are made of atheroma, a substance containing cholesterol, decaying cells, blood cells and fatty proteins. Narrowing of blood vessels raises blood pressure, which puts strain on the heart.

Chronic narrowing of the coronary arteries can lead to ventricular arrhythmia which can cause ventricular fibrillation. This is when disordered electrical activity causes the ventricles to quiver (fibrillate) instead of contracting, causing arrhythmia or abnormal heartbeats. This prohibits the heart from pumping blood, causing collapse and cardiac arrest. During ventricular fibrillation, an electric shock to the heart using a defibrillator can be vital for survival.

Example D.4.2.

Describe the use of defibrillation to treat life-threatening cardiac conditions.

Solution

An electrical impulse sent by the defibrillator is used to depolarize the heart muscle in order to re-establish the function of the natural pacemaker. Either two metal paddles or a few adhesive electrodes are placed on the patient's chest, close to the heart. Once on, the defibrillator detects whether the patient is having a dangerous heart arrhythmia or is in cardiac arrest. A series of electrical shocks is delivered through the electrodes and usually the patient is monitored through an ECG.

D.5 HORMONES AND METABOLISM (AHL)

You should know:

- ✔ endocrine glands secrete hormones directly into the bloodstream.
- ✔ steroid hormones bind to receptor proteins in the cytoplasm of the target cell to form a receptor–hormone complex.
- ✔ the receptor–hormone complex promotes the transcription of specific genes.
- ✔ peptide hormones bind to receptors in the plasma membrane of the target cell.
- ✔ binding of hormones to membrane receptors activates a cascade mediated by a second messenger inside the cell.
- ✔ the hypothalamus controls hormone secretion by the anterior and posterior lobes of the pituitary gland.
- ✔ hormones secreted by the pituitary control growth, developmental changes, reproduction and homeostasis.

You should be able to:

- ✔ describe hormonal secretion by endocrine glands.
- ✔ compare and contrast steroid and peptide hormones.
- ✔ explain hormonal control by the hypothalamus and pituitary gland.
- ✔ explain the control of milk secretion by oxytocin and prolactin.
- ✔ outline how some athletes take growth hormones to build muscles.

Hormones are not secreted at a uniform rate and exert their effect at low concentrations. Endocrine glands secrete hormones directly into the bloodstream. The endocrine glands are the pituitary, pineal, thymus, thyroid and adrenal glands, and the pancreas. In addition, women produce hormones in their ovaries and men in their testes.

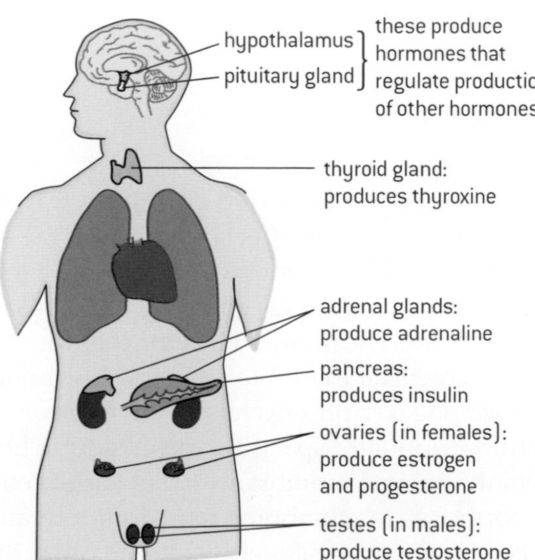

▲ **Figure D.5.1.** Main endocrine glands

In Topic 6.6 hormones, homeostasis and reproduction were studied.

The International Council for the Control of Iodine Deficiency Disorders includes a number of scientists who work to eliminate the harm done by iodine deficiency. Iodine is a mineral that is necessary for the correct functioning of the thyroid gland. This gland produces two similar hormones, triiodothyronine (T3) that contains three iodine atoms and tetraiodothyronine (T4) that contains four iodine atoms. These hormones are involved in metabolism. The inability to produce the thyroid hormones due to lack of iodine means that the hypothalamus and the anterior pituitary continuously stimulate the thyroid and this leads to goitre (enlargement of the thyroid).

- **Steroid hormones** act directly on receptors on cell membranes.
- **Peptide hormones** act through second messengers.
- **Second messengers** are molecules that are activated when the peptide hormone attaches to the receptor. These second messengers activate a series of reactions. Examples are cyclic AMP and calcium calmodulin.

Example D.5.1.

The thyroid hormones found in the bloodstream are tetraiodothyronine, also called levothyroxine (T4), triiodothyronine (T3) and reverse T3 (rT3). Reverse T3 is formed by the elimination of one iodine from T4, inactivating this hormone. The levels of rT3 have been recorded to increase in cases of cirrhosis.

The levels of T3 and rT3 were measured in the blood serum of men. The results are recorded in the scatter graph and the upper limit for the normal range of rT3 and lower limit for the normal range of T3 are shown.

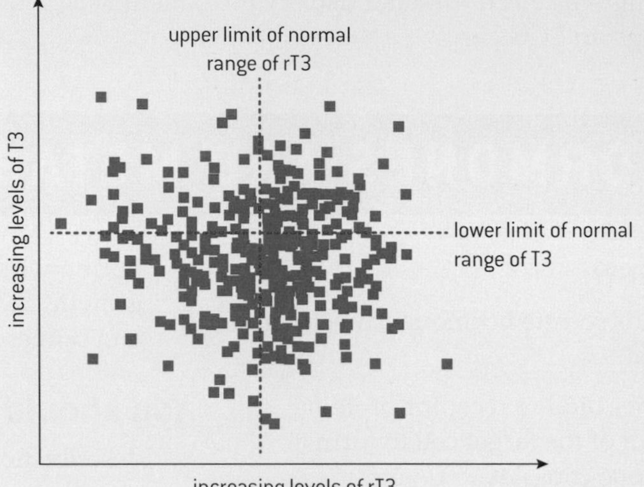

The data show that many of these men have thyroid and liver issues. Comment on the health information that can be provided by this data.

Solution

The data show that most of these men are not healthy. If you divide the graph in four, the men in the top left are healthy, and as can be seen, they are the minority. Those on the right upper and lower sections have elevated rT3 and are therefore sick (perhaps with cirrhosis). Those in the left bottom have correct levels of rT3 but low levels of T3 and therefore might have thyroid problems or lack of iodine in the diet.

Example D.5.2.

Compare and contrast the mode of action of steroid and peptide hormones.

Solution

Both peptide and steroid hormones travel through blood and act on target organs. The effect of both peptide and steroid hormones lasts for a long time. Both these hormones bind to specific receptors, but steroid hormones enter the cell while peptide hormones do not. Steroid hormones bind to receptor proteins in the cytoplasm of the target cell to form a receptor–hormone complex. The receptor–hormone complex travels to the nucleus where it promotes the transcription of specific genes. Peptide hormones bind to receptors in the plasma membrane of the target cell. Binding of the peptide hormones to membrane receptors activates a cascade mediated by a second messenger inside the cell which produces a series of enzymatic reactions. Peptide hormones require ATP while steroid hormones do not.

SAMPLE STUDENT ANSWER

Explain the role of receptors in mediating the action of both steroid and protein hormones.

This answer could have achieved 6/6 marks:

As receptors are specific for each hormone, the receptors allow both steroid and protein hormones to affect specific target cells and tissues.

Steroid hormone: Though steroid hormones can pass through the plasma membrane of their desired cells into the cytoplasm and/or nucleus, it is not until the hormone binds to the specific receptor for that hormone that the receptor–hormone complex binds directly to the DNA (once in the nucleus) and acts as a transcription factor for gene expression. (The complex can either promote or inhibit certain gene expression and thereby affect protein production.)

Peptide hormone: Requires plasma membrane receptors (which are coupled to internal proteins such as G-protein) to affect the gene expression of the cell through second messenger such as cyclic AMP.

▲ This is a very clear and complete answer. Three marks were given for steroid hormones cross plasma membrane and bind to receptors in the cytoplasm of the target cell to form a receptor–hormone complex. The other three marks were for peptide hormones bind to receptors in the plasma membrane of the target cell, acting through a second messenger inside the cell such as cAMP.
The answer obtained full marks, but other answers were that receptors are proteins; in steroid hormones the receptor–hormone complex promotes the transcription of specific genes; and that the binding of hormones to membrane receptors activates a cascade of reactions.

The hypothalamus controls hormone secretion by the anterior and posterior lobes of the pituitary gland.

Hormones secreted by the pituitary control growth, developmental changes, reproduction and homeostasis.

Example D.5.3.

Explain hormonal control by the hypothalamus and pituitary gland.

Solution

The hypothalamus controls hormone secretion by the anterior and posterior lobes of the pituitary gland. Hormones secreted by the pituitary gland control growth, developmental changes, reproduction and homeostasis. The cells in the hypothalamus secrete antidiuretic hormone (ADH) and oxytocin which are transported down to the posterior pituitary, where they are released into the bloodstream. The hypothalamus has neurons that control the endocrine functions of the anterior pituitary. The anterior pituitary gland produces prolactin, involved in milk production. It also produces follicle stimulating hormone (FSH) and luteinizing hormone (LH), involved in the growth of follicles and ovulation respectively in ovaries in females, and in sperm formation in testes in males. Thyroid stimulating hormone (TSH) is a hormone that induces the thyroid gland to produce thyroxine and other thyroid hormones. Somatotropin or growth hormone (GH) stimulates growth, cell reproduction and regeneration in bones and soft tissues.

Milk secretion is controlled by oxytocin and prolactin, two hormones produced by the pituitary gland. This gland is under the control of the hypothalamus.

- The **hypothalamus** is part of the brain that controls the pituitary gland.
- The **pituitary gland** is an endocrine gland.
- **Oxytocin** is a peptide hormone produced by the hypothalamus in charge of childbirth and milk secretion.
- **Prolactin** is a peptide hormone produced by the pituitary gland in charge of milk secretion.

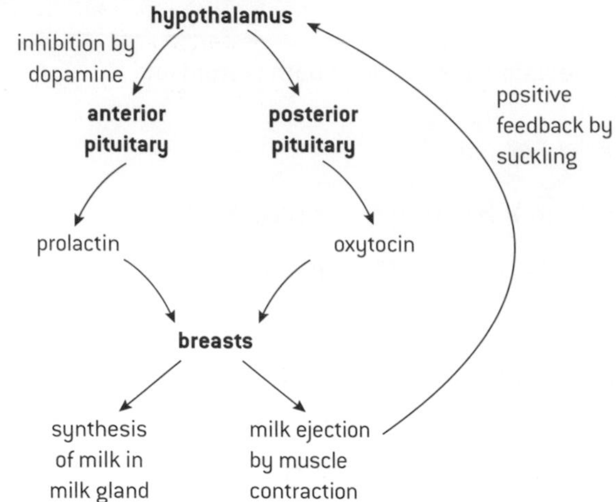

▲ **Figure D.5.2.** Control of milk secretion

Example D.5.4.

Outline how some athletes take growth hormones to build muscles.

Solution

Some athletes take human growth hormone (HGH) to build muscles. HGH increases body cell mass and decreases body fat mass. It also increases sprint capacity when administered together with testosterone. Anabolic steroid hormones have several adverse effects. They can produce sexual and mental problems, can increase the risk of coronary heart disease and are toxic to the liver.

D.6 TRANSPORT OF RESPIRATORY GASES (AHL)

You should know:

- ✔ oxygen dissociation curves show the affinity of hemoglobin for oxygen.
- ✔ carbon dioxide is carried in solution and bound to hemoglobin in the blood.
- ✔ carbon dioxide is transformed in red blood cells into hydrogencarbonate ions.
- ✔ the Bohr shift explains the increased release of oxygen by hemoglobin in respiring tissues.
- ✔ chemoreceptors are sensitive to changes in blood pH.
- ✔ the rate of ventilation is controlled by the respiratory control centre in the medulla oblongata.
- ✔ during exercise the rate of ventilation changes in response to the amount of CO_2 in the blood.
- ✔ fetal hemoglobin is different from adult hemoglobin allowing the transfer of oxygen in the placenta to the fetal hemoglobin.

You should be able to:

- ✔ identify pneumocytes, capillary endothelium cells and blood cells in light micrographs and electron micrographs of lung tissue.
- ✔ outline how pH of blood is regulated to stay within the narrow range of 7.35 to 7.45.
- ✔ explain causes and treatments of emphysema.
- ✔ analyse dissociation curves for hemoglobin and myoglobin.
- ✔ explain the consequences of high altitude for gas exchange.

Emphysema is a lung disease where the walls of the alveoli break down, so air sacs are fewer and larger. The enzyme elastase is a protease that digests elastin, a protein that gives alveoli their elasticity. It is caused mainly by smoking or pollution. Another cause of emphysema is the lack of alpha-1 antitrypsin, a protease inhibitor. This is a genetic disease. Treatment consists of the use of bronchodilators, corticosteroids to reduce inflammation, oxygen supplementation, and antibiotics if there are signs of infection. The most important step is to quit smoking. Surgery or even lung transplant is done in very severe cases.

Red blood cells (erythrocytes) are vital in the transport of respiratory gases. Oxygen dissociation curves show the affinity of hemoglobin for oxygen.

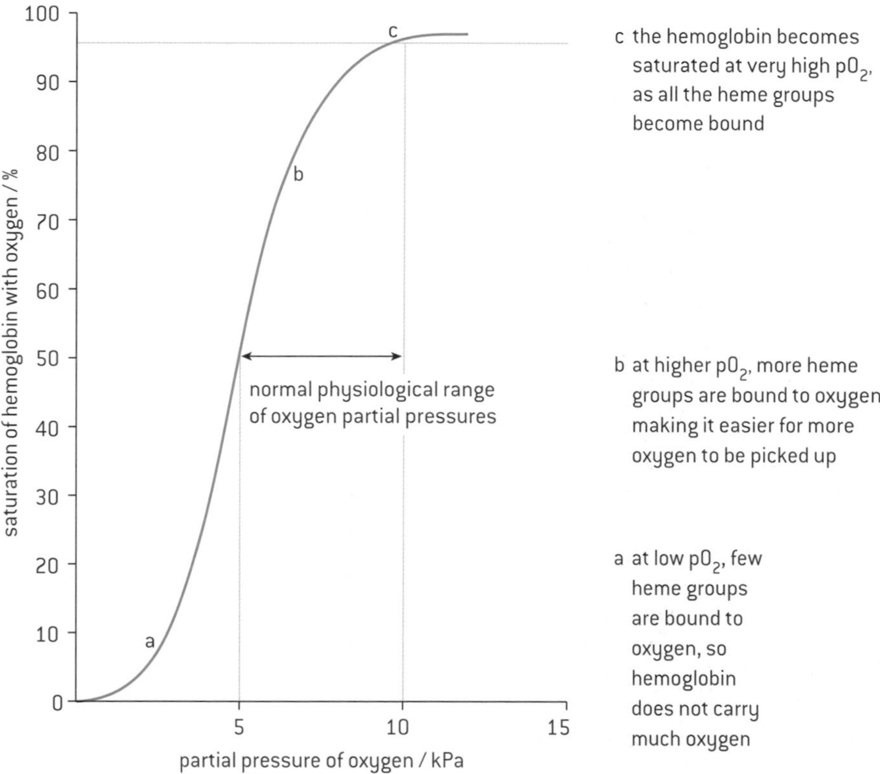
100
90
80
70
60
50
40
30
20
10
0
saturation of hemoglobin with oxygen / %
5
10
15
partial pressure of oxygen / kPa
a
b
c
normal physiological range of oxygen partial pressures
c the hemoglobin becomes saturated at very high pO2, as all the heme groups become bound
b at higher pO2, more heme groups are bound to oxygen, making it easier for more oxygen to be picked up
a at low pO2, few heme groups are bound to oxygen, so hemoglobin does not carry much oxygen

▲ **Figure D.6.1.** Graph showing oxygen dissociation curve for human hemoglobin

Example D.6.1.

Explain how carbon dioxide is transported in blood.

Solution

Inside red blood cells or erythrocytes CO_2 reacts with H_2O forming carbonic acid (H_2CO_3). This reaction is catalysed by the enzyme carbonic anhydrase. The carbonic acid breaks down to hydrogen ions (H^+) and hydrogencarbonate (HCO_3^-). Most of the carbon dioxide is therefore transported as hydrogencarbonate in blood. The rest of the CO_2 in the red blood cells binds to hemoglobin and is transported as carbaminohemoglobin.

Scientific research has led to a change in public perception of smoking. The carbon monoxide produced in smoking binds irreversibly to hemoglobin, decreasing gaseous exchange causing breathlessness. Nicotine increases heart rate, blood pressure and the chance of thrombosis. Tar coats the lining of alveoli, increasing the risk of emphysema. Infections and irritation of breathing pathways are increased because the cilia are burnt or damaged. This induces coughing. Scientists have proven that smoking affects lung function. There are increased chances of developing lung, throat or mouth cancer and cardiovascular disease.

- **Type I pneumocytes** are cells in the alveoli involved in the process of gas exchange.
- **Type II pneumocytes** secrete pulmonary surfactant, a fluid that decreases the surface tension within the alveoli. They are also capable of cellular division, giving rise to more type I pneumocytes.
- **Gaseous exchange** is the process by which gases move passively by diffusion across a surface. For example, the intake of oxygen and release of carbon dioxide between the alveoli and blood capillaries of the lungs.

Topic 6.2 covered the blood system and in Topic 6.4 gas exchange was studied.

- **Hemoglobin** is a protein in the erythrocytes formed by four peptides with four heme groups, each with an iron molecule.
- **Oxyhemoglobin** is hemoglobin that carries oxygen.
- **Carbaminohemoglobin** is hemoglobin that carries carbon dioxide.
- The **Bohr effect** is the shift of the oxygen dissociation curve to the right with the increased acidity or CO_2 in blood.
- **Emphysema** is when the walls of the alveoli break down, air sacs are fewer and larger.

The Bohr shift explains the increased release of oxygen by hemoglobin in respiring tissues. The oxygen dissociation curve is shifted to the right as the oxygen binding affinity is inversely related to acidity and carbon dioxide concentration. A shift to the right in the sigmoid dissociation curve means a decrease in O_2 affinity, therefore O_2 is released.

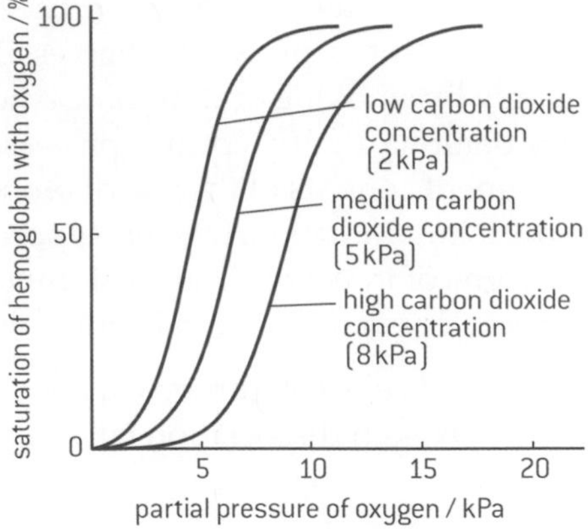

▲ **Figure D.6.2.** The Bohr shift

In the aorta and the carotid artery, there are special chemoreceptors that detect the levels of carbon dioxide in the blood. Additionally, the carotid receptors detect changes in the pH of blood. Blood pH is regulated to stay within the narrow range of 7.35 to 7.45. The respiratory control centre in the medulla oblongata controls the rate of ventilation in response to changes in pH by sending impulses to the external intercostal muscles and diaphragm. During exercise the rate of ventilation increases in response to the high amount of CO_2 in the blood.

SAMPLE STUDENT ANSWER

Explain how low blood pH causes hyperventilation (rapid breathing). [3]

This answer could have achieved 0/2 marks:

Low blood pH means that the blood has become more acidic. The binding of an acid substance can reduce the affinity of the hemoglobin. This means it is not able to carry as much oxygen. This means that an individual would have to breathe faster in order to obtain a normal amount of oxygen, therefore causing ventilation.

▼ This answer did not score any mark because it is not answering the question. In the first sentence the student mentions that blood becomes acidic but does not say that this is due to an increase in CO_2 lowering blood pH. The chemoreceptors, respiratory centre, intercostal muscles and diaphragm are not mentioned. Ventilation rate increase occurs to expel CO_2 rather than to obtain oxygen.

Example D.6.2.

a) Compare and contrast myoglobin and fetal hemoglobin.

b) Sketch curves to compare the saturation of myoglobin and fetal hemoglobin with adult hemoglobin.

Solution

a) Myoglobin is found in muscles. It consists of only one peptide chain with one heme group. Myoglobin can only bind one oxygen molecule, but this binding is stronger than that in hemoglobin, taking the oxygen from hemoglobin in respiring muscle cells. Fetal hemoglobin is different from adult hemoglobin. It has a greater affinity for oxygen, becoming saturated at a lower partial pressure of oxygen. This allows the transfer of oxygen in the placenta to the fetal hemoglobin.

b)

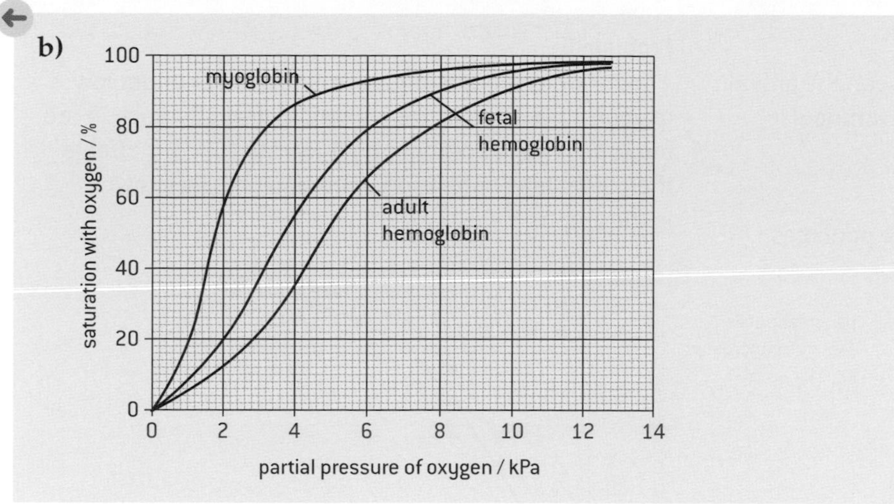

At high altitude there is less partial pressure of oxygen. A greater number of erythrocytes in blood are produced in response to this because there is an increased production of the hormone erythropoietin (EPO). EPO is produced naturally in the body, but has also been used for doping in some sports.

Assessment tip

In the sketch you do not need to be too precise with the values of saturation and oxygen partial pressures, but you need to clearly show the trends of saturation. In this case, the curves for fetal and adult hemoglobins are sigmoid, with the fetal hemoglobin above the adult. The curve for myoglobin is less sigmoidal and above both hemoglobins. The three curves plateau close to 100% saturation.

Practice problems for Option D

Problem 1

The graph shows the sodium, potassium and cholesterol consumption in a diet of 1,000 children in a school compared to the recommended dietary intake.

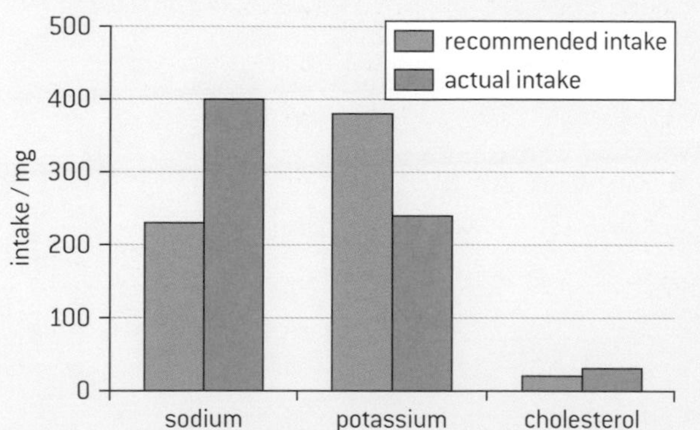

a) State the nutrient for which the diet of the children in this school is below recommended values.

b) Outline one problem these children might have in the future due to their cholesterol intake in diet.

c) These children have high chances of hypertension in the future if they do not change their diet. Comment on this statement.

Problem 2

Explain the hormonal mechanisms that control the secretion of digestive juices.

Problem 3

Explain the causes and consequences of jaundice.

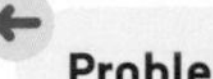

Problem 4

Explain the need for the time delay between the arrival and passing on of a stimulus at the atrioventricular node.

Problem 5

The diagram shows a section through the pancreas showing an islet of Langerhans.

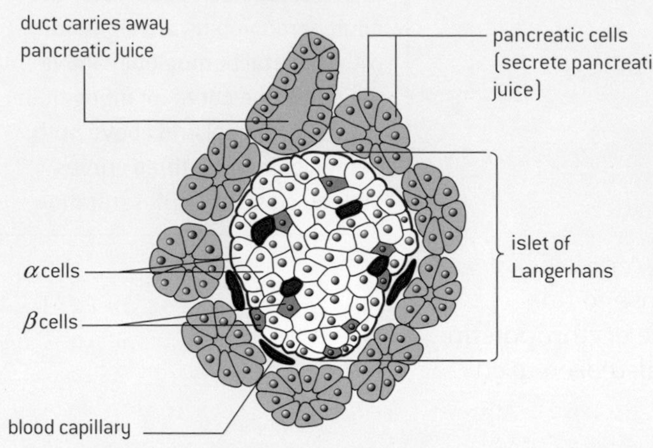

Identify the structures that show that the pancreas is both an exocrine and an endocrine gland.

Problem 6

A factor which adapts organisms to survive under low oxygen conditions is the presence of a substance in red blood cells called diphosphoglycerate (DPG). It affects the percentage saturation of hemoglobin as shown in the graph.

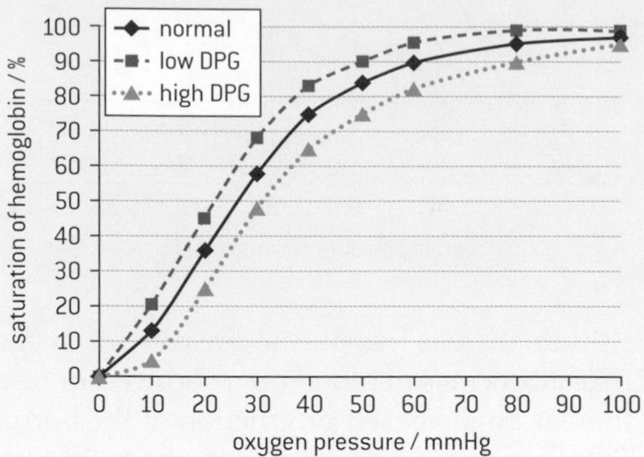

a) Identify the effect of DPG on hemoglobin saturation.

b) Suggest with a reason the DPG levels in organisms living at high altitude.

INTERNAL ASSESSMENT

For your internal assessment (IA) you will be asked to carry out an individual investigation and write a report. This report should be between six and 12 pages long and will form 20% of your final IB Biology grade.

Assessment criteria

Your investigation is assessed using the six criteria below, out of a total of 24 marks.

Criterion	Personal engagement	Exploration	Analysis	Evaluation	Communication	Total
Maximum marks	2	6	6	6	4	24
Percentage of mark / %	8	25	25	25	17	100

▲ **Table 1.** IA assessment criteria

Choice of research question

If you look at the assessment criteria in the table above, you can see that the most marks are available for exploration, data analysis and evaluation—that is why it is so important that the investigation yields good data. You can be adventurous in the choice of your investigation, but you must make sure it will provide you with enough data for analysis. Choosing an interesting research question that yields little data, or insufficiently useful data, will score very poorly.

>> **Assessment tip**

Research and try out your ideas to make sure you can generate sufficient useful data before you embark on a full investigation.

Not all IAs are based on primary data. You can use data generated from online databases. In cases where you work with a group of candidates collaborating to generate a database, you must generate your own unique research question.

>> **Assessment tip**

Collect sufficient data over a meaningful range, with adequate repeats, and use the most precise apparatus you have access to.

The independent variable is the variable that will be changed. The dependent variable is the variable that will be measured or observed. All other variables must be controlled. The research question may not necessarily include the actual dependent variable but a derived value. It is understood that for certain investigations there will be no independent variable (eg studying the relationship between the distributions of two species in a named ecosystem).

▼ Bad question	▲ Good question
How does the flow of water affect insects?	What is the influence of the flow rate of water in the Ibicui river in Brazil on the distribution of *Paracloedes leptobranchus* nymphs?

▲ **Table 2.** Good and bad research questions

>> **Assessment tip**

Make sure your research question is focused and precisely expressed. For example, "Measuring the effect of light on photosynthesis" is too vague as it does not indicate what the dependent and independent variables are. The independent variable could be light intensity, wavelength, photons, type of lamp, etc, while the dependent variable could be rate of photosynthesis, presence of starch, and many others.

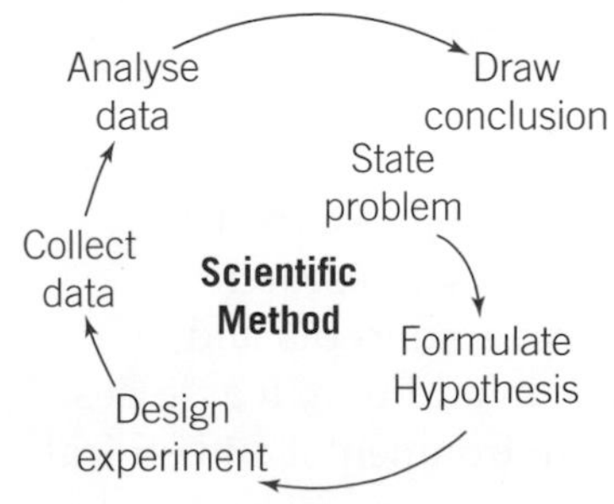

When formulating your hypothesis, consider:

- the independent variable or the two variables being correlated
- the range of the independent variable (try to keep it limited to a reasonable value)
- the dependent variable (or derived dependent variable) and how you are going to measure it
- the scientific name of the organism used
- a prediction of what you think is going to happen.

Once you have an investigative hypothesis, then you can add the statistical "null" and "alternative" test hypotheses:

- The null hypothesis will state a lack of correlation/association or impact between variables or groups.
- The alternative hypothesis will state a significant correlation/association or differences between variables or groups.

Criterion 1: Personal engagement

This criterion assesses your engagement with the exploration and how you have made it your own. For this criterion, make that sure you:

- communicate your interest in your investigation.
- pose a question to which you do not already know the answer.
- do not use a known method without making any modifications to it—examiners value originality in the design of your method
- explain why you are using the method you have chosen. Ensure that the method works, and if it does not, modify it and explain how and why you modified it in your report.

Assessment tip

The examiners will be looking for the personal significance of your research question.

Criterion 2: Exploration

This criterion is about establishing a scientific context, stating a clear and focused research question and using appropriate concepts and techniques in your method. In addition, it assesses your awareness of safety, environmental and ethical considerations.

To gain the exploration marks, include the following sections:

- your research question
- your hypothesis
- background information
- a statement of the variables
- apparatus
- method
- safety considerations.

Assessment tip

If you use secondary data (or data from models and simulations) instead of collecting primary data in the lab, you will need more than one source of data so that the variance between sources can be evaluated. This will make your error analysis and evaluation more successful.

After stating your research question you need to include suitable background information, which should focus on the specifics of your chosen research question and methodology. Describe the context for your question and existing knowledge.

Assessment tip

Remember, the background information provided for the investigation should be entirely relevant and enhance the understanding of the context of the investigation.

You will need to include a statement of the variables and how you will control them. You may find it helpful to use a table like this whilst planning your investigation to make sure you do not miss any out when writing your statement.

Independent variable (state the range)		
Dependent variable	*what will be measured (be precise)*	
Controlled variable	**Reason for controlling variable**	**Control method**

▲ **Table 3.** Planning around variables

Include a detailed list of the apparatus and materials used in your investigation. Choose apparatus that will enable you to collect reliable data, for example, by using graduated pipettes and measuring cylinders to measure volume rather than beakers.

▼ Bad naming of apparatus	▲ Good naming of apparatus
measuring cylinders	3 × 25.0 cm^3 ± 0.5 cm^3 measuring cylinders

▲ **Table 4.** Good and bad descriptions in an apparatus section

Make sure your method is written in sufficient detail for any reader to repeat the investigation and also so that an idea of the associated uncertainties can be gained. It is a good idea to describe in a paragraph how you chose and developed your methodology. This narrative will help explain the amount of data you are collecting and give insight into your decision-making, which is part of the evidence for the "personal engagement" criterion. You should plan in advance how many measurements you will need to take for a valid conclusion, and your method must allow sufficient repeats to ensure each measurement is valid.

After the method section, you should add in a short section about the safety, ethical and environmental issues relevant to your methodology. This might be quite basic, such as noting the need for gloves and safety glasses, but could also include the application of the IB animal experimentation policy, use of consent forms, disposal of waste and impact on field sites. If you cannot identify any safety, ethics or environmental problems, a statement to this effect will show that you have at least considered them.

Criterion 3: Analysis

The analysis criterion is about providing evidence that you have selected, recorded, processed and interpreted the data in ways that are relevant to the research question and can support the conclusion you reach.

In your analysis, you should include:

- raw data
- sample calculations for processed data
- qualitative and quantitative data
- statistical analysis of your data
- associated uncertainty table
- graphical analysis.

Assessment tip

Remember that bar charts are of limited analytical use and are used only if you have discontinuous variables.

You need to record enough data about the independent and dependent variables to enable meaningful processing and interpretation. You should also record qualitative observations, eg, fieldwork should always have some form of site description; this could take the form of maps, sketches or photographs. However, if you use photos make sure that that these are supporting and not replacing written qualitative data. Also, be careful not to include unnecessary photos, for example of the set-up of a respirometer under different conditions, where one picture will be enough.

▼ Bad graph	▲ Good graph
Units missing on the y axis Line of best fit is straight, but the data indicates a curve would be a better fit	Graph has a title Axes have labels Data points are clearly marked Scale of graph is appropriate

▲ **Table 5.** Good and bad aspects of a graph

Once you have collected the data, include the associated uncertainty in a table. You need to interpret and evaluate the uncertainty in the data and the size of any discrepancy between your results and those in the literature (if they exist). This gives an indication of how accurate your data is.

If you are familiar with statistical analyses, you can consider the impact of measurement uncertainty on your results using the following techniques:

- propagate the errors through numerical calculations
- calculate the standard deviation (if you have a sufficiently large data set)
- draw a line of best fit on your graph
- consider including error bars and/or the maximum or minimum slopes that can be drawn with the data you have, based on the range of values at each data point.

Statistical analysis of your data is expected. The layout of this will depend upon the tools used. Percentages, means, medians, standard deviations or other statistics can be presented at the end of the column or row of data they represent. For more complex processing of data using spreadsheets, such as correlation coefficient, t-test and X^2 (chi-squared) test, screenshots are acceptable. For other less orthodox processing such as analysis of variance (ANOVA) tests, a worked example may be necessary.

- **Mean:** the sum of the values divided by the number of values; ie, the average.
- **Median:** value found exactly in the middle of a data set.
- **Statistical significance:** these statistical tests are all about proving that any difference between sets of data is not random. The result of a test is said to be statistically "significant" when a test shows that your data sets are reliably different, and therefore not different by chance.
- **Standard deviation**: a measure that is used to quantify the amount of variation or dispersion of a set of data values.
- **Correlation coefficient (r):** a numerical measure of some type of correlation. It shows if there is a statistical relationship between two variables.
- **Coefficient of determination (r^2):** a statistic that will give some information about the goodness of fit of a model. An r^2 of 1 indicates that the regression predictions perfectly fit the data.
- **t-test:** a statistical test used to determine if there is a significant difference between the means of two groups of data. A t-test is most commonly applied when the data sets follow a normal distribution.
- **X^2 (chi-squared) test:** a statistical test used to determine whether there is a significant difference between the expected frequencies and the observed frequencies in one or more sets of data.
- **ANOVA (analysis of variance) test:** a collection of statistical models that help you calculate how much of the total variance comes from variance between groups of data and from variance within groups of data. ANOVA is useful for comparing three or more groups of data. It will show if there are any statistically significant differences between the means of three or more independent groups of data.

Assessment tip

When presenting your data, check you have considered:

- ✔ degrees of precision of instruments used
- ✔ variation in the material used
- ✔ standard deviations
- ✔ standard errors
- ✔ ranges (maximum–minimum)
- ✔ outlier data.

Outliers need to be recorded but marked as such and should not be used in the statistical analysis. You should check whether any data is really an outlier by repeating the experiment. You should not exclude outliers from processing just because they do not "fit well" in the general trend of the data. Their exclusion requires a justification.

Completing several trials will make it easier for you to decide which results are inconsistent. If you can, it would be good to compare the data from different data sources (secondary data examples) to evaluate reproducibility. If you are using data from a database you need to show you have done some research into the uncertainties associated with database data and applied it.

Assessment tip

In all of this section you need to keep an eye on your use of both significant figures and decimal places—they need to be consistent.

- A **random error** is a statistical fluctuation (in either direction) in the measured data due to the precision limitations of the measurement device.
- **Systematic error** is a problem which persists throughout the entire experiment that can be solved through a better experimental design.
- The **uncertainty** of a measurement is as accurate as the tool being used to make this measurement; it is one half of the smallest measurement possible with the device.
- An **outlier** is any value that is numerically distant from most of the other data points in a set of data.
- **Accuracy** refers to the closeness of a measured value to a standard or known value.
- **Precision** refers to the closeness of two or more measurements to each other.
- **Reliability** is the degree of consistency of a measurement.

Assessment tip

Using error bars and lines of best fit on graphs will help you to portray the data as realistically and honestly as possible.

Criterion 4: Evaluation

This criterion assesses your evaluation of your investigation and your results in relation to the research question and accepted scientific context. It also assesses your ability to identify limitations and suggest improvements.

Evaluation is the most challenging criterion. Drawing a conclusion that is consistent with the data is fairly straightforward, but correctly describing or justifying your conclusion through relevant comparisons to the accepted scientific context is much more difficult. However, you can do this by comparing your experimentally determined data to other available data and/or checking whether any trends and relationships that you identify are in line with accepted theory.

- A **correlation** indicates a predictive relationship between two variables.
- A **trend** is a pattern or general tendency of a series of data points.
- A **causal link** indicates that one event is the result of the occurrence of the other event; ie, the independent variable affects the dependent variable.

You must identify limitations and suggest improvements to your investigation. Try to identify and address any systematic and random errors in detail. Include a discussion of the likely effect of systematic errors; for example, if you measured the volume of an enzyme with a pipette and left a drop behind each time you will have a smaller volume than recorded. Do not be tempted to suggest things that are unnecessary, such as more repeats if you had sufficient results that agreed. You can also calculate the percentage error of your results.

Lastly, if you can see a good extension to your investigation, suggest it here.

Assessment tip

✔ Make sure you have a basis to your conclusion which you can then compare with secondary data or a scientific theory.
✔ Identify whether errors are random or systematic.
✔ Suggested improvements should address the errors identified.

Criterion 5: Communication

The communication criterion assesses whether your report effectively communicates the focus, process and outcomes of your investigation.

In general this will be the easiest criterion for you to do well in. Make sure your report is appropriately structured so that it presents the information well. Include enough detail in your description of the methodology, and at least one worked example of a calculation (if need be) so that the examiner can understand how you processed your data. It is important to include citations and references for anything that has come from another source.

Do not waste space by including unnecessary information or pictures. Do not include an appendix, as the examiner does not have to read it.

Report checklist

Complete this checklist before you submit your report.

Tasks
The report is within the 12-page limit
The title page has the candidate name and number on it
There is a contents page and all pages are numbered
The research question is specific
All background information included is relevant to the research question
The apparatus is listed precisely
The methods are described clearly (and are repeatable)
All illustrations have figure numbers
Raw data has been included
All graphs/tables have units and all graphs have labelled axes
Sample calculations have been included for processed data
Data is shown honestly, eg, using lines of best fit or error bars
Safety, ethical and environmental issues have been included
All sources are correctly referenced
There is a conclusion
The evaluation makes recommendations for improvements

PRACTICE EXAM PAPERS

At this point, you will have re-familiarized yourself with the content from the topics and options of the IB Biology syllabus. Additionally, you will have picked up some key techniques and skills to refine your exam approach. It is now time to put these skills to the test; in this section you will find practice examination papers, 1, 2 and 3, with the same structure as the external assessment you will complete at the end of the DP course. Answers to these papers are available at **www.oxfordsecondary.com/ib-prepared-support.**

Paper 1

SL: 45 minutes

HL: 1 hour

Instructions to candidates

- Answer all the questions.
- For each question, choose the answer you consider to be the best and indicate your choice on the answer sheet (provided at **www.oxfordsecondary.com/ib-prepared-support**).
- The maximum mark for the SL examination paper is **[30 marks].**
- The maximum mark for the HL examination paper is **[40 marks]**.

*SL candidates: answer questions 1–30 **only**.*

*HL candidates: answer **all** questions.*

1. Which property of stem cells makes them suitable for therapeutic use?

A. Their chromosomes are ideal for gene transfer and cloning
B. They can divide by meiosis to form gametes
C. They can differentiate into specialized cells
D. They can produce antibodies

2. Two epithelial cells were compared; one was larger than the other. Which statement about these cells is correct?

A. The small cell had less surface area relative to its volume
B. The big cell produced more waste per unit of volume
C. The small cell was more efficient at losing excess heat
D. Less waste diffused out of the big cell

3. What is a role of cholesterol in mammalian cell membranes?

A. Reduces membrane fluidity and permeability to some solutes
B. Clogs transport proteins disabling facilitated diffusion
C. Forms part of lipoproteins recognized by antibodies
D. Produces channels for transport through membranes

Questions 4 and 5 refer to the drawing at the top of page 229.

The drawing shows how proteins are secreted from a cell.

4. Which process is being shown in the diagram?

A. Diffusion
B. Exocytosis
C. Osmosis
D. Facilitated diffusion

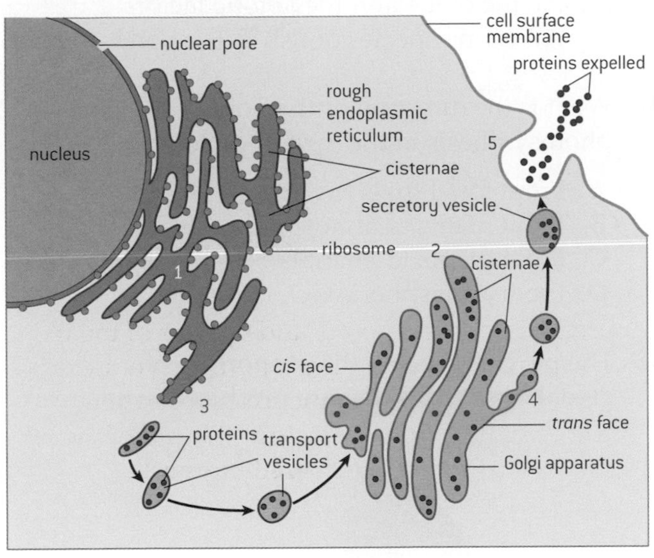

5. Which organelle(s) other than the nucleus show(s) that this is a eukaryotic cell?

I Golgi apparatus
II Ribosomes
III Rough endoplasmic reticulum

A. I only
B. III only
C. I and III only
D. I, II and III

6. Which diagram shows the interaction that occurs between water molecules?

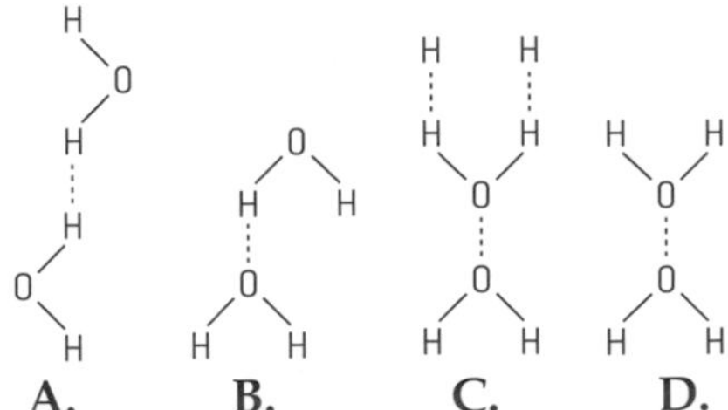

7. What is a difference between amylose and amylopectin?

A. Amylose is a larger molecule.
B. Only amylose is made of β-glucose.
C. Only amylopectin is made of glucose molecules.
D. Amylose is unbranched while amylopectin is branched.

8. Which diagram shows a peptide bond between two amino acids?

```
      R  O  H  R  O
H     |  ‖  |  |  ‖
 >N—C—C—N—C—C—OH
H     |        |
      H        H
A.

    H  O
H   |  ‖
 >N—C—C—OH
H   |
    R
    |
    R  O
H   |  ‖
 >N—C—C—OH
H   |
    H
B.

   O  R  H  H  R  O
   ‖  |  |  |  |  ‖
HO—C—C—N—N—C—C—OH
      |        |
      H        H
C.

  H  R  O  O  R  H
  |  |  ‖  ‖  |  |
H—N—C—C—C—C—N—H
     |        |
     H        H
D.
```

9. Which statement refers to DNA and RNA?

A. RNA usually comes in three forms and DNA only comes in one.
B. They always show complementary base pairing.
C. They are formed by the same nucleotides.
D. Only RNA is a helix.

10. Which statement is related to DNA replication?

A. It always requires Taq DNA polymerase to occur.
B. DNA polymerase is an enzyme that separates the DNA strands.
C. New nucleotide bases attach to the original sugar–phosphate backbone.
D. The new DNA contains one original strand and a new strand.

11. Which graph best represents the effect of increasing the substrate concentration on enzyme activity?

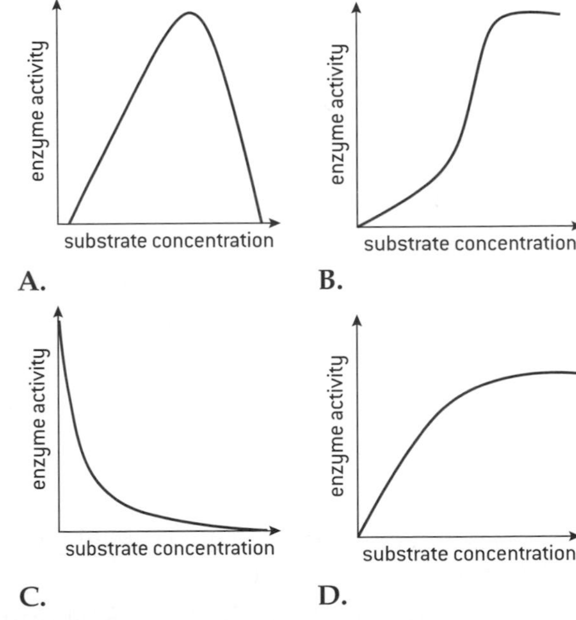

12. Which molecule is the source of the oxygen produced in photosynthesis?

A. Ribulose bisphosphate
B. Carbon dioxide
C. Glucose
D. Water

13. What is the relationship between Mendel's law of segregation and meiosis?

A. The separation of paternal and maternal chromosomes shows no pattern.

B. Only one of a pair of genes appears in a gamete.
C. Dominant and recessive alleles always separate.
D. Variation only results after two divisions.

14. A man has hemophilia, a condition caused by a recessive allele carried on the X chromosome. What is the possibility of having a hemophiliac child if his wife is homozygous normal for the gene?

A. 0% **B.** 25% **C.** 50% **D.** 100%

15. In goldfish (*Carassius auratus*) there is variation in colour pattern. The homozygous dominant (TT) are common orange goldfish and the homozygous recessive (tt) are transparent. When crossed, they produce mottled fish (Tt). What is this type of inheritance called?

A. Co-dominance
B. Non-mendelian
C. Non-disjunction
D. Dihybrid

16. In mice (*Mus musculus*) grey colour is due to a dominant allele (G) and the allele for albino colour (g) is recessive. Grey mice were crossed with albino mice. A total of 6 of the offspring were grey and 4 albino. The null hypothesis is that the alleles segregate independently (are unlinked) therefore the grey parent was heterozygous. The chi-squared value is 0.4. The table below shows the probability values:

	Probability			
Degrees of freedom	**0.99**	**0.95**	**0.05**	**0.01**
1	0.000	0.004	3.84	6.64
2	0.020	0.103	5.99	9.21
3	0.115	0.352	7.82	11.35

Which of the statements is correct about the null hypothesis?

A. It is accepted because the chi-squared value is less than 3.84.
B. It is rejected because the probability is less than 0.05.
C. It is accepted because the chi-squared value is less than 5.99.
D. It is rejected because the chi-squared value is greater than 0.352.

17. What constitutes a population in a pond ecosystem?

A. All the living organisms in the pond.
B. All the animals and plants in the pond.
C. All the biotic and the abiotic factors.
D. All of one species of fish in the pond.

18. What is the direction of the flow of carbon by photosynthesis in the carbon cycle?

A. From fossil fuels to atmospheric CO_2
B. From plants to atmospheric CO_2
C. From plants to animals
D. From atmospheric CO_2 to plants

19. The pie chart shows the proportion of greenhouse gases that are produced in an area.

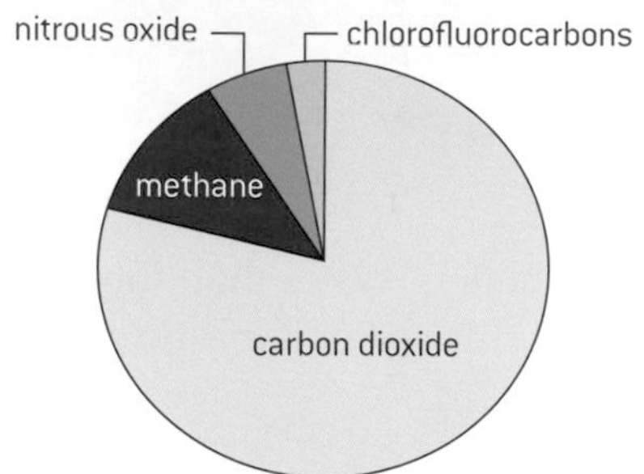

What does this graph suggest about the production of greenhouse gases?

A. Most production is probably due to cattle gut fermentation.
B. Most production is probably due to burning of fossil fuels.
C. Least production is due to burning of chlorofluorocarbons.
D. Least production is due to burning of biogas.

Questions 20 and 21 relate to the following data.

The number of snails (*Cepaea nemoralis*) eaten by birds was counted during late spring (days 1 to 10) and early summer (days 10 to 20), when the trees turn greener. More green snails are eaten when the background is brown and more brown snails are eaten when the background is green.

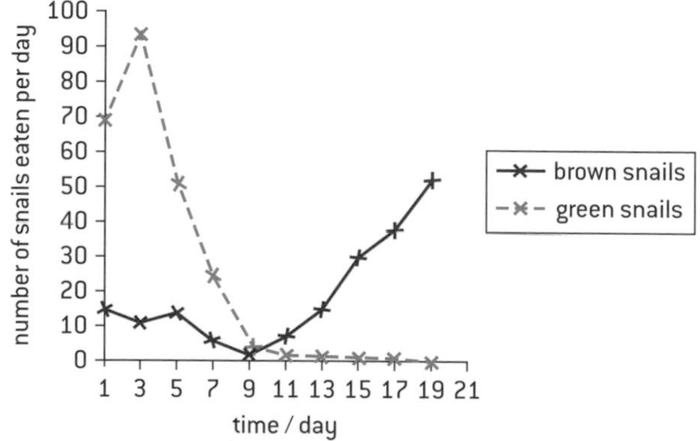

20. This pattern occurs continually throughout the years. What is a probable outcome?

A. Balanced polymorphism
B. Convergent evolution
C. Divergent evolution
D. Competition

21. What statement refers to green and brown snails?

A. They belong to a different genus.
B. They belong to the same phylum.
C. They belong to the *nemoralis* family.
D. They belong to the *Cepaea* order.

22. Studies from biochemistry provide evidence of the evolutionary relationships between primates. It was found that the sequence of hemoglobin in gorillas differs by one amino acid from hemoglobin in chimpanzees and humans, whose sequence is the same. Models were constructed showing the possible relationships between humans, chimpanzees and gorillas.

Which model best represents the relationship according to hemoglobin structure?

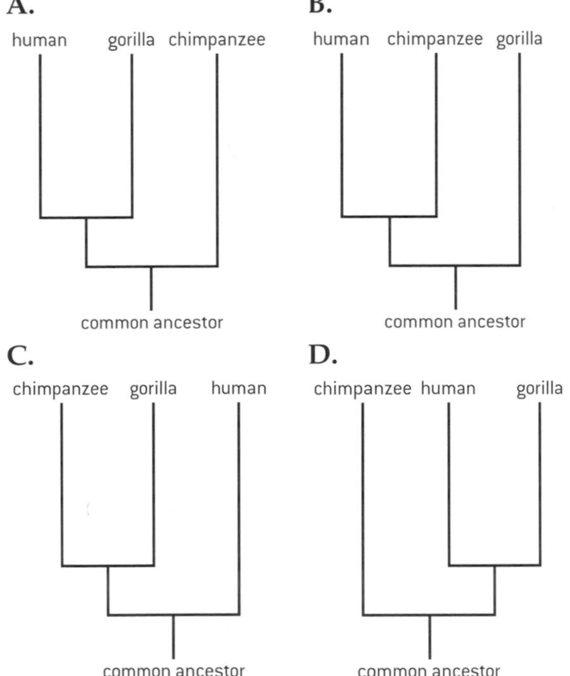

23. What characteristic is present in arthropods but not in molluscs?

A. Bilateral symmetry
B. Articulated legs
C. Calcium shell
D. Notochord

24. Blood cholesterol levels were measured in a group of teenage boys once a month during one year. Years later, when they were adults, they were measured again in the same way. The results were recorded in the scatter graph. Each dot represents the cholesterol level as a young teenager and as an adult for each man in the study.

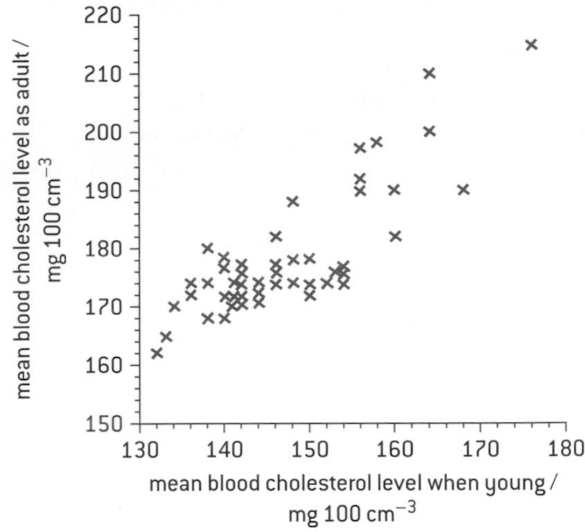

What does this graph show about blood cholesterol levels?

A. They increase as men grow older.
B. They are never below 160 mg in 100 cm^{-3}.
C. Low blood cholesterol levels in youth are healthier.
D. They remain constant throughout the lifetime of a man.

25. What is true about antigens?

A. They are produced in the bone marrow.
B. They are only found in white blood cells.
C. They may stimulate the formation of antibodies.
D. They can only be found on the walls of bacteria.

26. Where does gas exchange take place in humans?

A. Alveoli
B. Bronchi
C. Trachea
D. Diaphragm

27. During the menstrual cycle, what event occurs as the level of progesterone falls?

A. Growth of follicle surrounding the egg
B. Growth of uterine lining
C. Menstruation
D. Ovulation

28. What occurs during transmission of a nerve impulse?

A. There is a change in the permeability to ions of the membrane in the axon.
B. The resting condition is restored by Li^+ ions flowing out of the neuron.
C. During depolarization, K^+ ions flow out of the neuron.
D. In hyperpolarization, K^+ ions flow into the neuron.

29. What is true for type I diabetes but **not** for type II diabetes?

A. Glucose can be detected in urine.
B. High levels of glucose are detected in blood.
C. Blood insulin levels are lower than normal blood insulin levels.
D. Insulin receptors are blocked.

30. What is the sinoatrial node?

A. A blood vessel
B. The wall of the ventricle
C. A group of cells that initiate myogenic contraction
D. A node formed by sensory neurons

*The following questions are for HL candidates **only**.*

31. What forms a eukaryotic chromosome?

A. One DNA molecule and one large protein
B. Several DNA molecules and many proteins
C. One DNA molecule and many proteins
D. Several DNA molecules and one large protein

Questions 32 and 33 refer to the following diagram.

The diagram shows the formation of Okazaki fragments in the replication of DNA in prokaryotes.

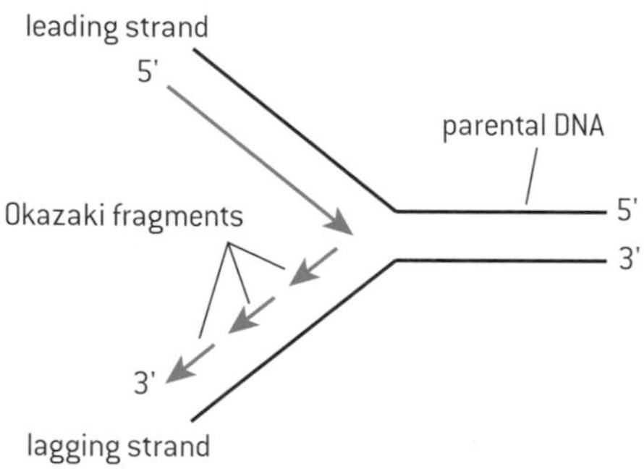

32. What enzyme links the Okazaki fragments together?

A. Gyrase
B. Ligase
C. Helicase
D. Primase

33. Why is the formation of Okazaki fragments necessary during replication?

A. DNA polymerase adds complementary nucleotides in the 5′–3′ direction only.
B. The RNA primer cannot be attached to the lagging strand.
C. A discontinuous replication is faster and more efficient.
D. Okazaki fragments only replicate non-coding regions.

34. What occurs during chemiosmosis?

A. Oxidation of molecules such as reduced NAD (NADH) and reduced FAD ($FADH_2$).
B. Electrons are passed from one carrier to another, liberating energy.
C. Protons move across a membrane, down a concentration gradient, releasing energy.
D. Transformation of energy stored in organic compounds into ATP.

35. The diagram shows two stems, one untreated and the other placed in ink for a day.

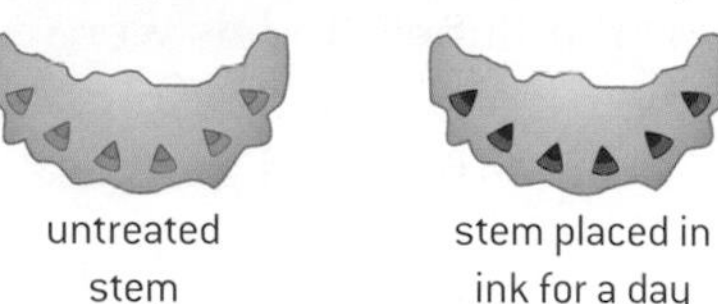

What causes the difference in these stems?

A. Coloured water transported through the xylem has passed to the phloem.
B. Coloured water has entered the phloem only.
C. The ink was transpired through the stomata of the stems.
D. The ink has stained all living tissues.

36. Scientists investigated the pollination of four species of plants. Each of the species was treated in three ways: some plants had their flowers covered to exclude pollinators, other plants were not manipulated, allowing birds to pollinate them, and a third group was pollinated by hand using a paintbrush. The percentage of the total flowers that developed into fruit was recorded.

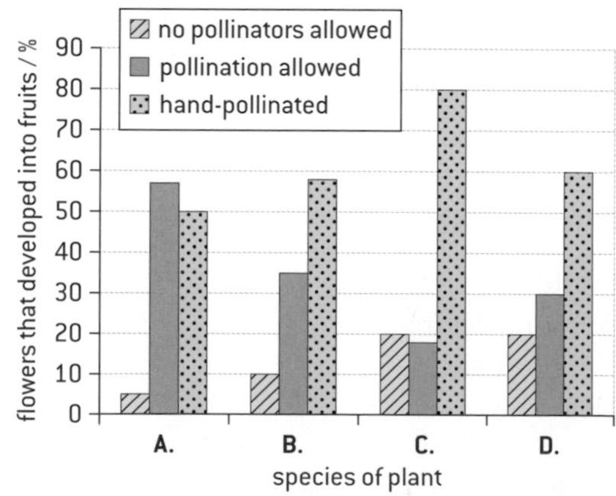

Which species of plant is most efficiently pollinated by birds?

37. A student carried out an investigation comparing the growth of 100 cress seedlings in each of two different environmental conditions. When processing the data collected the student carried out a t-test. Looking up the calculated value of t in the appropriate table gave a probability value between 0.15 and 0.25. What is the conclusion drawn from this result?

A. There is a significant difference between the two environmental conditions
B. The difference between the two samples is not significant
C. Cress seedlings prefer one environment over the other
D. There is insufficient data to carry out a t-test

38. What is punctuated equilibrium?

A. Long periods without appreciable change and short periods of rapid evolution
B. Evolution occurring uniformly at a steady state with continuous species transformation
C. Evolution with a long sequence of intermediate forms of a species
D. Evolution over long periods where a number of species are equilibrated

39. What is the source of glucose in the fluid flowing in the Bowman's capsule?

A. Filtrate in the distal convoluted tubule
B. Urine in the collecting duct
C. Blood in the glomerulus
D. Cells of the medulla

40. What happens immediately after the penetration of the sperm into the egg?

A. Development of the secondary oocyte
B. Development of the primary oocyte
C. Acrosome reaction
D. Cortical reaction

Paper 2

SL: 1 hour 15 minutes

HL: 2 hours 15 minutes

Instructions to candidates

- Answers must be written within the answers boxes (answer sheets are provided at **www.oxfordsecondary.com/ib-prepared-support**).
- A calculator is required for this paper.
- The maximum mark for the SL examination paper is **[50 marks]**.
- The maximum mark for the HL examination paper is **[72 marks]**.

*SL candidates: answer questions 1–4 from section A and **one** question from section B.*

*HL candidates: answer **all** questions from section A and **two** questions from section B.*

Section A

1. Transgenic plants are plants in which one or more genes from another species have been introduced into the genome, using genetic engineering. Some have been modified to resist pests. Bt corn is a variant of maize that expresses Cry protein from the bacterium *Bacillus thuringiensis*. This protein is poisonous to certain insects.

(a) Outline how the gene for the Cry protein is added to the Bt corn. [3]

The levels of uptake by the plant, transcription and translation of the gene for the Cry protein were measured. In order to do so, the amounts of DNA, mRNA and protein were detected. The darkness and thickness of the band is an indicator of amounts present. The results for leaves of three different plants (1, 2 and 3) are shown next to the respective ladders on the following page.

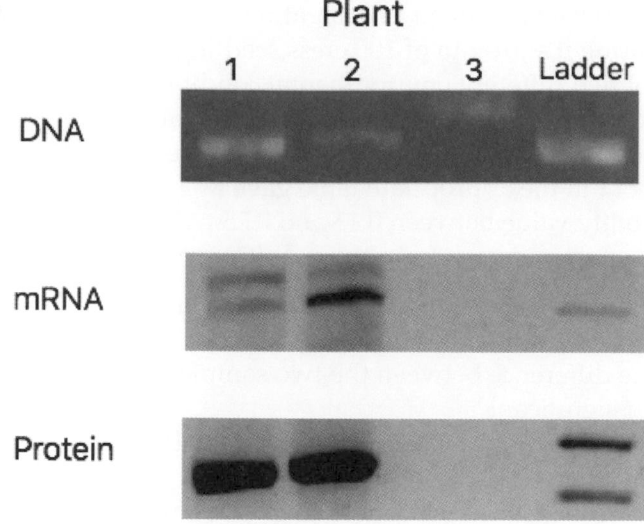

(b) In order to detect the presence of the gene for Cry, a PCR was performed. One plant presented a slightly larger gene for Cry due to an insertion (an extra sequence of DNA) in the promoter. Identify the plant that has the larger gene for the Cry protein. [1]

(c) Suggest one reason mRNA was not detected in plant 3. [1]

(d) Evaluate the data to confirm whether the genetic modification was successful in the three plants. [2]

Common milkweed (*Asclepiassyriaca*) is a plant that exists in cornfields across most corn-growing regions. The larvae of monarch butterflies (*Danaus plexippus*) live and feed on common milkweed plants. Bt corn is corn that has been genetically modified to produce a protein that is poisonous to certain insect pests.

A collaborative research effort by scientists produced information from 11 regions to develop a formal risk assessment of the impact of Bt corn on monarch butterfly populations. Information was sought on the percentage of corn that is genetically modified and the degree to which monarch larvae would be exposed to toxic amounts of Bt pollen on its host plant, the common milkweed, found in and around cornfields.

Region	Percentage of arable land planted with corn / %	Percentage of corn that is Bt / %	Percentage of corn area that is a breeding site for monarch butterfly / %	Probability of exposure of monarch butterfly to Bt corn pollen
1	42	26	19	0.0040
2	42	14	18	0.0026
3	38	7	10	0.0007
4	12	26	4	0.0002
5	13	11	2	0.0001
6	21	8	4	0.0002
7	29	30	8	0.0028
8	12	22	5	0.0002
9	39	26	7	0.0011
10	13	11	0	0.0001
11	26	37	9	0.0004
Mean	**26**	**23**	**8.6**	**0.0011**

(e) (i) State the region with the greatest percentage area with monarch butterfly breeding sites. [1]
(ii) Identify, with a reason, a region where pollen of Bt corn could be found on the leaves of milkweed. [1]

(f) Scientists concluded from this 2-year study that the impact of Bt corn pollen from current commercial Bt corn on monarch butterfly populations is negligible. Analyse the data to support this conclusion. [3]

In another study, monarch butterfly larvae were fed milkweed plants dusted with pollen from two brands of Bt corn (Bt corn 1 and Bt corn 2) and their survival rate was compared with that of larvae fed milkweed plants dusted with pollen from non-Bt corn.

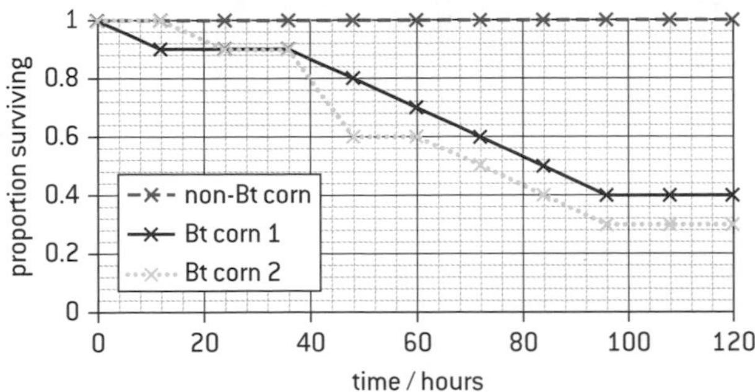

(g) Compare and contrast the survival of monarch butterfly larvae when using non-Bt corn pollen and Bt corn pollen. [3]

(h) Using **all** the data, discuss the effect of Bt corn on monarch butterflies. [2]

2. *Camellia sinensis* is a species of evergreen plant whose leaves are used to produce tea. The nucleotide percentages in the DNA and mRNA for *C. sinensis* are given in the table.

	Bases / %			
	Adenine	Guanine	Cytosine	Thymine or uracil
DNA	31.7	18.6	18.3	31.4
mRNA	25.7	29.8	23.9	20.6

(a) State the base found in mRNA but not DNA. [1]

(b) (i) Distinguish between the percentages of bases found in DNA and mRNA. [2]
(ii) Determine a reason for this difference. [1]

(c) Outline where DNA and mRNA can be found in the cells of the leaves of *C. sinensis*. [2]

3. The diagram shows the structure of a human heart.

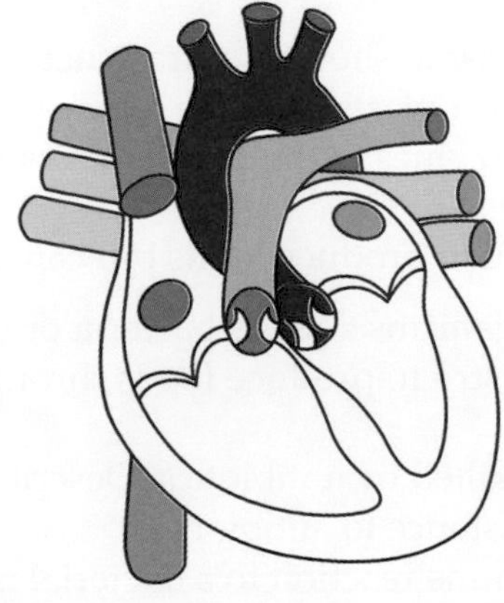

(a) On the diagram label the: [3]
   (i) left atrium
   (ii) semilunar valve
   (iii) pulmonary vein.

(b) Annotate the diagram to show the direction of deoxygenated blood through the heart. [2]

(c) Explain the difference in width in the walls of the ventricles. [3]

4. The photographs show parts of different plants (not to scale).

A

B

C

D

Determine the classification of plants shown in the pictures using the dichotomous key. [3]

1) No vascular tissue ....Bryophyta
   Presence of vascular tissue .... *go to 2*
2) Spores present.................................... Filicinophyta
   No spores .... *go to 3*
3) Flowers present......................... Angiospermophyta No flowers and presence of cones........Coniferophyta

*Question 5 is for HL candidates **only**.*

5. (a) List **two** structures present in mitochondria that support the endosymbiotic theory. [2]

(b) Explain chemiosmosis in mitochondria. [4]

Section B

*Questions 6 and 7 are for SL candidates **only**. Answer **one** of the questions. Up to one additional mark is available for the construction of your answer.*

6. Starch is produced in chloroplasts.

(a) Draw a chloroplast to show its ultrastructure. [4]

(b) Explain how chloroplasts produce starch by photosynthesis. [7]

(c) Describe how enzymes in the digestive system digest starch. [4]

7. Bacteria can be useful to produce foods but can also be pathogenic.

(a) Outline how organisms such as bacteria or yeast can be used in the food industry to produce foods through anaerobic respiration. [4]

(b) Bacteria can be killed by antibiotics. Describe how some bacteria can develop resistance to antibiotics. [4]

(c) Explain the immune reaction to a bacterial infection. [7]

*The following questions are for HL candidates **only**. Answer **two** of the questions. Up to one additional mark is available for the construction of your answers for each question.*

8. Starch is produced in chloroplasts.

(a) Draw a chloroplast to show its ultrastructure. [4]
(b) Explain how chloroplasts produce starch by photosynthesis. [7]
(c) Describe how enzymes in the digestive system digest starch. [4]

9. Bacteria can be useful to produce foods but can also be pathogenic.

(a) Outline how bacteria produce ethanol and carbon dioxide in anaerobic respiration. [4]
(b) Bacteria can be killed by antibiotics. Describe how some bacteria can develop resistance to antibiotics. [4]
(c) Explain the immune reaction to a bacterial infection. [7]

10. Waste matter needs to be eliminated to avoid harmful effects from toxic substances in the organism.

(a) Compare and contrast structural and functional aspects of excretion by Malpighian tubule systems and kidneys. [7]
(b) Outline the role of ADH in osmoregulation. [4]
(c) Uric acid is a normal component of urine. High blood concentrations of uric acid can lead to gout. This disease is characterized by inflammation of joints, such as the elbow. Draw a labelled structure of the elbow. [4]

Paper 3

SL: 1 hour

HL: 1 hours 15 minutes

Instructions to candidates

- Answers must be written within the answers boxes provided (answer sheets are provided at **www.oxfordsecondary.com/ib-prepared-support**).
- A calculator is required for this paper.
- The maximum mark for the SL examination paper is **[35 marks].**
- The maximum mark for the HL examination paper is **[45 marks].**

Section A

*SL and HL candidates: answer **all** questions.*

1. In an experiment, test tubes were filled with different glucose solutions and Benedict's reagent. The tubes were then heated in a water bath. A positive test results in an orange precipitate, which is seen in varying densities in the tubes.

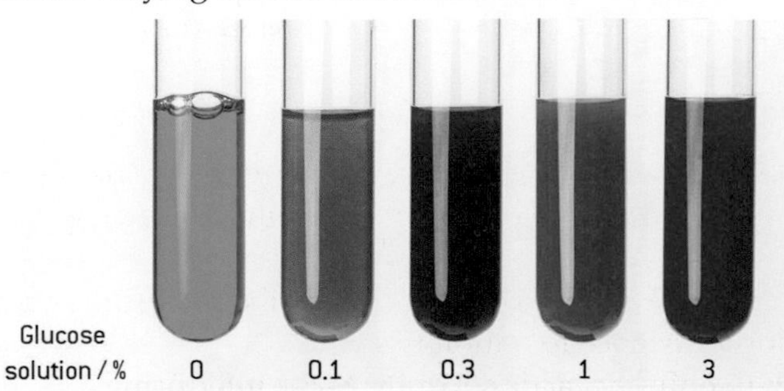

Two food samples were tested in the same way and the results are shown in the photograph.

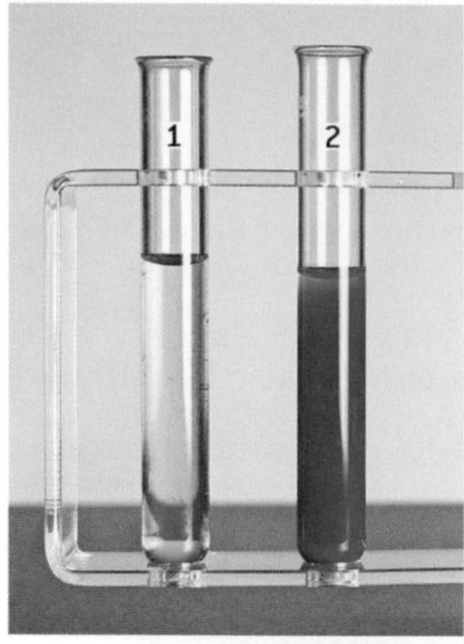

(a) Estimate the glucose concentration in foods 1 and 2. [2]

(b) Suggest with a reason whether this is a precise method of measuring the glucose concentration. [1]

(c) Describe one method to obtain the glucose solutions of 1% and 0.1% concentration. [3]

2. To test the effect of light intensity on the rate of photosynthesis, the following apparatus was set up in a laboratory.

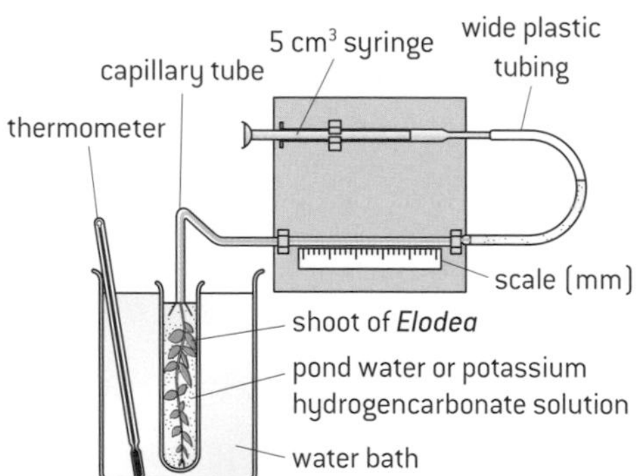

(a) Suggest one way light intensity could have been modified in this experiment. [1]

(b) Identify the reason a thermometer is placed in the water bath. [2]

(c) Explain how the rate of photosynthesis was recorded in this experiment. [3]

3. Polyacrylamide gel electrophoresis (PAGE) is a technique used to separate nucleic acids in forensic chemistry. The diagram illustrates the process.

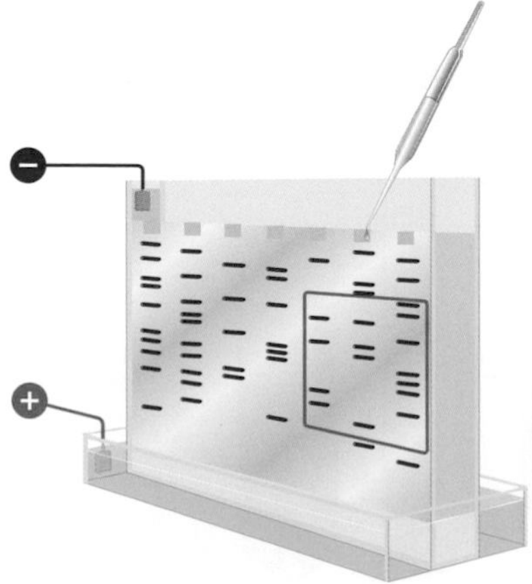

Explain how the nucleic acids separate in the gel during electrophoresis. [3]

Section B

*Answer **all** of the questions from **one** of the options.*

Option A: Neurobiology and behaviour

*SL candidates: answer questions 4–7 **only**.*

HL candidates: answer questions 4–10.

4. The graph shows the neurogenesis and dendritic growth in a mammal during embryonic stages and after birth.

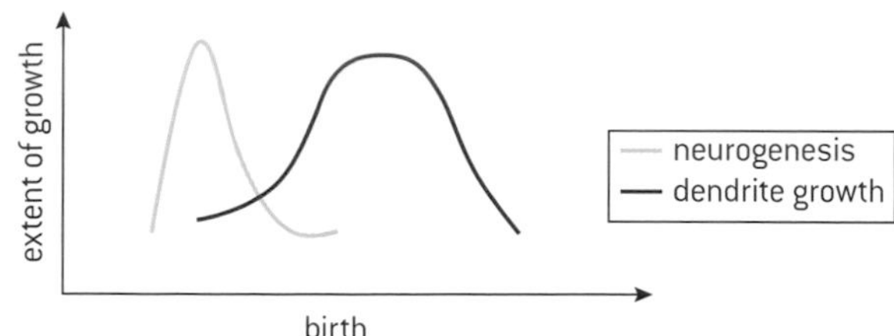

(a) Neurogenesis is the development of new neurons. Describe neurogenesis. [3]

(b) Neurulation needs to occur before neurogenesis. Outline the stages of neurulation. [3]

(c) State the name of the process by which neurons or synapses are eliminated if they are not used. [1]

5. It has been hypothesized that the evolution of increasing body mass in mammals has been accompanied by a reduction of the mortality rate and a corresponding increase in longevity. Another variable that could affect longevity is brain mass. Data on brain mass was analysed from many mammalian species.

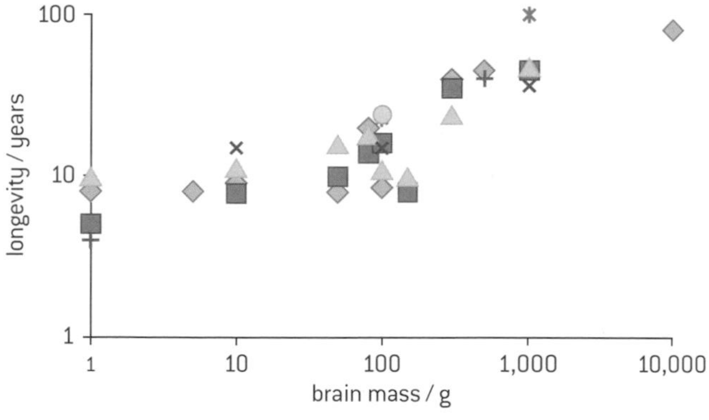

Analyse the data to see whether it confirms the hypothesis that increasing **body** mass can affect longevity. [3]

6. The diagram illustrates a transverse section of a human eye.

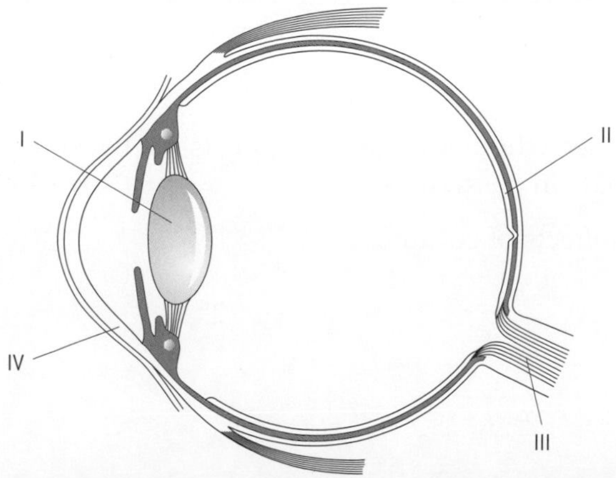

(a) Identify the label matching each function: [3]
   (i) light detection by photoreceptors
   (ii) sends impulses to the brain
   (iii) adjusts in order to bend the light rays to be focused on the retina.

(b) Draw an arrow showing the direction of the light rays when observing an object. [1]

(c) The ear is also a sensory organ. Distinguish between the hair cells in the cochlea and the hair cells in the semicircular canals. [2]

7. Explain the mode of action of the autonomic nervous system. [4]

*The following questions are for HL candidates **only**.*

8. The diagram shows the reflex arc. Label the parts 1 to 4. [2]

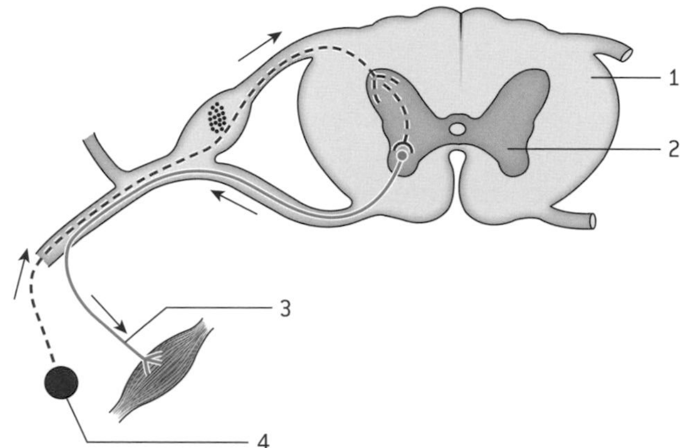

9. The number of different songs and lengths of these songs were recorded in two species of passerines, *Melospiza lincolnii* and *Melospiza melodia*. A group of birds was bred isolated from the rest of the passerines and the birds' songs were also recorded. The results are shown in the bar charts.

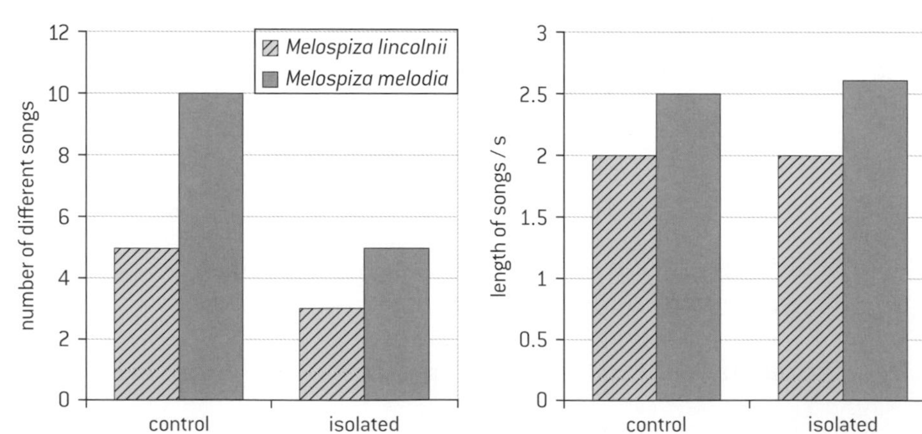

Analyse the data to see whether the data supports the hypothesis that birdsong is innate in passerines. [2]

10. Discuss the health effects of cocaine. [6]

Option B: Biotechnology and bioinformatics

*SL candidates: answer questions 11–14 **only**.*

HL candidates: answer questions 11–16.

11. Fermenters allow large-scale production of metabolites by microorganisms. The diagram shows the production of mycoprotein by fermentation. Mycoprotein is a fungus protein produced for human consumption. Mycoprotein is harvested every day.

(a) State **two** conditions that should be monitored by probes in this fermenter. [2]

(b) State **one** reason for the use of paddles. [1]

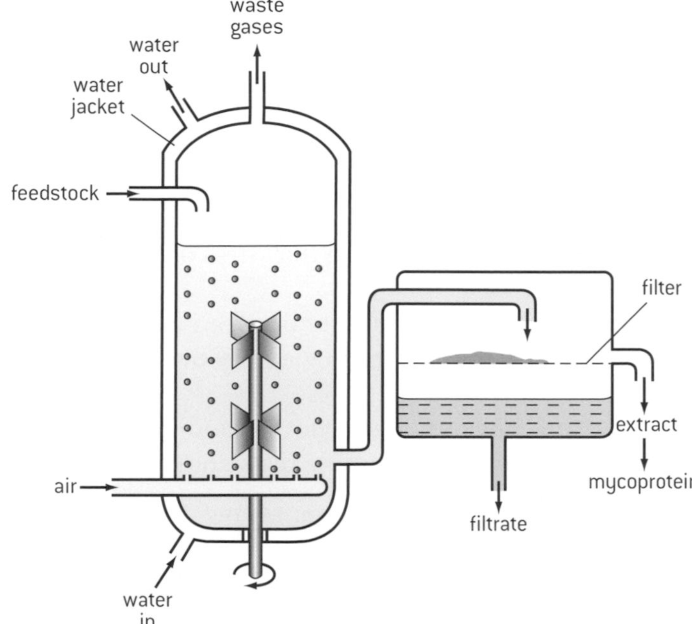

12. Bagasse is the fibrous matter that remains after the extraction of juice from sugar cane. Citric acid was obtained by fermentation of a nutrient solution containing different concentrations of bagasse using the fungus *Aspergillus niger*.

(a) State a use of citric acid in the food industry. [1]

Fermentation on trays of different depths was compared with an ordinary fermentation. The advantage of tray fermentation is that it is much cheaper than using a fermenter. Samples were collected for evaluation of citric acid production.

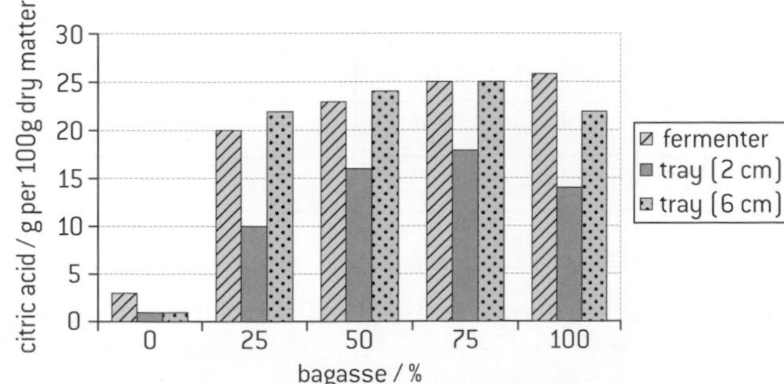

(b) Identify the best conditions to obtain citric acid in this experiment. [1]

(c) Based on the results of these experiments, comment on the effectiveness of tray fermentation of bagasse. [3]

13. The mixture of bacteria, saliva and carbohydrates found on teeth is known as plaque. It is a cause of tooth decay (caries). The most common bacteria in tooth biofilms are *Streptococcus mutans.*

(a) *S. mutans* are Gram-positive bacteria. Describe the method used to detect that the bacteria are Gram-positive. [3]

The thickness of a biofilm of *Streptococcus mutans* bacteria was measured for 3 weeks. The viability was measured as the percentage of live bacteria over the total after exposure to antibiotics.

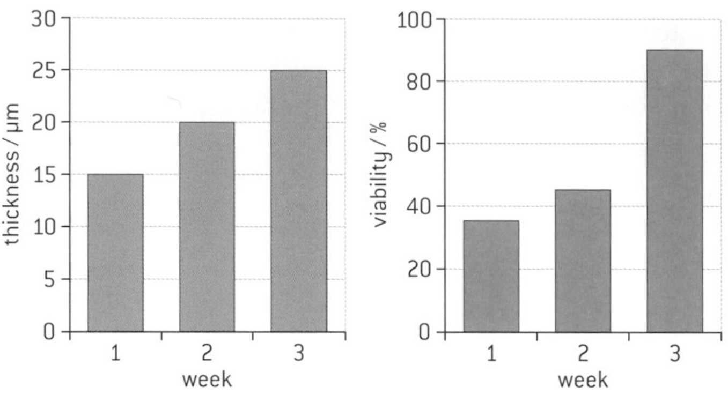

(b) Suggest reasons for the change in thickness of the biofilm over time. [2]

(c) Calculate the difference in viability between weeks 2 and 3. [1]

(d) Suggest reasons for the change in viability over time. [2]

14. Transgenic rice plants have been genetically modified by adding genes from different species to promote resistance to herbicides.

Explain **one** method by which these transgenic rice plants could have been genetically modified. [4]

The following questions are for HL candidates ***only****.*

15. Part of the sequence of the protein cytochrome c was compared in different organisms. The alignment is shown below. A dash shows the amino acid is repeated and a space that there is no amino acid.

(a) State the organism whose cytochrome c sequence is most similar to humans. [1]

(b) On the sequence, use a box to identify an amino acid that is conserved in all the species shown. [1]

(c) Outline how this sequence alignment was obtained using bioinformatics. [2]

Human	G	D	V	E	K	G	K	K	I	F	I	M	K	C	S	Q	C	H	T	V	E	K
Pig	–	–	–	–	–	–	–	–	–	–	V	Q	–	–	A	–	–	–	–	–	–	–
Fish	–	–	–	–	–	–	–	–	V	–	V	Q	–	–	A	–	–	–	–	–	–	N
Wheat		N	P	D	A	–	A	–	–	–	K	T	–	–	A	–	–	–	–	–	N	A
Yeast		S	A	K	–	–	A	T	L	–	K	T	R	–	Q	L	–	–	–	–	–	–

16. Using an example, explain the use of viral vectors in gene therapy. [6]

Option C: Ecology and conservation

*SL candidates: answer questions 17–20 **only**.*

HL candidates: answer questions 17–22.

17. Lettuce coral (*Undaria tenuifolia*) is the dominant coral species in shallow semi-exposed reefs in the Southern Caribbean Sea. Colonies of *U. tenuifolia* were classified according to the colony size into small, medium and large.

The number of colonies according to their size were recorded three times, starting in spring of the first year (1), then autumn of the first year (2) and ending in spring of the second year (3) in a location in the Southern Caribbean.

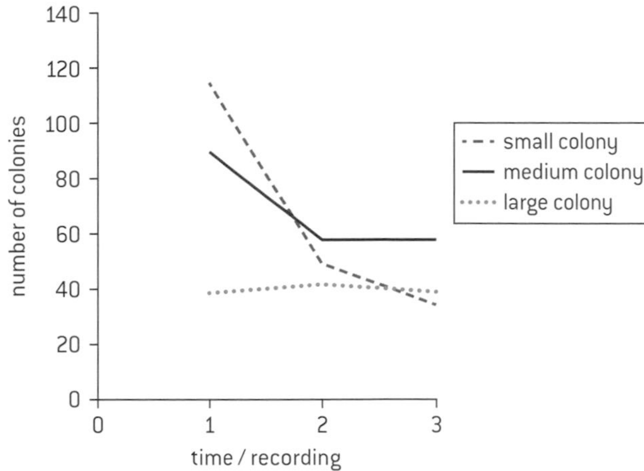

(a) State the number of colonies recorded in the first spring for each colony. [1]

(b) Describe the trends in numbers of colonies. [2]

(c) Suggest **one** reason for the change in population of small and medium corals. [1]

(d) Describe how bleaching is an indication of stress on corals. [2]

18. *Serratia marcescens* (first trophic level) is a bacterium eaten by *Paramecium aurelia* (second trophic level) which is eaten by a predator, *Didinium nasutum* (third trophic level). The effect of nutrients on this food chain was studied in three closed microcosms (small mesocosms). The microcosms were set up as follows at two different nutrient levels (low and high):

- microcosm 1: *S. marcescens* only
- microcosm 2: *S. marcescens* and *P. aurelia* only
- microcosm 3: *S. marcescens*, *P. aurelia* and *D. nasutum.*

The population density of the bacterium *S. marcescens* was measured after 7 days.

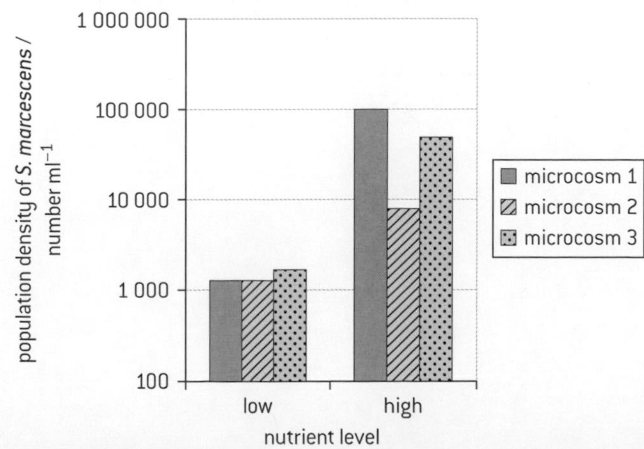

(a) State the effect of nutrient level on the density of population of *S. marcescens* in microcosm 1. [1]

(b) Explain the effect on the *S. marcescens* population density of adding a trophic level in each microcosm at a high nutrient level. [2]

(c) Suggest **one** reason for little change at low nutrient levels. [1]

(d) Using the data in this investigation, predict a relationship between nutrient levels and the length of a chain in a food web. [1]

19. The table shows the number of plants of each species counted in a sand dune.

Species	Number
sea holly (*Acanthus ebracteatus*)	3
sand couch (*Sporobolus virginicus*)	6
baby grass (*Sporobolus pungens*)	1
Total	**10**

(a) Calculate the Simpson reciprocal diversity index from the following data for this community, using the formula given below.

$$D = \frac{N(N-1)}{\Sigma n(n-1)}$$ [1]

(b) Comment on the richness and evenness of this sand dune community. [2]

The sand dune community was compared with that of a forest where 240 organisms were counted in a total of 10 species. The Simpson diversity index of this forest was 14.

(c) Distinguish between the biodiversity of the sand dune and the forest studied. [2]

20. Discuss the biological and ethical issues surrounding biological pest control. [4]

*The following questions are for HL candidates **only**.*

21. The study of species interactions in urban versus rural environments can improve the understanding of shifts in ecological processes due to urbanization. The leaves of the English oak (*Quercus robur*) in urban and adjacent rural areas were assessed to determine whether urbanization affected leaf nitrogen content.

(a) Outline **two** factors in the soil that could affect the nitrogen concentration in leaves of *Q. robur*. [2]

The graph shows the mean ± standard deviation of the nitrogen concentration in leaves of *Q. robur* growing in urban and rural habitats of large, medium and small cities.

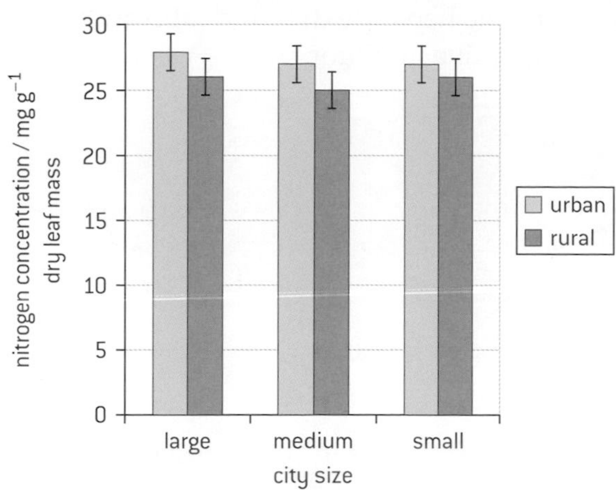

(b) Describe the effect of city size and urbanization on the *Q. robur* leaf nitrogen concentration. [2]

22. Explain bottom-up and top-down limiting factors in a pond environment. [6]

Option D: Human physiology

SL candidates: answer questions 23–26 ***only****.*

HL candidates: answer questions 23–28.

23. Phenylketonuria (PKU) is an inherited disease that results in decreased metabolism of the amino acid phenylalanine. It is caused by the deficiency of phenylalanine hydrolase, an enzyme that converts phenylalanine into tyrosine. Treatment is with a diet low in foods that contain phenylalanine and the addition of special supplements.

Serum phenylalanine levels are important indications for protein-restricted diet therapy. Brain damage might occur if the patient does not follow the diet. The scattergraph shows the incidence of brain damage and the plasma phenylalanine levels of patients who did not stick to the diet.

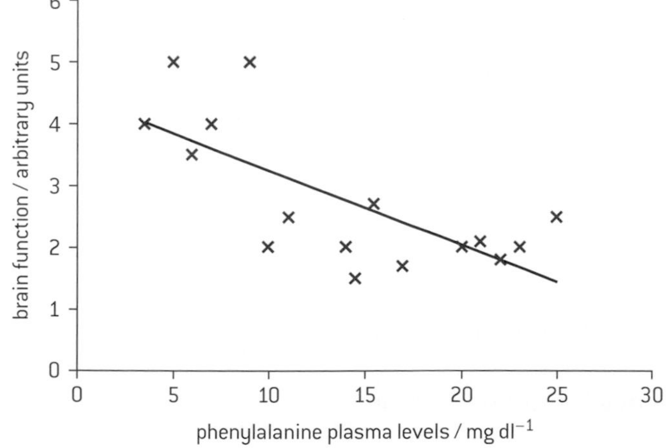

(a) Describe the relationship between phenylalanine plasma levels and brain function. [1]

(b) Discuss the use of a protein-free diet to improve brain function in patients with PKU. [3]

24. Orally administered drugs that are not excreted in urine must be completely eliminated by the liver by enzymatic reactions. The rate at which this happens is called the hepatic extraction ratio. A ratio of 1 means all the drug has been removed. Competitive inhibitors bind to the drug but do not catabolize it. The graph shows the relationship between the hepatic extraction ratio and percentage inhibition by a competitive inhibitor.

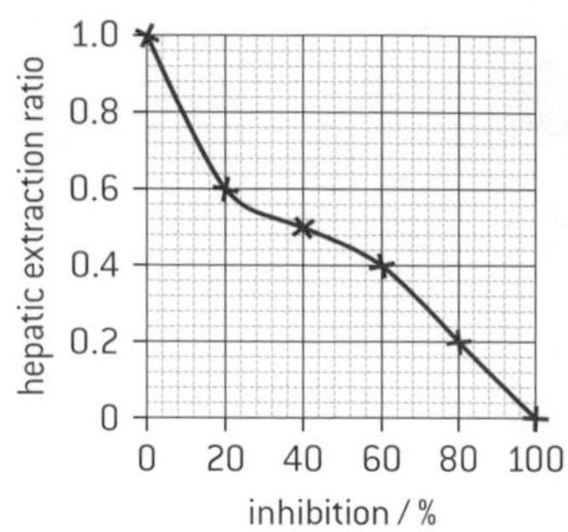

(a) It has been suggested that drug competitive inhibitors can cause drug plasma concentrations to increase, permanently damaging the liver. Discuss this hypothesis using the data provided. [3]

(b) List **two** functions of the liver other than detoxification. [2]

25. A study was performed on patients with hypertension to investigate the effects of exercise on heart functions.

(a) Define hypertension. [1]

For a period of 10 weeks a group of patients excercised using a treadmill. At the same time, control patients were kept under non-exercise conditions. The mean resting heart rate was recorded in beats per minute (bpm).

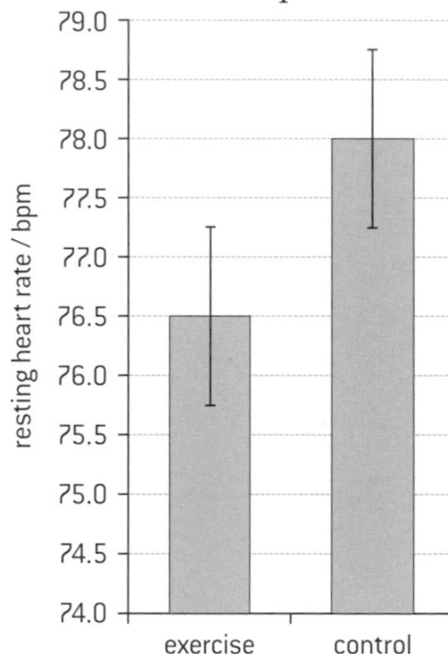

(b) Describe the effect of exercise on the resting heart rate. [2]

(c) Outline how heart rate is measured. [2]

The heart walls contract during the heart beats.

(d) Explain how the structure of cardiac muscle cells allows propagation of stimuli through the heart wall. [2]

26. Explain the nervous control of digestion. [4]

*The following questions are for HL candidates **only**.*

27. The human growth hormone, somatotropin, is a peptide hormone that stimulates growth, especially in childhood. It strengthens bone, increases muscle mass, reduces glucose uptake by the liver and promotes the breakdown of body fat.

Outline the mode of action of somatotropin on target cells. [4]

28. Fetal hemoglobin is different from adult hemoglobin. Comment on this statement. [6]

Index

Key terms are in **bold.**